RIT - WALLACE LIBRARY
CIRCULATING LIBRARY BOOKS

OVERDUE FINES AND FEES FOR <u>ALL</u> BORROWERS

- Recalled = $1/ day overdue (no grace period)
- Billed = $10.00/ item when returned 4 or more weeks overdue
- Lost Items = replacement cost + $10 fee
- All materials must be returned or renewed by the due date.

Practical E-Manufacturing and Supply Chain Management

Practical E-Manufacturing and Supply Chain Management

Gerhard Greeff ND Chem Eng, Dip Prod Man, Dip QA and QC,
Divisional Manager: Consulting for supply chain and E-manufacturing
at Altech Informatics (Pty) Ltd in Centurion, South Africa

Ranjan Ghoshal BSc (Chem Hons), MSc (Oils & Fats Technology),
PG Dip in Business Management (Manufacturing and Operations),
Dip in Bio-Process Engineering

Series editor: Steve Mackay

ELSEVIER

AMSTERDAM • BOSTON • HEIDELBERG • LONDON
NEW YORK • OXFORD • PARIS • SAN DIEGO
SAN FRANCISCO • SINGAPORE • SYDNEY • TOKYO
Newnes is an imprint of Elsevier

Newnes

Newnes
An imprint of Elsevier
Linacre House, Jordan Hill, Oxford OX2 8DP
200 Wheeler Road, Burlington, MA 01803

First published 2004

British Library Cataloguing in Publication Data
Greeff G.
    Practical E-manufacturing and supply chain management
    1. Production engineering   2. Business logistics
    I. Title   II. Ghoshal, R.
    658. ;

Library of Congress Cataloguing in Publication Data
A Catalogue record for this book is available from the Library of Congress

ISBN   0 7506 6272 7

For information on all Newnes Publications
visit our website at www.newnespress.com

Typeset and edited by Integra Software Services Pvt. Ltd, Pondicherry, India
www.integra-india.com
Printed and bound in The Netherlands

Working together to grow
libraries in developing countries

www.elsevier.com | www.bookaid.org | www.sabre.org

ELSEVIER   BOOK AID
            International   Sabre Foundation

Contents

Preface

Supply Chain Management involves the optimization of the way in which a company plans the production of goods or services, procures raw materials from various suppliers, manufactures the goods or services, delivers it to customers, and handles returns.

E-Manufacturing starts with an order for a product and then encompasses the entire manufacturing cycle of the product. Manufacturers need a highly responsive supply chain and manufacturing system to ensure that they meet the high expectations of their customers who, in today's economy, demand absolutely the best service, price, delivery time and product quality.

Manufacturing Execution Systems (MES) provide up-to-the-minute mission-critical information about production activities across the factory and supply chain via communications networks (e.g. Local Area Networks), resulting in the optimization of activities throughout all aspects of the manufacturing process. MES accomplish this task by guiding, initiating, responding to, and reporting on plant activities in real time, by using current and accurate data. This rapid response to changing conditions, together with a focus on reducing non-profitable activities, leads to more efficient plant operations and processes.

MES reduces cycle times, levels of Work in Progress (WIP), data entry time, paperwork and scrap. It also improves utilization of plant capacity, process control quality, arrangement of plant activities, tracking of orders, and customer service.

Research figures show that manufacturers have been able to achieve the following improvements through MES:

- An average reduction of manufacturing cycle time by 45%
- An average reduction of data entry time by 75%
- An average reduction of WIP by 24%
- An average reduction of paperwork between shifts by 61%
- An average reduction of lead time by 27%
- An average reduction of paperwork and blueprint losses by 56%
- An average reduction of product defects by 18%.

Implementation of MES invariably results in improved returns on production assets, on-time delivery, faster inventory turnover, larger net profits (through increased cost reduction) and improved cash flow.

Typical people who will find this book useful include:

- CEOs and CFOs
- Finance managers
- E-Commerce managers
- IT managers
- Business managers
- Strategy managers
- Operations managers and engineers
- Production managers and engineers
- Senior process engineers
- Network and telecommunications managers.

At the end of reading this book, you will be familiar with the following concepts:

- Enterprise and business automation system hierarchies
- Process and system design concepts and considerations
- The fundamentals of E-Manufacturing and Supply Chain processes
- Project motivation and benefit quantification
- Business and process modeling tools
- System integration concepts and models
- Managing software project partners, system integrators and consultants
- System implementation and behavioral change management considerations
- Product and vendor evaluation methodology.

As a result, you will be able to plan for the practical implementation of MES systems, and through that:

- Improve your company's financial performance
- Reduce inventories
- Improve cash flow
- Optimize productivity, the use of assets and interaction with suppliers
- Improve planning and forecasting.

An elementary understanding of business and manufacturing operations processes is expected to obtain maximum benefit from this book. No specific knowledge of software is required.

Disclaimer

Whilst all reasonable care has been taken to ensure that the descriptions, opinions, programs, listings, software and diagrams are accurate and workable, IDC Technologies does not accept any legal responsibility or liability to any person, organization or other entity for any direct loss, consequential loss or damage, however caused, that may be suffered as a result of the use of this publication. In case of any uncertainty, we recommend that you contact IDC Technologies for clarification or assistance.

Trademarks

All terms noted in this publication that are believed to be registered trademarks or trademarks are listed below:

IBM, XT and AT are registered trademarks of International Business Machines Corporation. Microsoft, MS-DOS and Windows are registered trademarks of Microsoft Corporation.

Acknowledgements

IDC Technologies and the author express their sincere thanks to all the engineers and software specialists in Altech Informatics who freely made available their time and expertise in preparing this manual. The author also wants to thank the management of Altech Informatics for allowing him to pursue this opportunity with IDC Technologies and Ranjan Ghoshal, who assisted greatly in the collation of the information and generation of graphics. A special thanks also to the family Greeff (Ida, Clariś and Cora-Lize) for their support and understanding through the rough editing and review cycles.

Who is Altech Informatics?

Altech Informatics assist customers to achieve operational excellence by combining its extensive experience in industrial processes and information technology to deliver high return yet low risk projects.

Altech Informatics consists of enthusiastic engineers, software specialists and support staff who are committed to deliver high value designs, high-quality solutions and exceptional project management services.

1

Introduction to e-manufacturing systems

Learning objectives

- To understand what e-manufacturing is all about.
- To understand where e-manufacturing is going and what is required to make this possible.
- To understand the implications of embarking on an e-manufacturing strategy.
- To understand how e-manufacturing integrates with supply-chain systems.

1.1 Preamble

To make e-manufacturing work, integration is needed between various disparate systems. To understand why this is such an issue, one needs to understand what the different systems or system components do, their objectives, their specific focus areas and how they interact with other systems. It is also required to understand how these systems evolved to their current state, as the concepts used during the early development of systems and technology tend to remain in place throughout the life cycle of the systems/technology. It is thus important to know how systems and technology evolved over time to understand their current operation, benefit and deficiencies.

The following chapters will take the reader through the evolution of systems and technology in order to establish a base-line understanding of the various kinds of systems and what lead to their development and evolution. It will then explore various standards, concepts and techniques used over the years to model systems and hierarchies in order to understand where they fit into the organization and supply chain. It will look at the specific system components and what they do, and at the ways in which they can be designed and graphically depicted for easy understanding by both information technology (IT) and non-IT personnel.

Without a good implementation philosophy, very few systems add any real benefit to an organization, and for this reason the ways in which systems are implemented and installation projects managed are also explored and recommendations are made as to possible methods that have proven successful in the past. The human factor and how that impacts on system success are also addressed, as is the motivation for system investment and subsequent benefit measurement processes.

Finally, the vendor/user supply/demand within the e-manufacturing domain is explored and a method is put forward that enables the reduction of vendor bias during vendor selection.

The objective of this book is to provide the reader with a good understanding regarding the four critical factors (business/physical processes, systems supporting the processes, company personnel and company/personal performance measures) that influence the success of any e-manufacturing implementation, and the synchronization required between these factors.

1.2   E-manufacturing definition

'The core of an e-manufacturing strategy is the technology roadmap for information transparency between the customer, manufacturing operations and suppliers. An e-manufacturing strategy takes e-business processes, such as build-to-order or reliability-centered maintenance and generates guidelines for implementing plant systems. The e-manufacturing strategy takes the e-business and manufacturing strategies and creates a roadmap for system development and implementation in the plant.' – AMR Research.

E-manufacturing can mean many things from e-procurement, B2B, B2C, industrial Ethernet, portals, TCP/IP, UDP, XML, collaborative manufacturing, wireless and embedded web servers, to supply chain management (SCM).

'E-manufacturing is the vertical (business) and horizontal (supply-chain) integration of systems to ensure the correct dissemination of information throughout the value-chain of a business, making use of appropriate technology like the Internet to ensure that real-time accurate information is available at all decision points throughout an organization and supply chain.'

1.3   Background

Earlier days, the plant floor was isolated from the rest of the enterprise, operating autonomously and out of the sight from the rest of the company as well as from the eye of the shareholders. Today everybody is driven by bottom line performance, and the shop floor is one area that shareholders and the financial analysts are becoming more interested in. Meeting or exceeding earnings forecasts and the return on net assets (RONA) are important criteria used to judge the performance of an organization and as such directly influence the share price.

In the case of a manufacturing company not only growing sales, but also the efficient operation of the unit are important criterion for success. With the removal of all non-value-added activities and making effective use of technologies like Internet, a company can operate better, faster and cheaper compared to its competitors.

The manufacturing plant, in a manufacturing company, is the heart where all value creation takes place. In the supply-chain model, connection of the plant with the entire chain is crucial and accurate, timely information is more critical than ever. All functions from planning to logistics are under review for potential cost saving opportunities. While many companies are facing the challenge of meeting consumer demands for e-commerce channels, most manufacturers are also trying to make sense of the maintenance, repair and operations (MRO) organization and understand the true potential of the plant floor.

The most strategic advantage of any organization today is information, and accurate information is critical for making the right decisions, whether the challenge is faster - to-market cycles, improved process yield, non-stop operations or tighter supply-chain coupling. The plant is the starting point for bigger information connectivity. Computer-based plant floor controls for manufacturing machinery, material handling systems and related equipment generate a wealth of information about productivity, product design, quality and delivery.

Today a company can have a single, complete set of operational capabilities including rapid plant design and deployment, real-time business management system connectivity, comprehensive asset management of people, products and process and a seamless coupling to the entire supply chain via the web. This is e-manufacturing, a concept much greater than the sum of its parts.

Keys for a successful e-manufacturing strategy:

- Integrated plant-floor automation
- Seamless connections to the enterprise systems enabled through software and services
- Comprehensive asset management and reliability-centered maintenance
- Tailored e-business strategies for supply-chain efficiencies.

1.4 E-manufacturing strategy

A company's manufacturing plants should be able to build-to-order and maintain non-stop operations. To achieve this four competencies are required of any manufacturer:

Design The ability to rapidly deploy and reconfigure manufacturing production capacity based on demand for goods. Companies should be able to accelerate and streamline the design and deployment of production processes. Fast product introduction in response to changing market demand is a critical competitive advantage and a key to growth.

Operate Optimization of process yield and consistency throughout the enterprise. Plant productivity has always been a focus in manufacturing. Initiatives like lean manufacturing drive out excess, achieving non-stop operations for maximum efficiency and throughput of production. Techniques like six-sigma reduce variability in processes to ensure peak quality.

Maintain Efficient management of all company assets – materials, processes, and employees to ensure non-stop operations and optimum asset productivity. It is not possible to sustain such a fast-paced environment where growth and profits are demanded simultaneously, without a solid information foundation.

Synchronize Tight coupling of a manufacturing operation into the greater supply chain, both up- and down-stream. This is best achieved only after the other three competencies are firmly implemented.

E-manufacturing requires a new approach to manufacturing and distribution systems throughout the design, operate, maintain and synchronize competencies. It is made possible through the complete integration of manufacturing control systems and enterprise applications using commercially developed, off-the-shelf-information technologies. E-manufacturing provides direct information exchanges between manufacturing, customer relationship management (CRM) systems and SCM systems.

Key requirements to successfully operate an e-manufacturing environment:

- 24×7 availability of information to the supplier/customer
- Scalability of systems – once information is available to people, many more people will immediately want to make use of it
- Variety of connectivity options – to handle many data sources
- Intuitiveness or organization of information – people should get the data easily
- Personalization of information needs
- Adaptability to constant change
- Rapid development – faster time to the market
- Legacy system integration – few people have the money and time for totally new systems on a frequent basis

- Application inter-operability
- Management
- Security
- Diagnostics.

1.5   E-manufacturing challenges

Changing attitudes of customers and the dynamic market environment increase the pressure on the organizations across the world whatever may be their field of operation, product/service and size. Following is only a few of the challenges facing manufacturing organizations in the e-manufacturing era.

1.5.1   E-commerce

E-commerce refers generally to all forms of transactions relating to commercial activities that are based upon the processing and transmission of digitized data, including text, sound and visual images. It extends beyond the boundaries of a single enterprise and can be applied to almost any type of business relationship. It is far more than business-to-customer interaction over the Internet, and those who deploy solid business-to-business Internet-enabled manufacturing technologies to fulfill the instant demand and mass customization expectations generated by e-commerce will have a definite advantage.

E-commerce and supply-chain collaboration through the Internet align and strengthen the outward view, impression and operations of an organization. An outward look, without an inward look to the operation itself, reduces the full potential benefits of e-commerce and supply-chain collaboration.

E-business customers want customized orders, more order information and faster response from the manufacturing supply chain, unlike other traditional customers who order from the already produced stock (refer Figure 1.1). It therefore requires an investment in a new generation e-manufacturing system, which provides speed, flexibility and visibility to the entire enterprise and connects e-business orders to real-time production processes.

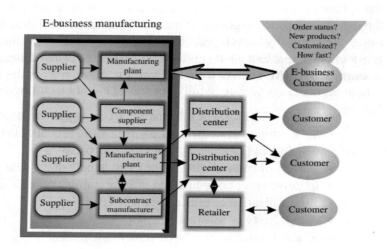

Figure 1.1
The e-manufacturing challenge

1.5.2　Industry drivers

Global market

- Customers now search the world
- Logistics systems are improving.

Technology is changing very fast

- Digital technology revolution
- Mobile technology exploring
- New economy trends are real.

Demanding and fickle customer

- Price, quality, delivery and service
- E-business and e-auction are changing the rules.

1.5.3　Company drivers

Cost efficiency

- Need for size and scale
- High global infrastructure costs.

Fast changing product lines

- Some are exploding and some dying
- Mechanical/electrical/electronic shift
- Software, increasingly the key driver.

Rising business complexity

- Convert to e-manufacturing
- Conversion to e-business.

1.5.4　Security

Security needs to be managed from the start. Protection mechanisms include network security, platform security, application security, client authentication and authorization. Many Internet applications do not work through firewalls or require special modifications to work, as security was not designed into the solution from the beginning. Another way to address the security issue is with dedicated fibre. Big companies may consider dedicated fibre while implementing system.

1.5.5　Seamless integration

E-manufacturing requires seamless 'sensor to boardroom and beyond' systems integration for maximum benefit delivery (refer Figure 1.2).

The challenge is that most enterprise systems do not integrate well with operations due to the barrier created by the MRPII (refer Figure 1.3).

Another problem is that most enterprise systems are patched together and not well integrated. 'Best of Breed' solutions are costly to implement, complex to manage and require never-ending integration. Businesses are demanding more value, less risk and better integration for a competitive advantage. They require affordable enterprise-wide business systems that really work, build-to-order manufacturing systems, supply chains that run at Internet speed and integrated, open-architecture systems that can be implemented quickly and applied seamlessly, producing higher productivity, less risk and higher returns from the software investment.

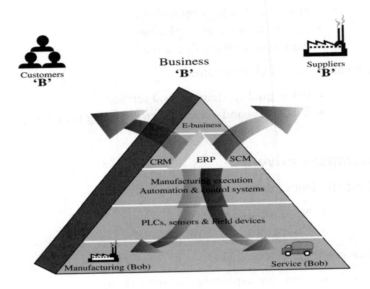

Figure 1.2
E-manufacturing – seamless system

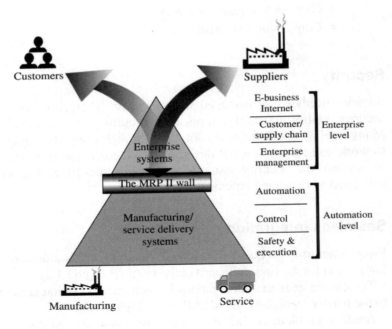

Figure 1.3
E-manufacturing challenge

1.6 E-manufacturing benefits

An 'E-manufacturing' survey report claims that organizations improved their performance by the following amounts through the implementation of an e-manufacturing strategy. This is compared to the historical 'lean manufacturing enterprise including six sigma' and classic 'MRPII enterprise' strategy implementations. This survey just indicates the potential benefits a well-implemented e-manufacturing system can provide to a business.

	Average Plants Classic MRPII Manufacturing	Best Plants Lean Enterprise and Six Sigma	E-Manufacturing Plants
Lead Time	3 months	3 days	12–24 h
Quality Level (Defects)	500 ppm	50 ppm	3 ppm
Inventory Turns	5 × per year	10 × per year	20 × per year
Delivery Performance	90%	98%	99.9%
Sales Growth	3%	8%	10%
RoS%	10%	16%	20%

1.7 E-manufacturing and supply chain

E-manufacturing systems enable collaboration within the ERP/SCM environment without overlap.

ERP/SCM	E-Manufacturing
Financial systems	Waste and downtime tracking
Procurement	Product tracking
Finished goods, raw materials	Production management
Customer service	Control systems integration
Customer orders	Process history
Capacity planning	Real-time quality management
Shipping and logistics	Shop floor metrics
Warehouse management	Decision support
Supply-chain planning	Shop floor user interface
Scheduling	

It has been stated that in the net age, the company with the best supply chain in terms of cost and quality is likely to be the winner. This is precisely why more and more manufacturing enterprises globally are outsourcing for the best quality materials.

The SCM focuses on those tasks that add real value to the product and give maximum profits to the organization. Global businesses are contributing toward the shift of the supply chain from a position where SCM was critical to cost and quality to one where it is becoming one of the most powerful ways for companies to offer greater and differentiated value to customers.

In the past, a great supply chain presupposed long-term relationships with vendors, with the enterprise and suppliers working together to improve design, boost quality, reduce costs and share benefits. Now, in the Internet market, every player is, necessarily, a global one and manufacturers are likely to discover a low cost, high-quality supplier who fits the bill. The Internet is changing the chain.

- Predetermined pricing is giving way to auction-based bidding for the best price.
- Sourcing is becoming global as suppliers all over the world sell on the Internet.
- Long-term partnerships with vendors are making room for deal-to-deal relationships.
- Buyers are being forced to compete with one another to secure the best and cheapest suppliers.

2

History of business automation

Learning objectives

- To provide a brief overview of business automation technology evolution.
- To obtain an understanding of what the various automation technologies are all about.
- To understand how the evolution of various automation technologies influenced others.

2.1 Introduction

If we look at the evolution of business automation, we will notice that it consists of control systems, accounting systems, execution systems, networks, the Internet and supply-chain systems. These, although they exist independently from one another, are still related and influence the development and improvement in others.

Most of these evolved over a period of time, getting more efficient and less maintenance intensive at a constant rate. Others moved in leaps and bounds, as some functions only became possible with the improvement of technology in one or more of the other related aspects.

Most of the above started development at a faster pace with the onset of the electronic age. It became possible to control processes without the need to manually open and close valves, and to keep track of income and expenses without having to manually record every transaction in a physical book.

The next big move was away from hardware toward software. Where control logic was historically built using circuit boards and switches, logic could be built using software code, making change easier. This also meant that control logic could be separated from hardware, making the hardware cheaper and easier to replace when it broke down, as long as the software logic was kept safe.

The move away from mainframe computers to PCs made it possible for almost every business owner, plant supervisor and manager (regardless of company size) to have a computer to assist in the management and control of their business.

More sophisticated communication and network technology, such as fiber optics and satellite communication made it possible for the transfer of large amounts of data over great distances at tremendous speeds. These assisted big corporations in planning and controlling remote and even international subsidiaries almost as if they were in the building next door.

The advent of the Internet made it possible for corporations to transfer and access information to and from subsidiary companies without having to invest in dedicated communication networks, as data could be posted through a multitude of different routes and servers, and still arrive at the intended destination intact.

This chapter briefly describes the evolution of the various components of business automation as the above influenced them.

2.2 Evolution of measurement instrumentation

The first instruments used by the mankind were mechanical in nature and the principles on which these instruments worked are still relevant today. The earliest scientific instruments used the same three essential elements as our modern instruments do.

These elements are:

- A detector
- An intermediate transfer device
- An indicator, recorder or a storage device.

If instruments did not evolve technologically, it would not have been possible for the evolution of manual control toward automated process control systems. The following instruments indicate the evolutionary steps in measurement instrumentation, although instruments using these principles are still in use today.

2.2.1 Mechanical instruments

These are unable to respond rapidly to measurements of dynamic and transient conditions. These instruments have moving parts that are rigid, heavy and bulky, and consequently have a large mass. Most of them are potentially noisy.

2.2.2 Electrical instruments

Electrical methods of indicating the output of detectors are more rapid than mechanical methods. An electrical system normally depends upon a mechanical indicating device.

2.2.3 Electronic instruments

Electronic instruments have higher sensitivity, faster response, greater flexibility, lower weight, lower power consumption and high degree of reliability.

Signals that vary in a continuous fashion and take on an infinite number of values in any given range are called analog signals.

Signals that vary in discrete steps and thus take up only finite different values in a given range are called digital signals.

Instrumentation exists for two purposes, indicating and recording. Indicating instrumentation exist to only show measurement values at a specific moment in time, while recording instruments record measurements over a period of time to make historical information and trends available to plant operators and management.

2.3 Evolution of control systems

2.3.1 Pneumatic process control

Some of the first attempts at process control were in the form of pneumatic control. Pneumatic process control made use of compressed air and multi-directional valves to affect process control (refer Figure 2.1).

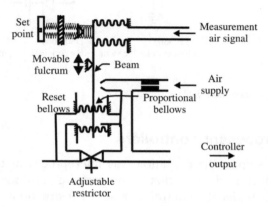

Figure 2.1
Pneumatic Controller

The effectiveness of this control method was very dependent on the original logic design of the control loops and the quality of the air supply. Maintenance and problem-solving on pneumatic control loops was also difficult, as it depended heavily on the version-control of loop-design documents. The valves were also very susceptible to the air quality, as oil in the air made them stick in partially open or closed positions. Today, some pneumatic control systems are still in use, but the skills required to operate and maintain them are very specialized and not easy to find.

2.3.2 Sequential relay control

Sequential relay control used electrical devices and switches to control processes. This method had most of the same maintenance and problem-solving problems (especially in the case of multi-relay systems) as pneumatic controls, except that it did not have to rely on compressed air.

Changes to processes also required changes to the control loops, as they were mostly hard-wired systems. As the switches were mechanical devices, they were also more prone to failure and had high replacement costs.

2.3.3 Electronic controllers

The first electronic controllers followed the principle of sequential-relay control, except that the control loops were embedded in a microchip. These were more reliable, smaller in size and easy to service and replace. Unfortunately, when the process needed change, the whole chip needed to be discarded and a new logic control had to be designed and embedded in a chip, sometimes at a very high cost (refer Figure 2.2).

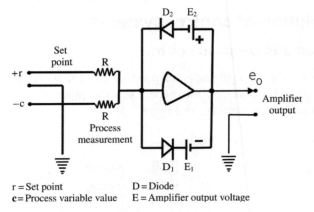

r = Set point   D = Diode
c = Process variable value   E = Amplifier output voltage

Figure 2.2
Typical two-position controller

2.3.4   Microprocessor controllers

The next step in the evolution was microprocessor controllers. A microprocessor is a semiconductor device that can perform arithmetic, logic and decision-making operations under the control of a set of instructions stored in memory. These systems used software as basis and were thus much more adaptable to changes in the process (refer Figure 2.3).

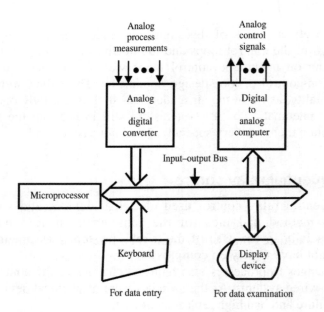

Figure 2.3
Microprocessor based controller

2.3.5   Programmable logic controller

Engineers at General Motors in the USA developed the original concept of the programmable logic controller PLC around 1968. At that time microprocessors were beginning to appear; however, they were difficult to use and required significant training and expertise. Most industrial control systems at that time were based on hard-wired circuitry constructed from electromechanical relays, timers and counters, etc. The PLC concept was developed as a microprocessor-based, programmable substitute for these hard-wired systems (refer Figure 2.4).

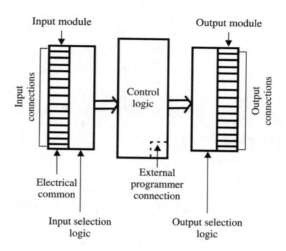

Figure 2.4
Programmable controller

2.3.6   Advanced process control

The next step in control systems is advanced process control (APC) or multi-variate control. It is used to control a collection of unit operations whose individual operations are coordinated to meet specified set of objectives i.e. each unit operation is controlled independently, with set points to optimize its operation.

Advanced process control is used to maximize the output of one or more of the unit operations by manipulating the set points of all the unit operations. In this way, the profit is maximized, as the products with the highest margins are maximized and the products with the lowest margins are minimized. This is normally done using some linear programing techniques and variables such as product specifications, energy consumption, product costs, product demand, etc.

An example of an APC system on a fractionator's complex is illustrated below, where only the APC outputs to the process are indicated. Without APC plant personnel has to determine the product specification of each column as well as the feed division everyday (total 15 variables, two product specifications per column and three feed distributions). With APC, only four specifications are required to optimize the operation (refer Figure 2.5).

Advanced process control improves performance of operations when conventional control systems have the following disadvantages:

- All controllable disturbances are not kept out of the system.
- For uncontrollable disturbances in related processes, there is no compensation, and manually adjusted controller set points are on variables of only indirect interest.

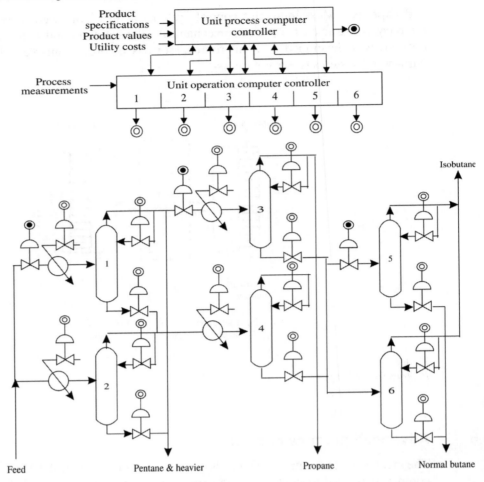

Figure 2.5
Advanced process control

The control of the complex continuous operations such as the one indicated above is divided into three levels: dynamic regulation, set-point control and performance optimization. Dynamic regulation includes the function of conventional controllers, which manipulate valves to regulate such variables as temperatures, flows, levels and pressures. Set point adjusts the set points of the first-level controllers to achieve the specified values for the process outputs. The unit-performance optimization, computes the specific process outputs, which results in optimum operation for the unit with maximum profit.

2.4   Evolution of process visualization systems

2.4.1   Supervisory control and data acquisition and distributed control systems

Supervisory control and data acquisition (SCADA) is not a full control system, but rather focuses on the human interface and supervisory level. It is basically a software package that is positioned on top of hardware to which it is interfaced, via PLCs to make information visually available to plant operators.

Systems similar to SCADA systems are seen in factories, treatment plants, etc. These are often referred to as distributed control systems (DCS).

Both are systems of process control automation that receive input from field devices in a process plant and directs output to such devices in order to control the process. They have similar functions, but the field data gathering or control units are usually located within a more confined area and more complicated with DCS. Communications may be via a local area network (LAN), and are reliable and of high speed. A DCS system usually employs significant amounts of closed loop control.

Supervisory control and data acquisition system can be used to monitor and control plant or equipment. The control may be automatic or initiated by operator commands. Supervisory control and data acquisition system read data from PLCs and other hardware and then analyzes and graphically presents that data to the user. Supervisory control and data acquisition systems are normally able to read and write from and to multiple sources of data. They can log incredible amounts of data for later review. This is helpful for solving problems as well as providing information to improve the process (refer Figure 2.6).

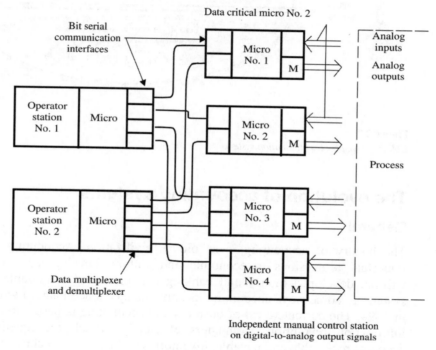

Figure 2.6
Distributed control system

2.4.2 Collaborative process automation systems

The latest concept in process automation is collaborative process automation systems (CPAS). This is not any specific system, but rather a concept of structuring and linking process automation systems with the rest of the organizational system infrastructure. The process industries are entering the collaborative manufacturing era as the existing process control systems cannot deliver the desired level of functional autonomy and coordination required to be competitive. The level of collaboration highlights the need for business performance requirements and emerging technologies to converge into a CPAS. The CPAS provides a strategic competitive advantage for both process manufacturers and their suppliers (refer Figure 2.7).

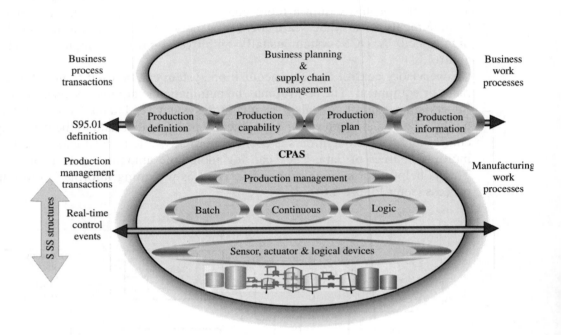

Figure 2.7
Collaborative process automation systems

2.5   The evolution of accounting systems

2.5.1   General

The history of accounting is as old as civilization. Accounting is among the most important professions in economic and cultural development. Accountants invented writing, developed money and banking and innovated the double entry book-keeping system. In some ways accounting has not changed since Paciolli wrote the first textbook in 1494. The current world of business and accounting is based on the computer and the information revolution. Computers efficiently crunch the repetitive transactions of accounts receivable and payable, inventories and payrolls. Technology exploded and new industries (and billionaires) created: personal computers, networks and the Internet, and 'killer applications' software such as the electronic spreadsheet. These evolved over time to the sophisticated systems we see today (refer Figure 2.8).

2.5.2   Spreadsheets

VisiCalc

Dan Bricklin is considered the 'father' of the electronic spreadsheet. VisiCalc was the first application for personal computers. A spreadsheet is a large sheet of paper with columns and rows that lays everything out about transactions for a business person to examine. It spreads or shows all of the costs, income, taxes, etc. on a single sheet of paper for a person to make a decision. An electronic spreadsheet organizes information into software-defined columns and rows. The data can then be 'added up' by a formula to give a total or sum. The spreadsheet program summarizes information from many paper sources in one place and presents the information in a format to help a decision maker see the financial 'big picture' for the company.

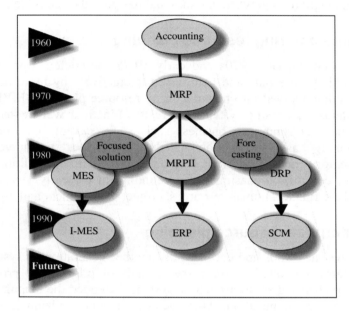

Figure 2.8
The evolution of manufacturing system

Lotus 1-2-3

Mitch Kapor developed Lotus and his spreadsheet program quickly became the new industry spreadsheet standard. Lotus 1-2-3 made it easier to use spreadsheets and it added integrated charting, plotting and database capabilities. Lotus 1-2-3 established spreadsheet software as a major data presentation package as well as a complex calculation tool. Lotus was also the first spreadsheet vendor to introduce naming cells, cell ranges and spreadsheet macros.

Microsoft Excel

The next milestone was the Microsoft Excel spreadsheet. Excel was originally written for the 512K Apple Macintosh in 1984–1985. Excel was one of the first spreadsheets to use a graphical interface with pull down menus and a point and click capability using a mouse-pointing device. The Excel spreadsheet with a graphical user interface (GUI) was easier for most people to use than the command line interface of PC-DOS spreadsheet products. Many people bought Apple Macintoshes so that they could use Bill Gates' Excel spreadsheet program.

2.5.3   Material requirement planning

Between the late 1960s and early 1970s, material requirement planning (MRP) was evolved from accounting systems, which was intended to help manufacturers for better planning of material availability.

Restrictions with MRP

Material requirement planning presented manufacturers with some restrictions. The MRP assumes deterministic lead-time and as such created over- or under-stock situations in some companies. It also assumes infinite capacity, resulting in material being purchased for orders that the company do not have the capacity to fill within the planning period. It also has limited lot-sizing capacity, resulting in a lot of re-planning on the factory floor. It also ignores safety stock issues and has poor data consistency.

2.5.4   Manufacturing resource planning

Between the late 1970s and early 1980s, computers were more powerful, capable of handling more data and being used interactively by more people. Material requirement planning evolved into manufacturing resource planning (MRPII) as shop floor reporting systems; purchasing systems and related functions were added to it. It plans all resources needed for running a business such as forecasting, customer order entry, production planning/master production scheduling, product structure/bill of material processor, inventory control and MRP. It also improved with the addition of capacity planning, shop floor control, purchasing and accounting. In the early 1990s, MRP-II was extended to include finance, human resources, project management and others.

2.5.5   Enterprise resource planning

In the late 1980s and early 1990s, another generation of systems was developed. These systems attempted to solve the 'islands of information problem' by providing broad comprehensive solutions. These systems extended the manufacturing oriented concept to the entire organization. Manufacturing resource planning systems became enterprise resource planning (ERP) systems, and are some of the most prevalent systems in use throughout all industries.

2.6   Evolution of computers

2.6.1   Mainframe computers

Mainframes were powerful machines that allowed organizations to begin to tap into the power of information. Companies were able to automate manual tasks, shorten the time to market for new products, run financial models that enhanced profitability, etc. However, these high-powered machines came with high-price tags. The cost of entry into the mainframe market was typically several hundred thousand dollars to several million. The minicomputer began to bring similar capabilities to a lower-price point, but the minicomputer configurations were often over a hundred thousand dollars as well. The mainframe model consists of centralized computers, usually housed in secure climate-controlled computer rooms. End users interfaced with the computers via 'dumb terminals'. These dumb terminals were low-cost devices that usually consisted of a monitor, keyboard and a communication port to talk to the mainframe. Initially these terminals were hard wired directly to communication ports on the mainframe and the communications were asynchronous.

Disadvantages of mainframes were:

- Character-based applications
- Lack of vendor operating system standards and interoperability in multi-vendor environments
- Expensive (high cost of entry)

- Potential single point of failure (non-fault-tolerant configurations)
- Timesharing systems – potential bottleneck.

2.6.2   Personal computers

The personal computer introduced the graphical user interface to users along with thousands of Windows-based applications that provided new levels of true employee productivity gains in a user-friendly environment. The low-cost PC has been widely adopted and along with it came distributed computing – for better or worse. Distributed computing simply means that a companys' central processing units (CPUs), and hence computing power, was extended out of the computer room to the users desktop. Now a user is not dependent on the availability or load of the mainframe because most of the functions are now performed at the desktop PC.

Advantages of PC Computing are:

- Standardized hardware
- Standardized operating systems (highly interoperable)
- Scalability
- GUI interface
- Low-cost devices (compared to mainframes), low cost of entry
- Distributed computing
- User flexibility
- High-productivity applications.

2.7   Evolution of networks

Exchange of information is required in many situations between two users. In such situations, data communication is needed. The concept of merging different computers and communication led to the first computer networks. A computer network is a collection of computers and peripheral devices (the network components) connected by communication links that allow the network components to work together. The network components may be located at many locations or within the same office.

Computer networks can be defined as an interconnected collection of autonomous computers. The term autonomous means it is an independent computer system. Normally, these computers are connected through a copper wire. Depending on the applications and the mode of processing, different types of connectors are used i.e. fiber optics, microwaves and cellular or satellite communication.

Computer network connections can be of two types:

1. *Broadcast networks:* These have a single communication channel which all the computers on the network share
2. *Point-to-point networks:* These networks have many connections between the source and the destination computers.

Classification of computer networks

Based on the size, capability, communication medium and cost, the computer networks are classified into two categories:

1. Local area networks
2. Wide area networks (WAN).

2.7.1   Local area network

A personal computer works in a stand-alone state with its own CPU. Organizations need a multi-user environment which allows for sharing of data as well as expensive resources i.e. printers and storage. It is an interconnection of autonomous computers within a single building or a small campus. Local area networks are privately owned computer networks. The range of LANs may vary from 10 m to 1.5 km. A LAN provides modularity, connectivity, superior performance, security and reliability in its operation. The LANs are normally small-sized networks. They work on the principle of 'load sharing' because the programs to be executed are downloaded into the personal computer's memory. In LAN transmission technologies, each computer usually contains a network interface device that connects the computer directly to the network medium such as a copper wire or coaxial cable.

2.7.2   Wide area network

Wide area networks are composed of a number of computers that are geographically spread over a wide area. The WANs emerged in the late 1960s. They operate usually at slower speeds (56 Kbps to 155 Mbps). Some examples of a WAN are the INTERNET, NICNET and COALNET.

2.7.3   Resurgence of the mainframe

As PC technology has improved dramatically, the power of the PC has risen to the point that they can perform enterprise level functions. In a distributed computing environment, however, these capabilities often go to waste. Companies cannot afford to put the latest and greatest technology on every desk, but in a distributed computing configuration, one could argue that they would have to in order to maximize employee efficiency. This became apparent and led to the invention of client/server computing.

On a PC LAN, there are usually desktop PCs (called clients) and one or more servers for file, printer and application sharing. True client/server computing is achieved when application processing (e.g. a database) is split between a client and a server. The server is the host of the database and it processes the query on behalf of the client. The results of the query are then sent back across the network to the client for final analysis and manipulation. This is in contrast to a non-client/server configuration in which the data itself, not just the query and the results, is sent across the network to the client machine for query processing. Client/server computing begins to tap into the potential power of a high end centralized PC, much like the mainframe model, but with GUI interfaces and state-of-the-art applications.

In the client/server environment we had all of the management and maintenance issues the PC LAN consists of, but we started to tap into the power of centralized systems while reducing network bandwidth requirements. In a mainframe environment the dependence is on a single computer or group of computers that can easily be centrally managed and maintained. This configuration has the additional advantage of being more secure not only because of the physical security of the computer room, but also because of the end-users lack of ability (not total inability) to introduce viruses into the system. The cost of virus protection and eradication today is costing companies hundreds of millions of dollars annually.

The technologies today are shifting more of the responsibility back to high-powered, highly reliable, centralized systems. This achieves significant reductions in the total cost of ownership by deploying low-cost, low-maintenance device at the desktop and centralizing the management functions once again.

2.8   Evolution of the Internet

2.8.1   Advanced research project network

Advanced research project network (ARPANET) was the first network created by the US Department of Defense in 1968. It was an experimental network designed to support military research – in particular, research about how to build a network that could withstand partial outages (such as a nuclear bomb attack) and still function. The goal was to work from any location on the network and exchange data. The ARPANET served as the basis for early network research and was the backbone during development of the Internet. In 1980, the National Science Foundation (NSF) developed the NSFNET network based on ARPANET IP technology (refer Figure 2.9).

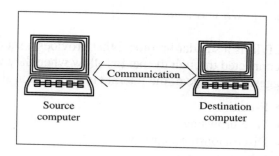

Figure 2.9
Arpanet model

Over time it became the largest network of networks: 'INTERNET' and uses the Internet protocol (IP) standards to enable communication.

2.8.2   Internet

The Internet is a massive collection of computer networks that connect millions of computers, through which people interact and exchange information.

> THE WORD 'INTERNET' IS BUILT FROM THE WORDS
> **INTER**CONNECTION + **NET**WORKS = **INTERNET**

All the systems in the networks are able to communicate with each other through a common set of rules called a protocol (refer Figure 2.10).

The Internet is not a single network, but a patchwork of networks run by different organizations using a system of packet switching for data transfer. The Internet is characterized by a set of protocols, known as the TCP/IP protocol suite that is fast becoming the world's most popular open-systems protocol suite. This is because it can be used to communicate across any set of interconnected networks and is equally well suited for LAN and WAN. The two best known communication protocols in the suite of Internet protocols, transmission control protocol (TCP) and the Internet protocol (IP), are not the only protocols within the Internet protocol suite, but it also specifies common applications such as electronic mail, terminal emulation and file transfer.

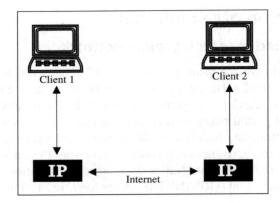

Figure 2.10
Communication between two clients

The TCP/IP is a standard protocol that provides a set of rules and regulations for all the systems connected through the net to follow when they want to exchange data.

The biggest advantage of the Internet is information sharing through:

- E-mail
- Web browsing
- Discussions/newsgroups
- E-business, etc.

2.8.3   Intranets

An Intranet is a self-contained, internal network within an organization. Intranets offer the means to be used as a business or organizational tool using the standard protocols of the internet (TCP/IP) and applying it to individual organizational requirements (refer Figure 2.11).

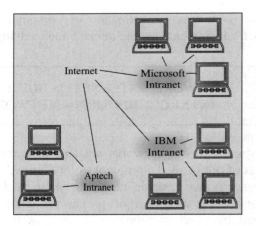

Figure 2.11
Intranet

Intranets offer a new and revolutionary way of communicating – both as a distribution channel and an information source. The most profound aspect of an Intranet is that it can provide everyone in an organization equal access to information.

2.8.4 Extranets

Extranets are business-to-business networks shared by two or more organizations, which accelerates the pace of business and movement toward an extra information economy. This extranet connection provides security to its business partners, suppliers or customers. The connection is typically made outside of the Internet backbone.

2.8.5 World wide web

The world wide web (www) was started at CERN, a particle physics laboratory in Geneva, Switzerland in 1989. To have a presence on the www, a website should have a web address.

2.9 Development of supply-chain management systems

In earlier times, manufacturers were the drivers of the supply chain, deciding the pace at which products were manufactured and distributed. Today, manufacturers are scrambling to meet customer demands for options/styles/features, quick order fulfillment, and fast delivery and manufacturing quality, a long time competitive differentiator. Meeting customers' specific demands for product delivery has emerged as the next crucial opportunity for competitive advantage.

The supply chain encompasses every effort involved in producing and delivering a final product or service, from the supplier's supplier to the customer's customer.

2.9.1 Internal supply chain

Today, SCM tools are used more as value chain management tools, managing and optimizing the internal value chain. This includes managing supply and demand, sourcing raw materials and parts, manufacturing assembly, warehousing and inventory tracking, order entry and order management, distribution across all channels, and delivery to the customer.

Supply-chain management (SCM) tools can address complex interdependencies due to its wide scope; in effect creating an 'extended enterprise' that reaches beyond the factory door. Today, material and service provider, channel partners and customers themselves, as well as SCM consultants, software product suppliers and system developers, are key players in SCM.

2.9.2 Extended supply chain

Supply-chain management easily integrates material planning, budgeting, requisitions, procurement, order handling, inventory handling, logistics and accounting. Organizations have felt the need for going beyond mere transaction processing and automation of business processes. They have misunderstood that the hierarchical organization structures, vertically integrated manufacturing and distribution processes, arm's length relationships with suppliers and customers, inflexible IT systems are inadequate for success. In a complex business situation, SCM can help in identifying and planning resources based on certain organizational constraints, which are dynamic in nature.

ERP packages are insufficient in meeting this specific need of the organization. Even though some supplier and customer information is captured by ERP systems, it is mainly for the internal needs of the organization. A supply-chain system can integrate the supplier and the customer with the organization. Overall, the top driver is improved supply-chain visibility – timely, accurate data about raw material requirement/consumption, production outages and finished products availability.

The Internet provides a simple, standard method for sending real-time dynamic electronic information between applications, over a commonly available, low-cost connection in such a way that both sender and receiver can understand the information.

The impact of this in business terms is that the distance between customers and suppliers has effectively shrunk. The first economic impact was the establishment of business-to-customer (e-commerce) via the creation of 24-h remote electronic storefronts on a global scale. This has led to a rise in customer expectations in terms of speed, personalization and quality of service, and these expectations have to be both managed and met.

The use of the Internet in business planning and control systems has thus become inevitable. The ERP and SCM vendors have been quick to entrench e-commerce principles into their software, and marketers and customers have quickly grown accustomed to only being a mouse click away from one another. The realization of the full potential of Internet technology on operations management, planning and control is, however, in its infant stage. With e-commerce and supply-chain Collaboration changing the nature of product demand, supplier relationships and customer demographics, it is inevitable that the nature of operations management, planning and control systems will have to evolve.

2.10   Evolution of manufacturing execution systems

The entire supply chain revolves around manufacturing thereby optimizing the supply chain requires good information about it. The management of assets on the plant floor requires tight coupling with the broader supply chain. Without adequate interconnections with the plant floor, the supply chain is just guessing about where orders are, what inventory is needed and what the most profitable use of resources should be.

It is no longer company vs company, but supply chain vs supply chain. Automation and control software, MES, plant portal and information systems, and other software aim to provide the interconnectivity with two-way information flow. The typical model for interfacing plant-floor systems with enterprise and supply-chain system is direct database access to a production database. Such a database maintains production information required for SEM, machine maintenance, quality control and management reporting. This information is also available to the enterprise via Intranet and Internet applications.

Material requirement planning was intended to help manufacturers achieve better planning of material availability. Between the late 1970s and early 1980s, computers were capable of handling more data and MRP evolved into (MRP-II) as shop floor reporting systems. Purchasing and related functions were also added. Manufacturing resource planning is the most widely used manufacturing management methodology in the world. Manufacturing resource planning did not address the requirement of forecasting and managing demand in distribution, nor did it do much of a job of managing the shop floor and the many disparate activities that took place there. To address the requirement in distribution, forecasting and distribution resource planning (DRP) were developed.

Manufacturing execution systems (MES) and some function-specific systems, e.g. quality management, evolved. Manufacturers were getting solutions for specific business problems with the help of these systems, but could not take the advantage of data from, or pass data to, other systems due to the lack of integration within the systems. Between the late 1980s and early 1990s MRP-II became ERP. The DRP became SCM and the shop floor solutions evolved into the integrated MES system (I-MES).

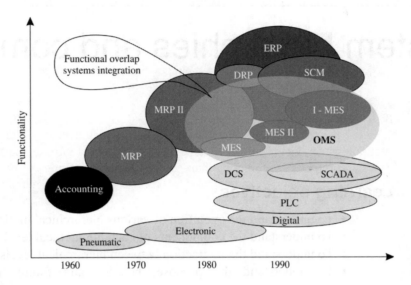

Figure 2.12
Manufacturing system evolution

During the same period, computers gradually replaced manual controls, and process and machine management became increasingly sophisticated. Controlled technology added new precision to real-time execution. Supervisory control and data acquisition systems also became available to manage these controls and to consolidate and evaluate the data they produced. Control specific interfaces became available with the advent of GUIs. During the late 1990s, SCADA systems began to acquire many of the functions of the traditional MES (refer Figure 2.12).

3

System hierarchies and components

Learning objectives

- To understand the evolution of various hierarchical models.
- To understand the components of each hierarchical level.
- To understand the interaction between hierarchical levels.
- To understand the purpose, benefits and future trends of hierarchical components.

3.1   Introduction

The ways in which system architectures have been viewed over the years were heavily influenced by the evolution of technology. As systems became more complex with more functionality, and more and more activities were systemized, the landscape of system architecture has changed. Some of the first architectures used to classify systems had three levels to indicate conceptual architecture layouts. AMR developed a three-tier model based on the plan, execute, and control philosophy (refer Figure 3.1).

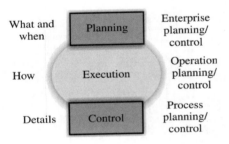

Figure 3.1
Three-level architecture

This model had the following advantages over the CIM model:

- Overlap of functionality between systems
- Information flow also includes horizontal flow

- Built around an integration architecture
- Built by operations people together with IT and engineering.

Supply-chain tools, however, did not quite fit this model. Manufacturing Execution Systems Association (MESA) then developed a four-tier model including supply-chain management (SCM) tools. These they placed above the Enterprise resource planning (ERP) level, as SCM tools are primarily planning tools, but at a higher level than conventional ERP (refer Figure 3.2).

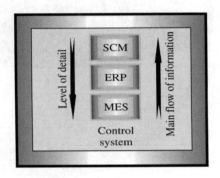

Figure 3.2
Four-level architecture

During this time, most companies (including instrumentation), especially for new manufacturing facilities, started using a five-level model. This model is still widely used in industry, especially by engineering, procurement and contract management (EPCM) companies. These companies use the five-level models to identify subcontractors for specific deliverables during project execution (refer Figure 3.3).

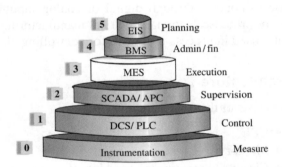

Figure 3.3
Five-level architecture

These two-dimensional models work well for most manufacturing companies, but do not address the needs of product development companies such as in the pharmaceutical industry. ARC developed a three-dimensional model called collaborative manufacturing management systems (CMMS). This model depicts systems on three axes, enterprise, value chain and life cycle (refer Figure 3.4).

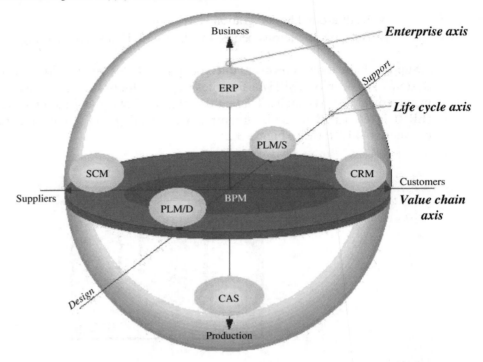

Collaborative manufacturing management

Figure 3.4
Simple Computer-integrated manufacturing model

3.2   Programmable logic controllers

A programmable logic controller (PLC) is a digitally operating electronic apparatus which uses a programmable memory for the internal storage of instructions by implementing specific functions such as logic sequencing, timing, counting and arithmetic to control, through digital or analog input/output modules, various types of machines or processes national electrical manufacturing association (NEMA). The PLC is a computer used in process industries for controlling electromagnetic devices.

3.2.1   PLC architecture

A PLC consists of the following:

- CPU
- Memory
- Circuits to receive and send input/output data.

A PLC microprocessor is reduced instruction set computer (RISC) based and is designed for high-speed, real-time control and is rugged and able to operate in industrial environments. The RISC is an architecture that reduces chip complexity by using simpler instructions. The RISC compilers generate software routines to perform complex instructions, which were earlier done in hardware by complex instruction set computer (CISC) computers. The PLC can perform the function of numerous relays, counters, timers and data storage locations, which physically do not exist but rather are simulated through bits (refer Figure 3.5).

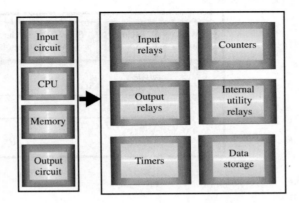

Figure 3.5
Components of a PLC

Input relay

Transistors physically exist and receive signals from sensors, switches, etc. and are connected to the outer world.

Internal utility relay

These are simulated relays with no physical existence. These types of relays do not receive signals from the outside world but perform one particular function at a time.

Counters

These do not physically exist. They are simulated and programed to count pulses up, down, or both up and down.

Timers

These do not physically exist but are simulated with increments varying from 1 ms through 1 s.

- On-delay and off-delay types
- Retentive and non-retentive types.

Output relay

These physically exist and are connected to the outer world sending on/off signals to solenoids, lights, etc.

Data storage

This is used to store data when power is removed from the PLC as well as for temporary storage, as well as for mathematics or data manipulation.

Input Devices	Output Devices
Push buttons	Solenoids
Limit switches	Contactors
Proximity switches	Motors

The PLCs work on continuously scanning programs and the scanning cycle consists of three important steps (refer Figure 3.6).

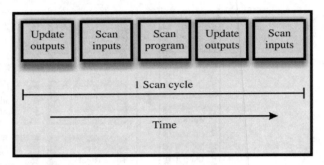

Figure 3.6
PLC operation

The time taken to complete one cycle is called the response time, which depends on the length of the programs and a PLC instruction execution rate. The response time is one important factor when choosing a PLC (refer Figure 3.7).

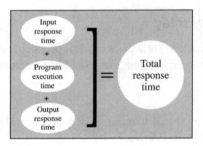

Figure 3.7
PLC response time

3.2.2   PLC trends

There is a stronger demand for nano- and microcontrollers, for competitive prices, enhanced functionality and flexible packaging in the PLC market.

Software-controlled systems provide an integrated configuration, operation maintenance and diagnosis environment, which are not available with conventional control systems. Control systems are embedded in the PC itself and deliver a higher level of flexibility at lower cost but PC reliability is a factor (PLC hardware is very robust), which hindered its inclusion in discrete manufacturing sectors.

3.3   Distributed control system

A distributed control system (DCS) is a control system that is used in complex manufacturing industries. It is an integrated system based on the concept of decentralization.

The DCS controls and monitors inputs and outputs to and from remote terminal units (RTUs). It handles both sequential and analog control, implements it and performs operations, e.g.

- Data gathering
- Data processing

- Data storing
- Monitoring of operational conditions to the operator
- Trend and analysis of data
- Historical record keeping (product/batch tracking)
- System maintenance from a single console
- Reporting of production status and information.

The purpose of the system is to control the process, and methods can differ:

- The control can be real-time based or event driven.
- The control can be of a close-loop feedback system or a specific feedback system.

Distributed control is based on a system which divides the plant or process control system into several nodes of responsibility and each of the nodes are managed by its own optimized processor. The entire system is interconnected with a single network together by communication buses of various kinds. The entire application is stored on one database. The system's functions and risks are divided with processors handling individual areas and tasks but the overall system still functions as one computer platform.

The DCS is based on dedicated hardware and process control and supervision software. Every part of the system, measurement and process controllers, communication links and even operators screens, is duplicated to minimize the risk of failure. If one module fails, the other one immediately becomes the primary module without losing any data. It ensures a high degree of reliability.

The DCS is typically used in large manufacturing establishments requiring significant amounts of control and a high degree of fault tolerance and redundancy. The DCS have developed systems for redundant processors and are designed to work in a single location with local area network (LAN) type links between the cabinets.

3.3.1 Emerging trends in the DCS industry

Open systems

Control system industries are leaning toward complete open systems i.e. hardware should be available with multiple suppliers and software should work on many different computer platforms. Some DCSs have completely open architectures, allowing RTUs to be interchanged between systems of different vendors.

System integration

Integration of the DCS includes the integration with other control systems and other types of systems. The DCS products include interfaces to other products as a standard product feature. System integration enhances the system functionality and value to a manufacturer. The DCSs can offer more than just process control when it is integrated with the enterprise's existing systems e.g. MIS/MES, etc.

Expansion of existing control systems

The DCS, scalable and expandable architecture, can easily accommodate/be expanded by adding work stations/servers and existing systems.

Lowering cost

New technology and architectures allow the DCS to function with the power and flexibility that is normally associated with supervisory control and data acquisition (SCADA) systems but retain the safety and reliability of a DCS.

Internet to leverage DCS

Internet-based systems enhance the ease of integration. Most new systems are Internet-enabled and therefore can act as a common platform between systems. Internet-based systems also enhance functionality since it makes remote access easy.

3.4   SCADA System

3.4.1   Supervisory control and data acquisition system (SCADA) architecture

Supervisory control and data acquisition system is an industrial measurement and control system consisting of a central host (Master Terminal Unit, MTU), one or more field data gathering and control units (RTU), and a collection of software modules (standard or customized) used to monitor and control remotely located data elements.

The SCADA helps with real-time visualization, monitoring, alarm logging, data logging, etc. from local and remote locations. Sensors and switches are connected to the PLC, which collect and accumulate data and transmit this to the SCADA computer. The computer then sends commands to the PLC to operate switches or valves, etc. The control may be automatic or initiated by operator commands.

The data acquisition is accomplished firstly by the RTUs (at a higher rate) and then the MTU scans the RTUs at a lower rate. Data are of three types:

1. Analog data
2. Digital data
3. Pulse data.

The primary interface to the operator is a graphical display of the plant/equipment in real time. Analog data can be shown as a number, or graphically. A high-level graphical overview would look something like Figure 3.8, and selection of any of the areas on the graphics will take the user to a more detailed version for that particular area.

An MTU consists of a base processing unit, operator interface, power supply, modem, I/O cards, relays, enclosure, battery backup, wiring terminals, lightning protection, programing and start-up software. The PLCs can be used in place of RTUs. The PLCs are used for routine and repetitive function whereas RTUs are used for remote control, communication and data acquisition purposes.

The SCADA is a multi-purpose, reliable and efficient management and control tool, which reliably monitors and controls a process without human involvement. The SCADA systems also gather valuable information about the process plant, which can be shared across the enterprise (refer Figure 3.9). It saves money on man-hours, maintenance and natural resources. The SCADA systems can reduce downtime, increase yield and improve quality, safety and record accuracy.

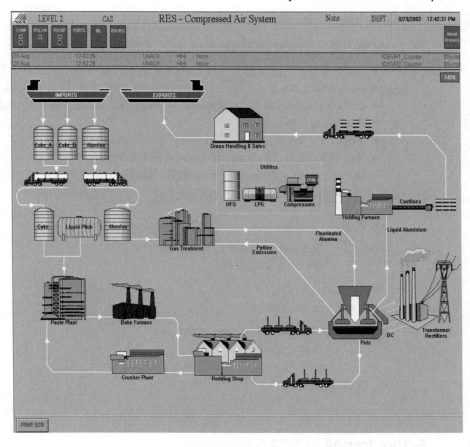

Figure 3.8
SCADA plant view

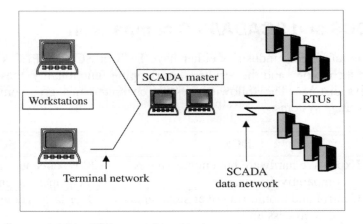

Figure 3.9
SCADA system

3.4.2 SCADA package

Standard SCADA system packages provide:

- User-friendly graphical user interfaces (GUI)
- Automatic control with high processing power and very quick response
- Data logging and historical trends recording with data manipulation
- Alarming functions
- Reporting tools
- Statistical analysis and off-line processing
- Integrated environments with SQL functionality.

Advanced SCADA system have the following features:

- System monitoring

    –Remote status (digital/analog)
    –Pulse accumulation

- System control
- System management

    –System data/status archiving
    –System analysis
    –Sequence of events analysis
    –System loss reduction
    –Activity analysis.

3.4.3 Future trends in SCADA systems

- Future SCADA systems will be based on open architectures
- Enable easy expansion and tailor made functionality
- Enable integration with other enterprise systems
- Have lower total cost.

3.5 DCS and SCADA/PLC comparison

Manufacturing industries either have DCS or SCADA/PLC systems. With the advances in technology and the resulting overlapping functionality, these systems became less and less distinct. The following system characteristics give some relevant features of the average systems.

DCS	SCADA/PLC
Expensive hardware but engineering is comparatively cheaper	200% cheaper than DCS Hardware is cheaper but engineering is expensive
Control and monitoring over small areas e.g. a process unit	Over large geographical areas
Data transfer via LAN infrastructure	Use leased telephone lines or radios
Large, extensive applications with many control and data transfer Analog control processing	Small medium-sized applications with majority open/close control

DCS	SCADA/PLC
Direct control, output directly to field actuators	Send set points to local controllers
DCS was a single vendor solution	With SCADA the connection with the field is done by third party hardware and software
Focused in process industry	Focused in discrete production industry
Inter-related continuous complex processes	Batch processing with low level of process interaction
Robust hardware and software are the part of the equipment on the shop floor and are fixed	Operating system software and the processor used in a SCADA PC undergo quick changes due to heavy competition
Process computer on the shop floor and PCs in the control room	PLC platform for process control and a PC platform (SCADA) for display
Data acquisition is event driven, rely on change. The RTUs can typically operate for extended periods of time w/o communications with the 'Host'	Data acquisition via one database with fixed scan cycle for each data point
Application stored on one database	Application data are divided over several databases
Fast with complex control	Fast when used in logical (on/off) application
Predictable, real time	Not completely predictable
Handle many controls	Limited number of controls

The goals of DCS and SCADA are quite different. It is possible for a single system to be capable of performing both DCS and SCADA functions, but few have been designed with this in mind, and therefore they usually fall short somewhere. It has become common for DCS vendors to think they can do SCADA because the system specifications seem so similar, but a few requirements paragraphs about data availability and update processing separate a viable SCADA system from one that would work OK but for the real world getting in the way.

The DCS is process oriented; it looks at the controlled process (the chemical process plant) as the center of focus, and it presents data to operators as part of its job. A DCS operator station is normally intimately connected with its I/O (through local wiring, fieldbus, networks, etc.). When the DCS operator wants to see information, he usually makes a request directly to the field I/O and gets a response. Field events can directly interrupt the system and advise the operator. The DCS is always connected to its data source; so it does not need to maintain a database of 'current values'. Redundancy is usually handled by parallel processing.

The SCADA is data-gathering oriented; the control center and operators are the center of focus. The remote equipment is merely there to collect the data, though it may also do some very complex process control. The SCADA must still operate when field communications have failed. The 'quality' of the data shown to the operator is an important facet of SCADA system operation. The SCADA systems often provide special 'event' processing mechanisms to handle conditions that occur between data acquisition periods. The SCADA needs to get secure data and control over a potentially slow, unreliable communications medium and needs to maintain a database of 'last known good values' for prompt operator display. It frequently needs to do event processing and data quality validation. Redundancy is usually handled in a distributed manner.

These underlying differences prompt a series of design decisions that require a great deal more complexity in a SCADA system database and data-gathering system than is usually found in DCS. The DCS systems typically have correspondingly more complexity in their process-control functionality. The SCADA database architecture is significantly different from the DCS data architecture. The SCADA system is event driven, while DCS is process state driven.

3.6   Hybrid control systems

A variety of disparate systems are supplied by different vendors. Hybrid control system (HCS) are component-based automation architectures, which can be tailor-made depending upon the requirements of the user. The HCS are easily replicated and modified. Vendors are also capable of assembling components (regulatory control, sequence control, advanced process control, and alarm management) for broader applications in many industries. Hybrid systems can utilize PLCs, DCSs, PC software, SCADA and man machine interface (MMI) software in the same architecture.

3.6.1   Advantage of integrated control systems

A PC controlled PLC has many advantages. Some of them are listed below:

- Easy troubleshooting
- Integration with other systems across the enterprise
- Easy to program and use, Windows interface
- Lower total cost
- Data manipulation and storage and reporting characteristics
- SCADA functionality and reporting.

3.6.2   PLC in a Windows environment

Software control on the factory floor has been made viable since the development of Windows NT combined with faster processor and expanded memory capacities. It is an integrated solution comprising operator interface, data processing, communication and control on a single PC.

3.6.3   Advantages of HCS

- Lower I/O cost
- Easy programing
- Troubleshooting is easy
- Better interfaces
- Higher functionality and performance
- Easy maintenance
- Modularity.

3.6.4   Automation of business transformation

PLC prices and margins are falling, as vendors are moving toward commodity products (refer Figure 3.10). Successful business models are:

- Delivery of systems based on low-cost automation components
- High value-added assembly of complete automation systems with integrated applications.

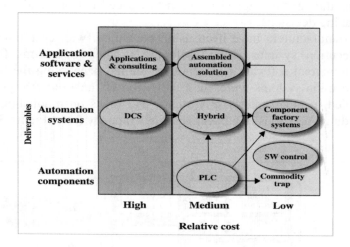

Figure 3.10
Automation business transformation

The different types of products compete based on a different combination of features.

Automation value table

Key Elements	DCS	Hybrid	PLC+MMI	SW Control
Price	4	3	2	1
Functionality	3	4	1	2
Configurability	4	3	1	2
Diagnostics	4	4	1	3
Connectivity	2	4	1	3
Service	4	4	2	1
SW applications	4	4	2	1
Field Acceptances	4	2	3	1

Scale: 1(low) to 4(high).

3.7   Manufacturing execution systems

With the continuous change in customer behavior and requirements, manufacturers are forced to produce or deliver a tailor-made product or service for an individual customer in a shorter period of time. It is often a question of survival in a cut throat competitive market.

To maintain competitiveness in this fast-paced business environment, manufacturers are always looking for ways to increase productivity to a higher level by reducing inventories further. That is to run the plants with the best possible scheduling at the lowest cost. Every manufacturing organization is a complex web of activities and information flow.

Most of the manufacturing plants' shop floors are managed inefficiently where work is piled up due to improper coordination of factory resources (people, information, materials and tools). It causes confusion amongst the workers on the shop floor whilst the entire production control is under constant challenge to meet customer requirements. The manufacturer is trapped between fast-paced demand and lagging information useful for making decisions.

Manufacturers have been using computers with custom-built specific Manufacturing execution systems (MES) software in their operations for a long time. However, commercial software products have started penetrating the MES market. Manufacturing execution systems represent a collection of functions that provide unique benefits when added to other enterprise systems, and focus on production activities – the core value adding operations of a manufacturing company (refer Figure 3.11).

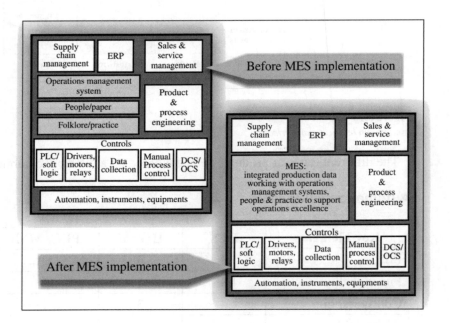

Figure 3.11
Plant information model

3.7.1   What is MES?

'Manufacturing Execution Systems (MES) deliver information that enables the optimization of production activities from order launch to finished goods. Using current and accurate data, MES guides, initiates, responds to, and reports on plant activities as they occur. The resulting rapid response to changing conditions, coupled with a focus on reducing non-value-added activities, drives effective plant operations and processes. MES improves the return on operational assets as well as on-time delivery, inventory turns, gross margin, and cash flow performance. MES provides mission-critical information about production activities across the enterprise and supply chain via bi-directional communications.' MESA (Manufacturing Execution System Association founded in 1992)

'MES provides facility-wide views of critical production processes and product data in such a way that it is usable by supervisors, operators, management and others in the enterprise and the supply chain. MES guides production activities to meet global standards. MES provides a current view of what is possible in production, providing real-time key information for supply chain management and sales activities.

Manufacturing Execution Systems (MES), are information systems that reside on the plant floor, between the planning system in the offices and direct industrial controls at the process itself.' AMR (Advanced Manufacturing Research founded in 1994)

'A manufacturing execution system (MES) is an on-line integrated computerized system that is the accumulation of the methods and tools used to accomplish production.' McClellan, M. 1997.

3.7.2   MES functionality

The MES is not a single function, it has functions that support, guide and track each of the primary production activities (refer Figure 3.12).

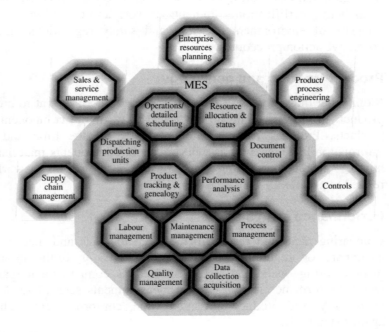

Figure 3.12
MES functional model

MESA International has identified eleven principal functions of MES.

Operations/detail scheduling

Sequencing and timing activities for optimized plant performance based on finite capacities of the resources. It provides sequencing based on priorities, attributes, characteristics, recipes associated with specific production units at an operation e.g. shape of color sequencing which, when scheduled in sequence properly, reduce set-up. It recognizes alternative and overlapping operations in order to calculate exact time or equipment loading and adjust to shift patterns.

Resource allocation and status

Guiding what people, machines, tools, and materials should do, and tracking what they are currently doing or just have done. It provides a detailed history of resources and ensures that equipment is properly set up for processing, and provides status in real time. The management of these resources includes reservation and dispatching to meet operation-scheduling objectives.

Dispatching production units

Giving the command to send materials or orders to certain parts of the plant to begin a process or step; it can alter as well as control the prescribed schedule or the amount of work in process at any point on the factory floor. Rework and salvage processes are available, as well as the ability to control the amount of work in process at any point with buffer management.

Document control

Managing and distributing information on products, processes, designs or orders, as well as gathering certification statements of work and conditions. It includes the control and integrity of environmental, health and safety regulations, and ISO information like corrective action procedures.

Product tracking and genealogy

Monitoring the progress of units, batches or lots of output to create a full history of the product. Product tracking allows traceability of the components and usage of the end products. Provides the visibility of work all the time and its disposition. Status information includes, who is working on it, components materials by supplier, lot, serial number, current production conditions, and any alarms, rework, and other exceptions related to products.

Performance analysis

Comparing measured results in the plant to goals and metrics set by the corporation, customers or regulatory bodies. It provides up-to-the-minute reporting of actual manufacturing operations results as well as comparison to past history and expected result. Performance results include measurements such as resource utilization, resource availability, and product unit cycle time, conformance to schedule and performance against standards.

Labor management

Tracking and directing the use of operations personnel during a shift based on qualifications, work patterns and business needs. It provides information of personnel in an up-to-the-minute time frame. It interacts with resource allocation to determine optimal assignment. It tracks time and attendance reporting, as well as indirect activities such as material preparation or tool room work as a basis for activity-based costing.

Maintenance management

Planning and executing appropriate activities to keep equipment and other capital assets in the plant performing to the overall goal. It maintains a history of past events or problems to help in diagnosing problems. It tracks and directs the activities to maintain the equipment and tools to ensure the availability for manufacturing and gives alarm to immediate problems.

Process management

Directing the flow of work in the plant based on planned and actual production activities. Process management monitors production and corrects automatically or provides decision support to operators for correcting and improving in-process activities. These activities are inter-operational and focused mainly on machines.

It tracks the process from one operation to the next and may include alarm management to inform the factory personnel. Process management provides interfaces between intelligent equipment and MES through data collection/acquisition.

Quality management

Recording, tracking and analysing product and process characteristics against engineering ideals. It provides real-time analysis of measurements collected from manufacturing to ensure/assure correct product quality control and identifies problem. It recommends action to correct the problem, correlating the symptom, actions and results to determine the cause. Quality management include SPC/SQC tracking, management of off-line inspection operations and analysis of laboratory information management system (LIMS).

Data collection/acquisition

Monitoring gathering and organizing data about the processes, materials and operations from people, machines or controls. The data are collected from the factory floor manually or automatically from equipment in an up-to-the-minute time frame. It provides an interface link to obtain the intra-operational production and parametric data which populate the forms and records related to production unit.

3.7.3   MES and future trends

The MES is a collection of different types of software products; hence technologies used and planned are also varied a lot, but some general technology trends are evident. The MES is a reasonably low-overhead set of applications. The requirement for MES is a high number of users distributed on a network even in a plant. Many MES functions have run on PCs and workstations (refer Figure 3.13). The client-server technology, robust PCs and fail-safe LAN have been a major benefit to MES (refer Figure 3.14). Almost all commercial products provide application programing interfaces (APIs) to connect to other applications.

MESA International has identified four business characteristics driving MES implementation architectures:

1. *Low capital expenditures:* This factor drives users who change systems in the plant more slowly than the technology evolves, PC-based architectures, and thin-client architecture, in which application logic resides mostly on the server.
2. *High degree of change:* Data capture and archiving becomes important to track the rapid change in a plant. How much data to store actively vs put into an archive becomes a trade-off between operational analysis capabilities and storage and processing burdens.

3. *Short cycle time:*   The speed at which products move through a plant also dictates how rapidly transactions must be processed to measure operational performance.

4. *Functionality flexibility:*   MES products will specialize based on which attributes they can handle and how configurable they are because plants vary widely.

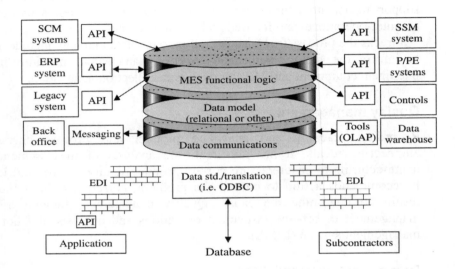

Figure 3.13
MES current technology model

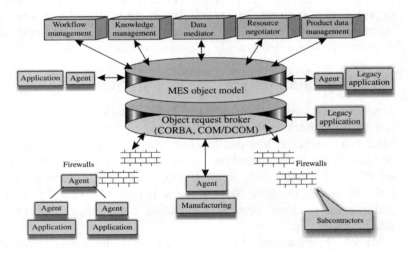

Figure 3.14
MES future technology model

3.7.4   MES for future consideration

MES is now considered by many manufacturing organizations as it addresses business drivers in the following ways:

- To respond rapidly in production to changes in the marketplace
- To increase profits

- Operational personnel are empowered by MES
- As new technologies increase the sensitivity of production processes, MES can provide structure to execute them properly
- Short product life cycle make the MES assistance essential for achieving full yield and plant productivity
- Regulatory and customer compliance demands are becoming difficult without MES
- Improved ROI on assets.

3.8   Enterprise resource planning systems

Enterprise resource planning (ERP) systems comprise a commercial software package supported by a structured management approach, that promises the seamless integration of all the information flowing through the company – financial, accounting, human resources, supply chain and customer information. An ERP system uses, or is integrated with, a relational database system. The database often forms the information backbone of the enterprise.

The ERP is a software mirror image of the major business processes of an organization, e.g. customer order fulfillment and manufacturing. It is an industry term for the broad set of activities supported by multi-module application software. The integrated software helps a manufacturer or service provider manage the important parts of their business.

These parts of business include:

- Product planning
- Parts purchasing
- Maintaining inventories
- Interacting with suppliers
- Providing customer service
- Tracking orders.

3.8.1   How does ERP work?

The ERP system integrates and maps the business processes of an enterprise. It resides on an enterprise's computer network. The users only see the computer screens and perform their duties accessing the system from their desktop computers. The logic and business processes reside in the system database and link all transactions and data according to the business processes while the transactions and records are all logged on the system. The administrator can update the system according to the processes and rules.

As ERP systems evolved out of accounting systems, it is transactional focused with the company financial model at the center. Around this model, the modules that support or influence the financial health of the company were developed. As most organizational activities influence finances, these modules cover most of the organization on a management level. The ERP systems also focus very much on 'best practice', often requiring major organizational change during implementation.

This is different from MES in that MES has a specific manufacturing process focus, and adapts to the organization's operational processes more readily and with less effort than ERP systems (refer Figure 3.15).

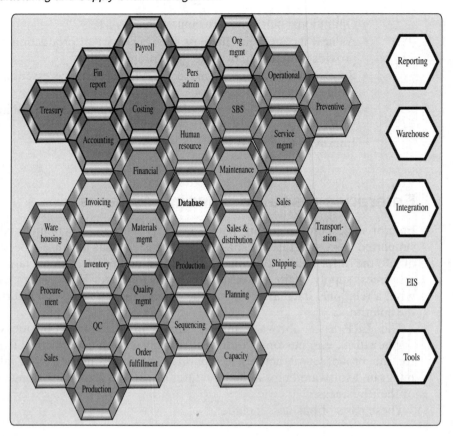

Figure 3.15
ERP Functionality

3.8.2   ERP functionality

Core Subsystems of an ERP System	Key Functions of ERP Systems
Sales and marketing	Finance/Accounting
Master scheduling	Sales and distribution
Materials requirement planning	Budgeting and planning
Capacity requirement planning	Human resource and Personnel
Bill of materials	Fixed assets
Purchasing	Material management and inventory control
Shop floor control	Master scheduling
Purchasing	Work order management
Shop floor control	Logistics and warehouse management
Accounts payable/receivable	Purchasing/sourcing
Logistics	

ERP systems provide the following to an enterprise.

Supply-chain visibility

The entire supply chain revolves around manufacturing; so optimizing the supply chain requires good information about it. Of the various applications of supply-chain solutions

(SCS), available-to-promise, manufacturing planning and production scheduling are most closely tied to both plant operations and to ERP.

Plant decision support

The most frequently cited benefit area is improved decision support within the plant. Most of the information used in plant decision-making comes from the plant itself, not from ERP. However, most plant-level systems deal more with detailed data (temperatures, pressures, flow rates) instead of higher-level business information (pricing, shipment schedules, production orders) that is valuable in decision support.

Integrating plant data into ERP first requires some type of transformation of that data to be more meaningful – production orders into set points, flow rates into production totals, etc. The ERP integration is often the catalyst for automating that transformation.

Better data

Another important area is improved cost and financial accuracy. Improved accurate data may really be the result of improved consistency. Prior to integrating with ERP (refer figure 3.16), it is quite common for companies to implement multiple, disparate systems for capturing similar information for different uses.

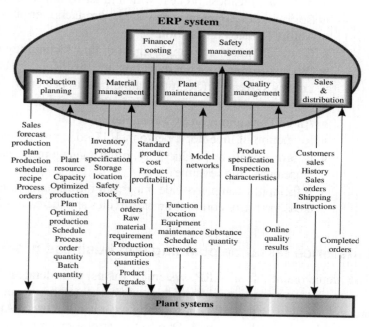

Figure 3.16
ERP/plant messages

Technological

ERP provides data of higher integrity, as disparate systems often cause poor quality of information. These systems are often also not integrated and hence the difficulty in data acquisition and capture. A lot of these systems may also be obsolete and unable to support organizational growth.

Operational

The integrated nature of ERP also assists organizations to sort out poor performance and very high cost structures. As information is more readily available throughout the organization, ERP assists in increasing the responsiveness to customers. It also supports the organizational strategy especially in the light of globalization. Enterprise resource planning also assists in the standardization of complex processes and inconsistent business processes throughout the organization.

Benefits of integration

ERP systems are used to perform business functions that focus on the entire business's transactions. Benefits of an ERP system include:

- Access to an expertise database
- Automatic introduction of latest technologies
- Better customer service, customer satisfaction
- Better project management
- Integration of the system across all departments in a company as well as across the enterprise that enables the enterprise to operate as a unit
- Perform corporate activities in its functional areas
- Automate business process which in terms improves overall business
- Business development – new areas, products, services
- Ability to face competition
- IT, development and employment of new technology
- Other software does not meet business needs
- Legacy systems difficult to maintain, euro currency
- Obsolete hardware/software difficult to maintain
- An ERP system is an SCM enabler
- Reduction of inventory, personnel, IT cost, procurement cost, etc.
- Improvements of productivity, order-management, cash management, financial close cycle and revenue
- It facilitates company-wide integrated information systems covering all functional areas, performs core corporate activities and increases customer service.

3.8.3   Enterprice resource planning and business process re-engineering

The deployment of an ERP system involves considerable business process analysis, employee training and new work procedures. Enterprise resource planning and business process re-engineering (BPR) historically went hand in hand. All the ERP implementations coincided with a BPR exercise. Due to this strong historical connection, it is important to mention the relationship between ERP and BPR.

The BPR is aimed at improving business processes, not removal of people (a common myth). A BPR exercise is a dramatic process with questionable results, and an ERP implementation is not commonly executed together with BPR these days.

Traditionally, BPR used to be implemented to provide a launching pad for an ERP implementation. This led to mismatches between the proposed model and the ERP functionality, which caused customization, extended implementation time frames, higher costs and loss of user confidence. A BPR running parallel to an ERP implementation was one reason for long implementation times.

Today, it is too costly to wait 2 years for the ERP implementation and a BPR process that accompanies an ERP implementation is an exception.

Advantages and disadvantages of BPR in conjunction with an ERP implementation

Advantages	Disadvantages
Functionality and solutions may not undergo changes	Stability of the system is known only in long run
Can be completed within time frame	Long-term process changes are not addressed
Cost will be within control	Best business practices may not be in practice
Level of integration is moderate	May not achieve corporate wide systems integration
	New technologies have to be studied

3.9 ERP and SCM relationship

The functionality of ERP systems is very close in functionality to that offered by SCM products. The level of detail at which each system operates differentiates these systems. Supply-chain management (SCM) goes beyond traditional planning solutions such as manufacturing resource planning (MRP) and distribution resource planning (DRP) by simultaneously considering demand, capacity and material constraints. The ERP provides planning capability and flow of information through an organization whereas SCM provides flow of information up and down the supply chain (refer Figure 3.17).

The ERP and SCM systems that share certain functions share information in a different form and at different levels. Manufacturers develop current SCM systems to be installed on top of an ERP system. The ERP system then feeds the SCM system with required information. That is precisely the reason why ERP vendors are moving toward SCM products (as they know how ERP functions and how to integrate it with other systems).

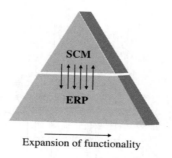

Expansion of functionality

Figure 3.17
Expansion of functionality

Functionality is the main driving force behind SCM systems, but besides the functional reasons there are also other reasons that can play an important role in the software market and offer some insights about the future trends. These are:

- Competitive strategies
- Lower cost
- Product differentiation and premium price

- Create barriers to entry
- Reduce power of suppliers
- Lock in customers, switching costs
- Value-added services
- Change basis of competition (collaborate with competitors).

ERP	Supply-Chain Planning (SCP)
The planning systems within ERP are elementary	SCP are more detailed and complex in nature
ERP provides a snapshot of time w/o supporting continuous planning	SCP support continuous planning by refining and enhancing the plan up to the very last minutes
ERP is not effective for optimal plan	SCP is a complement to ERP systems to provide intelligent decision support
ERP systems offer limited planning or decision support functionality	An SCP system is designed to overlap existing ERP systems, and pull data from every step of the supply chain
The ERP systems indicate about the costs, the financial performance of the company	Providing a clear, global picture of where the enterprise is heading
ERP process the order but not providing information about the profitability of the order or the best way to deliver it to the customer	SCP system allows companies to quickly assess the impact of their actions on the entire supply chain, including customer demand

Figure 3.18 illustrates a typical relationship of plant floor automation systems with ERP and supply-chain systems.

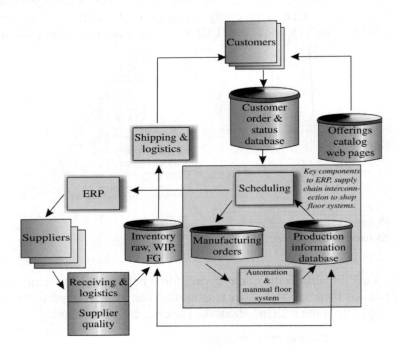

Figure 3.18
Supply chain at plant level

3.10   Supply chain management

The concept of the supply chain has evolved from the idea of the value chain. It is also important to understand and define what a value chain means in the context of SCM.

3.10.1   Value chain

Value chain is a method of dividing a business into a number of linked activities, each of which may produce value for the customer. The value chain is a framework for analysing the contribution of individual activities in a business to the overall level of customer value the enterprise produces and ultimately to its financial performance. The value chain concept helps identifying processes that add value and eventually bringing an organization into an integrated supply chain.

Customer value is a function of the following factors:

- Those that differentiate the product
- Those that lower its cost
- Those that allow organization to respond to customer needs more quickly.

3.10.2   What is a supply chain?

A supply chain is the process of creating products for customers as efficiently as possible. The supply chain is a portfolio of business entities or assets involved in the flow of products from the raw material, including raw data and concepts, procurement through manufacturing and logistics to delivery of the finished goods to customers.

It is a business process that links manufacturers, retailers, customers and suppliers in the form of a 'chain' to develop and deliver products as one 'virtual seamless' organization of pooled skills and resources, that reaches far beyond the factory doors. The objective is to obtain benefits by streamlining the movement of manufactured goods from the production line into the customer's hands, by providing early notice of demand fluctuations and coordination of business processes across a number of cooperating organizations (refer Figure 3.19).

3.10.3   Supply-chain models

There are two types of supply-chain models – the push and pull. In the push model, manufacturers produced/delivered in terms of size, quality, price, etc. keeping in mind their own interest but not that of the actual customers. In this model, the focus was on process efficiency, plant throughput and economies of scale. Most companies operated on a make-to-stock principle with long production campaigns. Quality and product customization were counter-productive to profitability, and so not easily entertained. And the customers were also helpless as there were no alternative suppliers.

Today, the pull model is more prevalent, as customer demand determines what needs to be made and when. Markets have become global, with a greater number of suppliers competing for a customer with high quality, competitively priced products. The customer–supplier relationship is becoming more like a partnership. Organizations have become highly flexible and produce exactly what the customer wants in a fraction of the time at a fraction of cost compared to that of the old manufacturers.

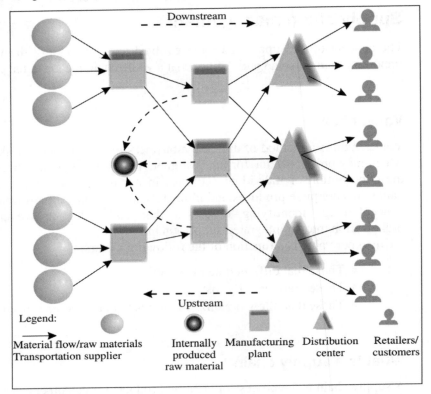

Figure 3.19
Supply-chain network

Out	In
Mass production	Customization
Product driven	Customer intimacy
Incremental change	Discontinuous change
Gradual improvement	Re-engineering/optimization
Functions	Processes
Hierarchical structure	Self-organizing teams
Do it all	The virtual enterprise
Customer vs supplier	SCM

Organizations are focusing on the required processes to meet the specific customer demands and looking to improve their performance at all phases of the demand fulfillment process. To achieve this, there was a need to manage information and systems in different ways. Without real-time systems, it is difficult for the manufacturer, to plan, schedule and execute the operations strategy effectively and efficiently. The old method of manufacturing resource planning (MRP) is no longer enough in this competitive market.

These two models refer to the different flows of the information that drives the supply chain. The dotted arrows indicate the flow of information. The downstream flow of information refers to the push model and the upstream flow refers to the pull model. A supply chain is assumed to be a pull type of system.

Products flow downstream, toward the customers that requested it, and information flow up and downstream, keeping every partner informed along the way, providing insight into the future (refer Figure 3.20). This is the most dynamic type of supply-chain system. Where the supply-chain functions really well, it delivers the greatest benefits to the end customer, the driving force behind the entire concept.

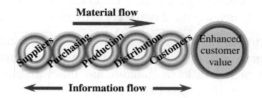

Figure 3.20
Supply-chain management

3.10.4 SCM systems

Supply-chain management is the center of a major business revolution to get products to market faster at lower costs. Supply chain management is a generic term for the coordination of order generation, order taking, order fulfillment and distribution of products, services and/or information. Material suppliers, channel supply partners (wholesalers/distributors, retailers) and customers are all key players in SCM. Supply-chain management extends beyond narrow functional areas, taking into account the needs of the customer. There has been a shift from a purely functional view to a more process-oriented view. SCM tools are supplied to manage the total process with different players integrated as one. It is a concept that requires a holistic view of the supply chain from start to finish.

3.10.5 Elements of SCM

Supply-chain management can be divided into four business activities:

1. *Planning:* The ability to anticipate the future and respond to changing situations.
2. *Optimization:* The ability to automate the supply chain and execute optimized plans with predefined constraints.
3. *Execution:* The ability to automate the supply chain and execute optimized plans within predefined constraints.
4. *Performance measurement:* The ability to define key performance/benchmark indicators and continually monitor performance against them.

3.10.6 Key functions of SCM

- *Customer asset management:* Managing information about demand to enable a better understanding of the markets and customer needs.
- *Integrated logistics:* Managing the flow of physical goods from suppliers.
- *Agile manufacturing:* Managing the manufacturing process to ensure low production costs.
- *Financial and accounting management:* Managing the financial flows with suppliers and customers through financial intermediaries.

These activities are applicable horizontally across every business process: design, planning, sourcing, manufacturing, order management, delivery and service/support. The terminology of SCM is broken down into related terms.

3.10.7   Supply-chain planning

Setting up an effective supply-chain plan (SCP) consists of two activities:

1. Structuring the supply chain through supply-chain analysis
2. Managing the day-to-day activities of the supply chain.

Structuring the supply chain requires an understanding of the demand patterns, service level requirements, distance considerations, cost elements and other related factors. These factors are highly variable in nature and this variability needs to be considered during the supply-chain analysis process.

3.10.8   Supply-chain optimization

The primary concern of supply-chain optimization (SCO) is the management of stocks and flows of material and information such that specified service performance is achieved at acceptable cost. The ROA (Return on Assets) value is used to measure the profitability/ productivity of the supply chain. The ROA value indicates the degree of optimization achieved.

SCO increases the benefits while reducing time and cost by:

- Reducing stock
- Producing actual quantity and supplying the same on time
- Increasing customer satisfaction that leads to increased market share, contributing toward higher profits
- The economy-of-scale in production
- Accurate scheduling of resources, improving resources productivity.

3.10.9   Supply-chain execution

Supply-chain execution (SCE) is the execution of the supply chain plans. The SCE focuses on the daily operations of the distribution process, which is one of the most important critical links in the supply chain. It mainly deals with warehouse, transportation, logistics, inventory and order management.

The ERP applications do not address the dynamic and complex requirements of SCE. Advanced planning systems (APS), part of SCM, was developed from resource and capacity planning processes for manufacturers and are concerned primarily with the planning of production, not distribution. It controls the process to ensure that the right goods are delivered in the right quantity at the right place, at the right time, at the minimum cost and according to a valid schedule.

The SCE addresses the distribution-related issues of the supply chain, it includes:

- Order management
- Enterprise-wide inventory control
- Advanced planning and replenishment
- Purchasing
- Warehouse management
- Transportation management
- Equipment management.

3.10.10   Supply-chain solutions

The SCM software packages consist of different modules that can be integrated to form a 'complete' solution. Many of these modules are industry specific. There is a move to develop standard ready-made software, which can be updated as per the requirements of individual customers by extending the desired functionality. These types of ready-made software modules are cheaper and easier to change while implementing.

The resistance to packaged software sometimes emerges when organizations believe a package supplier could never have a good appreciation of their business process needs. In reality, the best package suppliers develop solutions based on the needs and best practices of a wide range of organizations.

Much of the functionality of SCM, ERP, operation management system (OMS) and MES systems overlap. The distinguishing aspect is the level of detail and the time frame in which these systems operate. The level where it is used corresponds with the time the information is valid for, the longer the time validity, the higher the level of use.

3.10.11 SCM and the influence of technology

The backbone of a supply chain is the information channel. The technology that is currently enabling the supply chain is electronic data interchange (EDI) and electronic mail (e-mail). Web-based information sharing and data transfer technology is being used increasingly. An Intranet is a corporate network that is fenced in with firewalls on internal web servers. Where the Intranet is extended for use to outside users, customers, partners, suppliers or others outside the company, it is referred to as an Extranet.

The Intranet makes the modern supply chain possible and very cost-effective. This communication and data transfer infrastructure is one of the most important enabling technologies. It enables the sharing of data with the correct interpretation, as needed by the different users, that could be scattered around the globe, up and down the supply chain to work as one organization toward a common goal.

3.11 Operation management systems

In today's competitive market, manufacturers have to take care of every element of the operation processes for business excellence, and it has to be in real time. Operation Management systems is a management strategy based on information utilization. With an integrated computer-based system that is based around a single common repository of data, control can be achieved online and in real time.

An OMS is in many ways a strategic decision. There is a shift in attitude toward 'management by information' instead of 'management by guessing' in operations management. High manufacturing costs, long product lead-times, unreliable service or unpredictable quality generates a high degree of customer dissatisfaction. Manufacturers have discovered that most of their competitors have solved their SCM problems. Implementation of OMS can avoid erosion in market share and profitability.

The realities of systems integration requires logical division or separation of systems, data and applications according to:

- Amount of process and operations data to be handled
- Frequency of data capturing and processing required
- Focus of functionality (plan, execute and control or ready, execute, process, analyze and coordinate)
- Needs of users (management, engineers, operators), and the number and functionality of interfaces.

3.11.1   System integration vision

System integration, a conceptual systems integration paradigm based on enterprise-wide information sharing and business process integration. It divided into three sections (refer Figure 3.21).

- Business planning and control cycle
- Operational planning and control cycle
- Process control cycle.

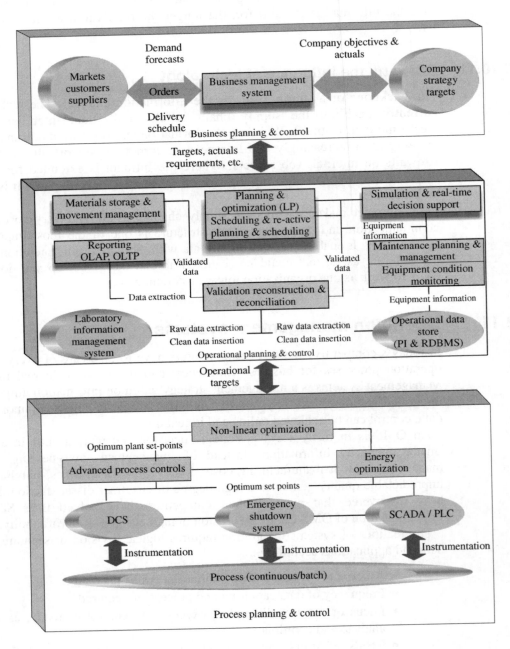

Figure 3.21
Conceptual architecture

3.11.2 Business planning and control cycle

An ERP system from an operational point of view must provide plant targets that are used to develop operational plans such as a budget, forecasted market demand, actual orders, etc.

This cycle is performed by an ERP system and has the following objectives:

- Financial and transactional visibility and automation of the enterprise is supported by the system
- Budgeting on enterprise and operational levels
- Long-term, medium-term and short-term operational targets
- Capturing of forecasts, actual orders and financial reconciliation.

3.11.3 Operational planning and control cycle

A large number of stand-alone or custom-developed software products historically used resulted in data duplication, functional silos and loss of synergy.

All the functions listed below are based on sharing a common operational data store (ODS) containing relational and real-time data.

- Production planning and optimization using related technology producing long-term, medium-term and short-term plans. Blending schedules and recipes as well as dispatch plans can be generated as well.
- Productions scheduling and re-active scheduling for handling upsets in the production process or supply chain.
- Production simulation and real-time decision support for test-runs and decision support.
- Integrated laboratory information management system, which supports the quality assurance and controls business process by providing timely analysis results and documentation for shipment of products.
- Availability of plant is an important issue in capital intensive industries as it directly influences the yield and should be managed as an integrated part of operations using a maintenance management information system, which is used to plan maintenance activities as well as exploit opportunities such as trips and breakdowns.
- Yield and availability prediction and management of equipment is of great importance for pro-active maintenance.
- Data used in operational decision-making must be reliable and there is a need for data validation on process data contained in the ODS. Reconciliation is an essential next step in ensuring data reliability.
- Yield accounting and product reconciliation is essential for effective production accounting and financial reconciliation.
- Considerable money is wasted on non-value adding activities such as storage, movement and dispatching of raw materials, intermediate and final products, and process material such as catalysts and additives, without an effective logistic execution function.

3.11.4 Process control cycle

The operational cycle produces operational targets for the process control cycle, which includes functionalities such as:

DCS, PLC and SCADA are the basis of this layer. Basic regulatory and logic control is provided from distributed control rooms and process data is passed on the operational layer ODS.

Non-linear optimization, which receives targets from production planning and converts it into set points for advanced process controls.

Advanced process controls have the objective to minimize deviations from optimal operating conditions and to ensure that transitions between different production modes are stable.

Energy is one of the major operating costs of industrial plants and is therefore necessary to monitor and manage energy consumption to optimize the production plan for an optimum energy profile on a continuous basis.

Emergency shutdown, fire and gas detection and alarming functions are the protection against abnormal plant situations. These functions are integrated into an abnormal situation management system to ensure that plant upsets have minimal effect on production.

3.11.5   System functionality

System functionalities are integrated around the ODS and divided into six sections (refer Figure 3.22) that should be evaluated in terms of business benefit:

- Production management
- Quality management
- Maintenance management
- Logistics management
- Human resources management
- Safety, health, environment and risk (SHER) management.

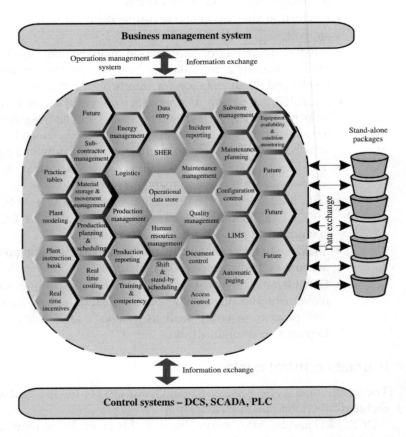

Figure 3.22
OMS conceptual functionality

3.11.6 OMS and SCM

Both OMS and SCM systems are concerned with relevant and accurate information. An OMS provides that information and an SCM system performs its functions based on that information. An OMS provides the right information, in the right form, on the right time, and the right information at the right time ensures a successful supply chain. An OMS feeds the supply chain with the information that it needs. An OMS system is almost a mini SCM system that is limited to an organization or plant.

An OMS system differs from an ERP system in that it uses real-time information from the control systems on the plant floor, whereas ERP use OMS information for longer-term planning without a direct link with plant control and manufacturing.

An SCM system normally operates on information that lies on an ERP system, although SCM systems can be implemented without having an ERP system in operation. In that case, all information will be provided by the OMS. An OMS can fulfill the role of an ERP system in providing information to an SCM system, since both systems use the same information (provided the information is processed according to the specified system's requirements).

An SCM system can therefore be implemented on top of an OMS instead of an ERP. Each link in the chain is linked with the other via information, whether it is from an ERP or OMS, which communicates the link's status. The status of each link is communicated to the other links, by means of Internet-based technology. The accurate situation at a given moment is known at each link and each of them can plan their business according to the situation of their suppliers and their buyers, up and down the supply chain. A link can only communicate its status if it is known to itself; this will be the purpose of an OMS in an organization. An OMS enables a link to function as part of a supply chain.

The benefits of an SCM and OMS are much the same; both systems strive toward:

- Improved product and process quality
- Better accountability
- More accurate and realistic production planning and scheduling
- Better accountability of time
- Improved work in process tracking
- Improved handling of production changes and exceptions
- Better customer order tracking and supplier accountability
- Reduction in cycle times, data entry, WIP, paperwork, defects and lead-times.

These two systems share common information and are based on the same principles visibility through information that leads to effective management. The integration of these two systems is natural, provided that the level of detail can be changed for the different roles of each system. Some functions of these two systems can use the same platform for its information.

3.12 Holonic manufacturing system

In January 1990, an intelligent manufacturing systems program was initiated to develop manufacturing systems, which are flexible, adaptable and reusable in different environments.

An HMS (refer Figure 3.23) integrates the entire range of production activities in a manner that provides a dynamic, agile manufacturing system. An HMS is built on autonomous, intelligent, cooperative building blocks (Holons) that are capable of reorganizing themselves and rescheduling the resources of the system to deal dynamically with external and internal change. Scheduler's and data collection agents are developed to operate when some parts of the systems are not functional and can resynchronize and reestablish cooperation.

Data collection on the shop floor gets affected due to multiple reasons e.g. cable failure, power problems, files damaged, etc. In the case of a malfunctioning network, data collection modules repair the damaged data files and resume the operation. The HMS can operate, self-diagnose and cooperate as is possible without constant human intervention.

Global competition and socio-economic trends are constantly forcing change on the manufacturing companies and the workforce they employ.

- Shortage of skilled workers for sophisticated manufacturing units
- Need for product and process standardization in a globalized market
- Consumer demand for customized higher-quality products at lower prices with shorter design and delivery time.

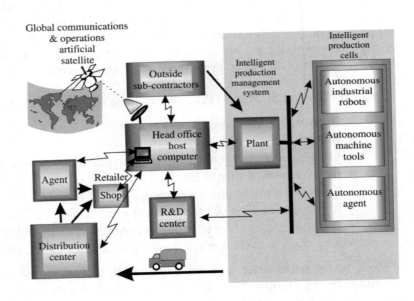

Figure 3.23
HMS vision

3.12.1   Obstacles to the holonic system architecture

- Islands of automation without standardized interfaces
- Manufacturing systems and technologies are not reusable, and the cost of technology transfer is too high
- The operational systems are inflexible and factories are unable to change quickly in terms of production volume, product design, etc.

3.12.2 The holonic model

Conventional manufacturing system architectures are based on a command–obey relationship whereas holonic architectures use the whole–part relationship.

An HMS is composed of holons. Each holon is a system building block – autonomous and cooperative. It consists of an information processing part and a physical processing part and can be part of another holon.

An HMS is organized as a holarchy, which defines the basic rules for cooperation of the holons and thereby limits their autonomy.

An HMS is not organized in a fixed way, but organizes itself dynamically to meet its goals and adapts itself to internal and external environmental changes.

An HMS integrates the entire range of production activities from order booking through design, manufacturing and marketing to realize the agile manufacturing enterprise (refer Figure 3.24).

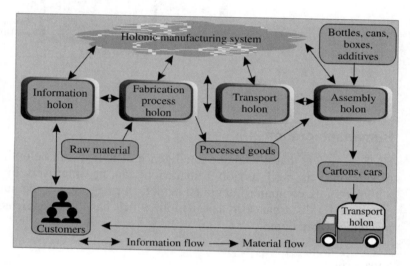

Figure 3.24
The general concept of HMS

3.12.3 Operational benefits of HMS

- HMSs are capable of rapid self-recognition in response to the change and uncertainty in the manufacturing environment.
- The role of the human is explicitly taken into account in the HMS architecture, thereby enabling enterprises to maximize the use of human intellectual skills and flexibility.
- The incorporation of human and machine intelligence into holons, and their inherent cooperative behaviors, enabling the formation of 'virtual companies' both within and across enterprise boundaries.
- Holons can be introduced in an incremental manner into the current manufacturing environment.

3.12.4   What holonic implies

Integration and decentralization

Holonic implies integration and decentralization of diverse resources of the enterprise into the HMS. The holonic philosophy (refer Figure 3.25) generates the rules for this integration.

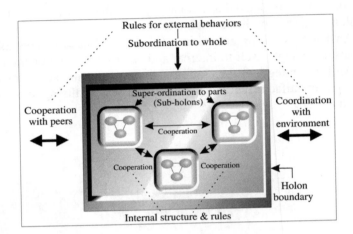

Figure 3.25
'Social contract' in a holarchy

Human integration

HMS considers that the unique intelligence of human can be optimally utilized, and both productivity as well as job satisfaction can be improved by a suitable degree of automation. In computer-integrated manufacturing systems, the human operator was often considered a disturbance in automation, which has to be restricted to defined inputs and mechanical functions.

Synergy

Holons must preferably cooperate with other holons to fully exploit all their capabilities.

Modularity

Each holon has interfaces for interacting with a range of other holons, upgrading the HMS by a simple replacement of holons.

Improvement

An HMS is improved from external sources and equally from the holons' built-in capabilities such as self-learning and adaptation without external support.

Fault tolerance

An HMS has the ability for each of its modules to work independently, or within the web of systems. It provides a highly reliable and fault tolerant architecture. When implemented with messaging and transaction protocols, the ability to provide strong fault tolerance is inherent in the architecture.

3.12.5   Characteristics of a holon or subsystem

Autonomy

Once an order is given by its master, a holon will behave correctly for a period of time. In normal conditions, a holon functions under its own initiatives, following its own set of rules:

- Basic functional goals to be accomplished: processing, assembling, testing, conveyance, calculation, data storage and communication.
- A holon has the capacity to control its conditions continuously. It analyzes the problem immediately and in extreme conditions separates the problematic subholon without affecting any other holons.
- Self-monitoring, autonomous, self-diagnosis, self-repair and self-restoration for unmanned operation or fault tolerance.
- Capacity to metamorphose, changing components or internal organization to change or improve its capabilities.
- A holon is a self-learning and self-organization function.
- Capacity to support a job other than its original, or to replace its master/neighboring machine, when its behavior is incorrect.
- The ability to change the order of jobs and steps to accommodate new priorities or current problems.
- Understanding the targets or goals of the total system and principally of the holon of which it is a part.

Cooperation

- Recognize the conditions of neighboring holons and total system and conceptual targets at any time
- Support of neighboring holons of the same level
- Capacity to support the neighboring holon in problems of its own initiative
- Capacity to report its condition to the total system and to neighboring holons
- Each holon communicates bidirectionally with holons of one rank above and with neighboring holons
- Easy entry and separation from the total system.

Intelligence

Figure 3.26 depicts a typical HMS.

- Capability to acquire and use knowledge, to develop or refine goals, and to develop strategies and make decisions of how these system goals may be achieved.

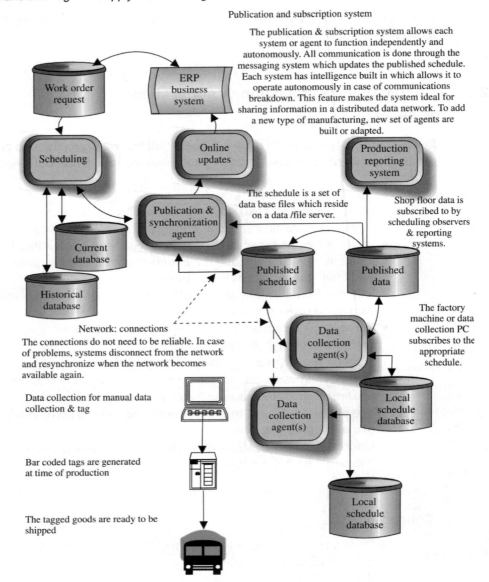

Figure 3.26
Publication and subscription system

3.12.6   Development of HMS

Incremental development

A strategy to develop the system step-by-step (refer Figure 3.27) beginning with its core and its most important functions.

- More 'learning cycles' in a defined period
- Faster response to committed errors and wrong assessments
- Better control of investments.

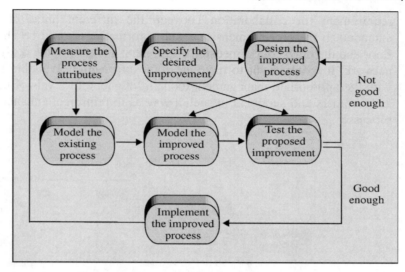

Figure 3.27
Incremental HMS engineering process

3.13 Collaborative manufacturing management systems

Collaborative manufacturing management (CMM) is formally defined as 'the practice of managing by controlling the key business and manufacturing processes of a manufacturing enterprise in the context of its value networks' (ARC Advisory Group – October 2001). It builds on a collaborative infrastructure, business process management (BPM) services and real-time strategic business management tools, together with critical applications, production systems and enterprise information, to maximize the responsiveness, flexibility and profitability of the manufacturing enterprise, together with the overall effectiveness of the value networks. At a conceptual level CMM is a new model for business management which shows the relationship between the real time and transaction paradigms of reality by expanding the traditional two-dimensional model into a third dimension.

The CMM model (refer Figure 3.28) has three dimensions or axes onto which the systems can be mapped. The enterprise axis is the linear dimension where the system sits in relation to the real-time/transaction distinction. The value chain axis defines the position in the process that starts with the supplier's supplier and ends with the customer's customer. The life cycle axis looks at the traditional product design-launch-growth-maturity-support continuum. The nodal sphere defines the areas of the business and this sphere does not exist in a vacuum either as there are linkages to the outside world, from every possible point on the surface of the sphere. The Internet, exchanges and portals can be used to link the enterprise to the outside world.

'The essence of collaboration is the ability for individual plants to synchronize their work in real time based on accepted orders, and to coordinate the production and delivery of component materials at the production level in a highly distributed manner' (ARC – October 2001).

The functions of CMM have been divided into seven areas.

3.13.1 Synchronize business processes with manufacturing processes

Coordinating the two sides of the business-production axis in the sphere is one of the most important, and difficult, tasks in systems and business process design. Collaboration between the two, in real time, is essential for successful modern business. The requirement for collaboration between the different areas of the model should be extended to the other dimensions, and information should be shared between the plant floor and the other four dimensions: suppliers, customers, the design team and the support network. It is important to have the right information available at the right time, along with the appropriate management tools, throughout all levels of the organization, as well as customers and suppliers in such a way as to reinforce, enhance and optimize business processes.

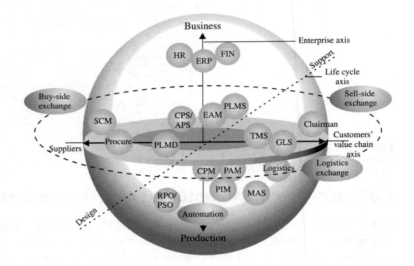

Figure 3.28
Detailed Computer-integrated manufacturing model

3.13.2   Optimize the supply-side value chain

The concept of real-time business control through the SCM system, and optimizing the information sharing processes allow process decisions based on actual demand for products rather than forecasts, leading to a competitive advantage.

3.13.3   Automate business processes across departmental and business boundaries

Manufacturers having vision and cultural adaptability will be able to capitalize on the agility and responsiveness of real-time business process control by automating the entire supply chain.

3.13.4   Generate value by empowering people and measuring results

Managers in CMM environments will be able to use portals and other web-based tools to view information from a variety of systems in a single interface.

3.13.5   Implement collaborative design and engineering

The collaborative manufacturer must design and manage change in a transparent, mutually beneficial manner across the product value chain. Collaboration between any two of the axes in the model are to be ensured.

3.13.6 Link operations with customers

Collaboration with customers, via EDI and vendor managed inventory (VMI) has been around for a long time. Close collaboration throughout the value chain ensures communication on a much more real-time basis than before and instantaneous information transfer, in both directions, is essential for a high degree of confidence in the value network.

3.13.7 Enable collaborative maintenance and manufacturing support

The implementation of an enterprise asset management (EAM) system can integrate equipment conditions into production commitments and ensures customer satisfaction. Internet technologies enable one to monitor equipment at a customer's site thus ensuring optimal maintenance and providing the best possible after-sales support.

4

Business process design models and concepts used in operations systems

Learning objectives

- To understand the models, concepts and theories used in the design of software solutions.
- To understand the standards and regulations influencing system functionality design.
- To obtain an understanding of the basic assumptions behind the models and standards.
- To understand the correct usage of models and standards.

4.1 Introduction

Various design concepts, standards and models have been put forward over the years to assist in the design of business processes and systems. Depending on the specific requirements of the client and the type of industry they operate in, some would be more applicable than others. There is thus no 'one correct way' in all circumstances, as most models/concepts were developed for a specific set of industry standards, production philosophy or level of technological capability.

The best way therefore is to approach each design project starting with a clean slate. Understanding of the business drivers, the internal and external markets and the related business processes is paramount to enable a good design. Only then can a decision be made as to the appropriate concept/model to use during the design phase. In most cases, what will be used is a combination of concepts and models that best fit and describe the unique requirements of a specific company.

Various research institutions have attempted to produce generic models to assist in the analysis and design of specific industry solutions. The supply chain operation reference (SCOR) model from the supply-chain council for instance is a good tool that can be used to analyze logistical processes of a company and compare them to world best practice. The ready, execute, process, analyze, coordinate's (REPAC) model developed by AMR Research in turn can be used for a more detailed analysis of the 'make' part of the SCOR model.

As the technological capabilities of software as well as manufacturing companies increased over the last two decades and more systems are installed, both are increasingly looking for better and more effective ways to integrate disparate software systems. This task is daunting if one considers that large numbers of platforms, programing languages, databases, database management systems, programing styles/methods and tools are available today. Various standards have been developed in an attempt to make this integration (as well as system maintenance and safety) easier to manage. The Instrumentation, Systems and Automation Society (ISA) has developed (amongst others) the S 88 Batch control and S 95 Enterprise-Control System Integration standards in an attempt to standardize on the structure used to design and configure these systems. The International Electrotechnical Commission (IEC) has developed the IEC 61131 Programmable Controllers standard for PLC programing.

Other concepts and business philosophies such as just in time (JIT) manufacturing, total productive maintenance (TPM), total quality management (TQM), statistical process control (SPC) have been around most of the past three decades. One of the more recent concepts is the theory of constraints, which turns some of the past paradigms regarding how businesses should be run and measured on their heads. These have all been implemented around the world with varying degrees of success. This is due in part to the unique circumstances of individual companies, the available technology and resource skills, the technological and logistical abilities of suppliers and customers, but mostly the level of success depended greatly on the readiness of the company employees and management to make and accept the changes.

The rest of this chapter will detail some of the above concepts, models and standards.

4.2 Theory of constraints

The theory of constraints (TOC) is an overall management philosophy that has its basis in the manufacturing environment, was developed by Dr EM Goldratt. TOC recognizes that organizations exist to achieve at least one goal. A factor that restricts company's ability to achieve its goal is referred as a 'constraint'.

A manufacturing company can be considered as chains of dependent events that are linked together just like a chain. The activities of one 'link' are dependent upon the activities of the preceding 'link'. According to the TOC, management needs to identify the week link in the chain since 'a chain is only as strong as its weakest link'. Strengthening other links is going to be waste. Therefore a company needs to focus on 'chain strength' by strengthening the weakest link – the constraint.

4.2.1 Five steps of TOC

In case of easy identification of the constraint, five processes of ongoing improvement steps are to be followed.

1. Identify the constraint.
2. Exploit the constraint.
3. Subordinate and synchronize every thing else to the above decisions.
4. Elevate the performance of the constraint.
5. Go back to step 1.

The TOC provides a theoretical framework and the tools to continually identify the constraints. For an organization to have a process of on-going improvement, certain basic questions need to be answered faster and more effectively.

The fundamental questions are:

- What to change?
- To what to change?
- How to cause the change?

4.2.2   Thinking process

There are five thinking process tools that allow executives to identify what to change in the organization, what to change it into and how to implement the change. The five tools are as follows:

Current reality tree It identifies the root causes and the core problems of an organization.

Evaporating cloud It uncovers the conflict that brought about the core problem. It identifies and uncovers the assumptions that allow the core-conflict to persist.

Future reality tree   It helps to evaluate and improve the solution before implementation. It constructs a future reality tree that lays out the complete solution/strategy that:

- Resolves all of the undesirable effects by making their opposites the desired effects
- Ensures alignment with strategic objectives of the bigger system that the subject is a part of
- Ensures that no negative side effects occur from implementing the solution/strategy
- Identifies the necessary changes required in the culture of the system
- Measures the bottom-line value of achieving those desirable effects and objectives.

Prerequisite tree   It identifies all the intermediate steps that are needed to reach the chosen solutions.

Transition tree It identifies those actions needed to achieve the intermediate objectives that were identified earlier with the prerequisite tree.

4.2.3   Applying TOC to an organization

The TOC process is used to improve the health of an organization by solving problems using the following fundamental questions.

What to change?

Cause and effect analysis is used to identify the common cause, the core problem, for all the observed symptoms. In an organization the core problem is an unresolved conflict that distracts the organization constantly – management vs market, short term vs long term, centralize vs decentralized, process vs results, etc. This conflict is called the core conflict and the effects of core conflicts can be devastating. In order to treat these conflicts, every organization has to develop policies and behaviors to remove, modify or replace such negative effects.

What to change to?

A solution to the core conflict has to be identified by challenging the logical assumptions behind the core conflict. A strategy has to be developed to completely and irrevocably resolve all the initial symptoms. The strategy must also include the related changes which are required to ensure that the solution to the core conflict works better. This often leads to changes in the policies, measurements and behaviors identified in earlier stages. The strategy is complete only when all potential negative side effects of the strategy have been identified. The ways that help prevent the negative effects constitutes the main element of the strategy.

How to cause the change?

A plan consisting of what action to be taken, by whom and when has to be developed for the successful implementation of the strategy.

4.2.4 Resistance to change

TOC has developed a process based on the psychology of change that systematically address the questions frankly and effectively with both the people who must implement the change, and those who will be affected by it. Without this, the proposed change will not be successful. The following are some examples of resistance natural to all humans.

- Is the right problem identified?
- Is the solution leading to the right direction?
- Is the solution really solving the problem?
- What may go wrong with the solution? Is there any side-effect?
- Can this solution be implemented?
- Are we all really up to this?

4.2.5 Drum-buffer-rope tool

Drum-buffer-rope (DBR) is the TOC production planning methodology. The traditional DBR model is designed to regulate the flow of work-in-process (WIP) through a production line at or near the full capacity of the most restricted resource in the manufacturing chain. To achieve this optimum flow, the entry of work orders into production is synchronized with the current production rate of the least capable part of the process, referred to as the capacity-constrained resource (CCR). The production rate of this CCR is typically likened to the rhythm of a drum, and it provides the pace for the rest of the system. The rope is essentially a communication device that connects the CCR to the material release point and ensures that raw material is not inserted into the production process at a rate faster than the CCR can accommodate it.

The purpose of the rope is to protect the CCR from being swamped with WIP. To protect the CCR from being 'starved' for productive work to do, a time buffer is created to ensure that WIP arrives at the CCR well before it is scheduled to be processed.

- *The drum:* The drum in DBR is the constraint. According to DBR, non-constrained resources should be subservient to the constrained ones.
- *The buffer:* A time/material buffer is used to avoid disruptions in the production process. These disruptions can be because of breakdowns, longer setup times, delays by suppliers and so on. Some companies also use shipping buffers to enhance on-time deliveries.
- *The rope:* A schedule is executed to release either materials or jobs into the system. The basic aim of the schedule or the rope is to ensure that all workstations are stretched to perform at the speed of the drum.

An underlying principle of TOC is that manufacturing to firm orders with defined due dates is the most desirable situation possible, and preferable to manufacturing to stock. To that end, applying traditional DBR starts with some desired master production schedule (MPS) that includes firm customer orders with delivery due dates. Next, the existence of an internal physical resource constraint is verified. The identification of such a constraint (CCR) can be supported by computerized capacity analysis, but should be validated by production management.

Immediately, there are two distinct possibilities:

1. There is no capacity constraint currently active, or
2. A definite capacity constraint is identified.

The DBR method strives to achieve the following:

- Very reliable due-date performance
- Effective exploitation of the constraint
- As short response time as possible, within the limitations imposed by the constraint(s).

Conceptually, the three main process steps of DBR are:

1. Identify the constraint – the plan for exploiting the capacity constraint (the 'drum').
2. Determine the amount of buffer to ensure that the drum is not idle on the basis of past experience – Protection against 'murphy' (the 'buffer' expressed in time rather than in things that are stocked somewhere).
3. The schedule is determined by working backwards from the due date of the job – A material release schedule (the 'rope') that protects the shop floor from excess WIP and priority confusion. The DBR assumes that true material constraints are very rare and proper inventory management should ensure material availability as required.

The DBR scheduling results in flexible batch manufacturing, less investment in inventory and a reduction of work-in-process inventory.

DBR when no CCR is active

When no CCR is active, there is no reason why all the firm orders should not be delivered on time. The list of those orders constitutes the 'drum', which is really the master production schedule (MPS).

DBR when a CCR is active

When a CCR is confirmed to exist, a finite capacity schedule is generated for it, based on the preliminary MPS. The MPS is subsequently revised, based on the limitations imposed by the CCR. The new MPS and the detailed schedule for the CCR constitute the Drum.

In this situation, three buffers are established as a protection mechanism against variability ('murphy'):

- *A shipping buffer:* This is a liberal estimation of the lead-time from the CCR to the completion of the order or the lead-time from raw materials to completion.
- *A CCR buffer:* The CCR buffer is a liberal estimation of the lead-time from raw material release to the site of the CCR.

- *An assembly buffer:* This is a liberal estimation of the lead-time from the release of raw materials to a process step where parts that do not use the CCR are assembled with parts that do.

The rope is the schedule for release of materials as dictated by the three buffers. The three schedules are the typical output of DBR planning. The rest of the resources are not specifically scheduled. They are directed to process any order arriving to their site as fast as possible. The rope ensures that no order is released to the manufacturing floor until the CCR or shipping buffer times.

A simplified DBR (S-DBR)

To apply S-DBR, the presumption is that the company is not currently constrained by any internal resource. In other words, the market is the overarching constraint for the company.

When the market is clearly the constraint, the combination of the simplicity of DBR planning with the highly focused control afforded by buffer management, results in full subordination of operations to sales (the constraint). However, when a CCR begins to emerge the following significant changes are observed:

- The decreasing capacity of the internal resource constraint may limit the company's ability to respond to the market. Some orders may not be delivered on the required dates. To keep this condition from deteriorating even further, either some of the market demand must be reduced, or capacity must somehow be increased.
- The actual lead-time from raw material release to order completion and shipping increases significantly.
- Every unit of product needs to pass through two buffers covering various non-constraint operations (assembly and shipping) rather than just one.
- Buffer management now includes three buffers, each of which must be monitored and managed. This can create conflicts when a single resource must expedite different orders for different buffers.

The basic assumption underlying S-DBR are as follows:

Basic assumption No. 1 The market dictates certain requirements that a company must meet, otherwise, demand for the company's product or service will diminish and perhaps vanish completely in the future. These requirements imposed by the market sometimes conflict with full exploitation of an internal constraint (CCR).

Basic assumption No. 2 A small change to the actual processing sequence at an internal constraint does not have much impact on overall system performance.

Differences between traditional DBR and S-DBR

There are some key differences between traditional DBR and S-DBR:

Level of throughput Traditional DBR is capable of squeezing more throughputs out of the CCR in certain peak demand periods, due to the detailed CCR schedule.

Customer satisfaction The role of the shipping buffer in S-DBR is more dominant than in traditional DBR. When the shipping buffer is added to the CCR buffer, the protection of promised due-dates is less effective. This is essentially the same phenomenon that causes a critical chain project completion buffer to be more effective than a number of buffered intermediate points.

Focus Traditional DBR is usually focused on the internal resource while S-DBR is focused on the market demand.

Lead-time Having one buffer, rather than three, enables S-DBR to achieve shorter lead-times. The accumulation of protection is always more effective than spreading it.

In most situations market demand fluctuates, and so dealing with peak and off-peak demand is frequently required. Assuming we cannot fully level the load on a CCR throughout the year, we can conclude that the CCR is active only around the peak period, and the market demand is the sole active constraint in any off-peak periods.

Shifting from three buffers to one buffer and then back to three buffers again represents a huge policy change for traditional DBR, with significant ramifications for management. For this reason, in most cases organizations using traditional DBR would regard the CCR as a constraint even in a period of low demand. This produces suboptimal results (longer delivery times) in off-peak periods. The S-DBR is able to shift smoothly between peak and off-peak periods, as the main focus of planning and control have not changed (satisfaction of the market).

Support from common information technology (IT) packages: S-DBR is much easier to plan and control with common MRP systems. In fact, specialized DBR software packages are not really needed, since MRP systems can be adjusted to support S-DBR. This can be a real benefit to companies that already have MRP systems but might be unable or unwilling to invest in specialized DBR software.

4.3   The supply-chain operation reference model

The supply-chain council (SCC), a non-profit, an independent global organization has developed this model for companies/organization interested in advanced SCM. The SCOR model was developed to achieve customer satisfaction, the aim of most businesses these days. It describes all phases of customer demand and is divided into five primary management processes – plan, source, make, deliver and return.

By describing supply chains using these process building blocks, the model can be used to describe supply chains that are very simple or very complex, using a common set of definitions. As a result, disparate industries can be linked to describe the depth and breadth of virtually any supply chain. The model has been able to successfully describe and provide a basis for supply-chain improvement for global as well as site-specific projects.

Span of the SCOR model (refer Figure 4.1):

- Entire customer interaction i.e. order booking to payment invoicing.
- Entire physical transactions i.e. supplier's supplier to customer's customer.
- Entire market interactions i.e. understanding the demand to fulfillment process of every order.

The model does not attempt to describe every business process or activity. Specifically, the model does not address sales and marketing (demand generation), product development, research and development, and some elements of post-delivery customer support.

The SCOR model is designed in a manner that can support a variety of supply-chain complexities of multiple business houses. The SCC focuses on process levels but does not suggest the method of conducting business in any particular organization. Organizations using the SCOR model should extend it down to level 4 using organization-specific processes, systems and practices.

Process type Level 1 of the SCOR model defines the scope and content of the model. It sets and fixes a basis for competition performance targets.

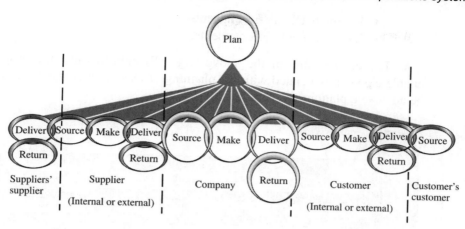

Figure 4.1
SCOR is organized between five major management processes

Process categories In level 2 of the SCOR model the operations strategy of a supply chain can be configured. A company's supply chain can be 'configured-to-order' from 26 core 'process categories.' Companies implement their operations strategy through the configuration they choose for their supply chain.

Process Element level Level 3 defines a company's ability to compete successfully in its chosen markets, and is used to fine-tune the operation strategy. It consists of:

- Process element definitions
- Process element information inputs and outputs
- Process performance metrics
- Best practices, where applicable
- System capabilities required to support best practices
- Systems/tools.

Implementation level (decomposed process elements) – Level 4 defines practices to achieve competitive advantage and to adapt to changing business conditions. Companies implement specific SCM practices at this level.

The SCOR model focuses mainly on activities and not on any organizational element or person performing the activity. The reference model links process elements, metrics, best practice and the features related to supply-chain execution. The successful implementation of the SCOR model is based on these four elements.

The model distinguishes three processes: planning, execution and enable besides the five basic management processes (plan, source, make, deliver and return). There is a common internal structure within the source, make and deliver process elements, and the model focuses on three possible business environments – make-to-stock, make-to-order, and engineer-to-order.

Any supply chain is characterized by performance attributes, without which it is highly difficult to compare an organization that prefers to be a low-cost provider, against an organization that competes on reliability and performance (refer Figure 4.2).

Throughout the model, a standard notation is used:

- P depicts PLAN elements
- S depicts SOURCE elements
- M depicts MAKE elements

- D depicts DELIVER elements
- R depicts RETURN elements.

An E preceding any of the above (e.g. EP) indicates that the process element is an enable element associated with the planning or execution element (in this case, EP would be an enable planning element).

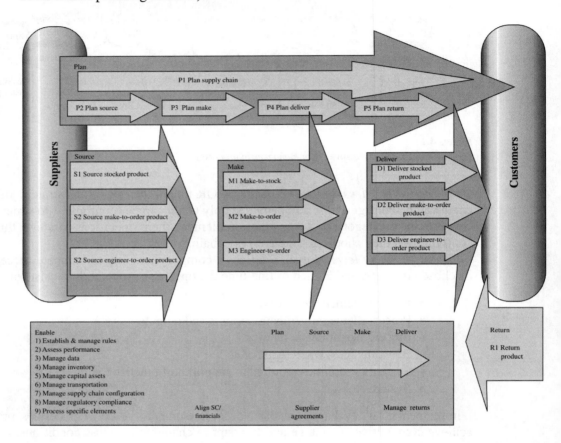

Figure 4.2
SCOR Toolkit

The model is hierarchical with three levels. P1.1 is a notation that indicates a third-level process element. In this case, it is a plan (P) element that is concerned with supply-chain planning (.1) and is specific to identifying, prioritizing and aggregating supply-chain requirements.

The SCOR model contains six basic sections: introduction, plan, source, make, deliver and a glossary. The plan and execution (source, make, deliver, return) sections are the heart of the model, while the glossary provides a listing of the standard process and metrics terms that are used within the Model.

The plan, source, make and deliver sections are organized similarly. At the beginning of each section, there are graphics that provide a visual representation of the process elements, their relationships to each other and the inputs and outputs to each process element. Following the graphics are text tables that identify:

- The standard name and notation for the process element
- SCC's 'standard' definition for the process element

- Performance attributes that are associated with the process element
- Metrics that are associated with the performance attributes
- Best practices that are associated with the process (not necessarily an exhaustive list), and features (generally technologically related) that can contribute to heightened performance of the process.

Within the source, make and deliver process elements, a common internal structure is used. As the model focuses on three environments, make-to-stock, make-to-order, and engineer-to-order, S1 becomes source-make-to-stock product, S2 becomes source-make-to-order product and S3 becomes source-engineer-to-order product. This same convention is used for make (M1 – make-make-to-stock) and deliver (D2 – deliver-make-to-order).

Metrics are used in conjunction with performance attributes for, supply-chain reliability, supply-chain responsiveness, supply-chain flexibility, supply-chain cost and supply-chain asset management as indicated in the table below.

Performance Attribute	Customer Facing		Internal facing		
	Reliability	Responsiveness	Flexibility	Cost	Assets
Delivery performance	•				
Perfect order fulfillment	•				
Order fulfillment performance		•			
Fill rate					
Order fulfillment lead-time					
Supply-chain response time			•		
Production flexibility			•		
Total supply-chain management cost				•	
Cost of goods Sold				•	
Value-added productivity				•	
Warranty cost or returns processing cost				•	
Cash-to-cash cycle time					•
Inventory days of supply					•

Performance attributes

The performance attributes are characteristics of the supply chain that permit analysis and evaluation against other supply chains with competing strategies. Without these characteristics it is extremely difficult to compare organizations and supply chains. The performance attributes and their associated definitions and level 1 metrics are provided in the table below:

Performance Attribute	Performance Attribute Definition	Level 1 Metric
Supply-chain delivery reliability	The performance of the supply chain in delivering: the correct product, to the correct place, at the correct time, in the correct condition and packaging, in the correct quantity, with the correct documentation, to the correct customer	Delivery performance Fill rates Perfect order fulfillment
Supply-chain responsiveness	The velocity at which a supply chain provides products to the customer	Order fulfillment lead-times
Supply-chain flexibility	The agility of a supply chain in responding to marketplace changes to gain or maintain competitive advantage	Supply-chain response time Production flexibility
Supply-chain costs	The costs associated with operating the supply chain	Cost of goods sold. Total supply-chain management costs Value-added productivity Warranty/returns processing costs
Supply-chain asset management efficiency	The effectiveness of an organization in managing assets to support demand satisfaction. This includes the management of all assets: fixed and working capital	Cash-to-cash cycle time Inventory days of supply asset turns

4.4 The ready, execute, process, analyze, coordinate model

During the modeling of a business it is important to address both business and operations business processes, to ensure alignment and completeness of any solution. Interaction is required with and between business processes. Five business processes are necessary to describe operations processes and improve the plant performance (Figure 4.3) the REPAC processes. The ready, execute, process, analyze, coordinate (REPAC) model is used to describe in more detail the 'make' portion of the SCOR model. (refer Figure 4.3).

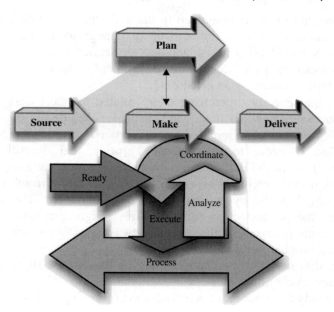

Figure 4.3
SCOR and REPAC Model

In the REPAC model each category is divided into a set of subcategories and users can further expand the different sub-subcategories as per their plant requirement.

4.4.1   Ready

The ready business process prepares production for the introduction of a new product line or modification of old products by transforming product designs or specifications. This process enables the most suitable plant configuration for the manufacturer of new and old products. It also administers process and product improvements.

Process definition outlines the process characteristics that are to be implemented in the new process (machinery layout, setup, material flow, etc.) and describes the functionality to be implemented in the definition i.e. What system do we have to perform inventory or resource checking at lot/piece level? Do we have electronic access to supplier and customer production system? Or are data always available to us? Do we manufacture as per the customer order or depending upon the warehouse stock level? etc.

The process definition function should be integrated with the production engineering system, document management system, shop floor control system, quality management system, SPC, purchasing, etc. Data to be interchanged may be the latest design documents, material specifications, cycle time, etc.

Product definition outlines the product characteristics (size, weight, performance, material, etc.) of the new product and describes the functionality to be implemented in the definition i.e. what system do we have to administer the Bill of materials? Do we have routing information? Do we have labor instructions, quality instruction? etc.

Data that may be interchanged may be the latest design documents, material specifications, pricing, etc.

Engineering product change identifies the additions and modifications made to production equipment and procedures, coordinates their implementation and describes the functionality to be implemented in the definition i.e. Which system allows start-up with these changes? How is the start-up of change to be managed, by predefined date of inventory level? What new machinery or service is required? Is there an SPC package, maintenance package? document management system? Are there any external interfaces? What type of databases is in use? etc.

Data that may be interchanged are related to latest design documents, schedule for introduction of new machinery and changes, schedule for services installation, etc.

Improvement definition defines the process characteristics (speed, accuracy) to be modified in order to improve any process and describes the functionality to be implemented in the definition i.e. Which system do we have to manage the requests for process improvement? Where does the request come from? How much time is required for testing? Is the cycle time to be decreased? etc. The improvement function also outlines the product characteristics (size, weight, performance, etc.) to be modified to improve the product. Requirement of new package? new raw material? Will the new price list, new brochure be necessary? etc.

Data that may be interchanged may be the latest design documents, capability study data, new cycle times, etc.

The ready business process includes the following high-level functionality:

Administer quality and legal procedures

Administering quality consists of product quality, data quality and work processes.

Product quality includes quality control procedures, specifications for all process materials and inspection/testing methods, raw material assessment, quality inspections of equipment, process samples and control samples, quality inspection of final products and quality reporting.

Work processes include standard operating procedures (SOP), implementation of engineering change procedures, production standards and metrics, set-up, switchover and training.

Data quality includes instrumentation configuration, calibration record and reporting.

Product data management

Product data management includes the maintenance of recipes as well as defines and compiles 'batch' data such as components, process material used, process metrics, quality, etc.

Plan-do-review

Plan-do-review comprises data collection and comparison with metrics, setting-up corrective and preventive actions (CAPA), management and review of the actions.

Modeling

Make provision for the definition, building and validating of models, and the analyses of problem root causes to enable the generation and implementation of process improvements.

4.4.2   Execute

The execute business process executes an optimized production plan to ensure that products comply with their specification and to manufacture with high efficiency. It also provides process with suitable machinery configuration (set-up, bill of materials, routings, labor and quality instructions) needed to make specific products and to reduce start-up time.

Work order management (WOM) is a core function of the execution business processes that accept planned orders by an MRP/ERP system or optimized sequences by a finite capacity scheduler. It also releases orders to production and establishes a current order priority list based on sequencing rules or other schedule-production methods, such as simulation or constraints management techniques.

The role of WOM is to accept automated or manually entered information, and to manage the plant accordingly until and unless it is necessary to respond to unplanned events. It is normally directly joined with the scheduling section to enable information to be passed on to the production system.

The WOM defines the functionality to be implemented in the definition i.e. How are the orders on the plant floor prioritized and with what time horizon? How is material allocated to orders? Is alternative routing available? How can orders be combined or split? How frequently will WOM be required to manage production plan changes in response to new order priority, changes in quantity, etc.? WOM also defines integration with ERP systems, scheduling packages, inventory systems and plant automation and control systems, etc.

Work-in-process (WIP) is another core function of the execution business process that provides visibility to where work is at all times and the deposition of this work. Work-in-process provides real-time views of production goals and status information and may include personnel working on it, component material by supplier/lot/serial number, current production conditions and any alarms, rework and other exceptions related to the product.

It is a very important function within the discrete and batch process industry where defined actions are performed on the lot at each point in the process. Additionally the operator may take some manual actions.

Data collection is an important aspect of WIP, used to build an information base for future analysis and comparison. It allows routing as well as location definitions (e.g. intermediate storage, assembly station, etc.) and describes which materials and information are relevant to the WIP tracking.

Resource management implements the planning, the scheduling, the loading and/or the coordination of the resources. Resources may include machinery, tools, labor skills, materials as well as other equipment and entities (such as documents) which must be available for work to start on an operation. Resource management defines the functionality to be implemented in the definition i.e. How is resource management handled presently? Is the information passed to WOM? Is a logical model, listing departments, resources and operations in the facility, available? etc.

Inventory tracking develops, stores, and maintains detail of each lot or unit of inventory, including the current location. All materials used in the production process are appropriately tracked and accounted for by means of the inventory tracking system. These materials include tooling, fixtures, raw materials, WIP, drawings, specific labor skills, and any other items that could be listed in bill of materials and routings. It is implemented at two levels – via the planning systems and the MES on the plant floor. The inventory tracking system should be integrated to the planning system, automatic storage/retrieval system, CAD system, plant control system, etc.

Material management manages and schedules the manual or automated movement of material. It is a non-value adding function and determines when to move what, from where, to where and issues the appropriate instructions.

Data collection allows the system to remain current by retrieving all information that is needed and/or generated within the production facility, which include direct input of data being generated at an upstream supplier location within and/or outside the company.

It should be integrated to bar code scanners, radio frequency transmitters, control systems, quality assurance system, WOM, WIP and allow for manual data entry, etc.

Exception management provides the ability to respond to un-anticipated events that affect the production plan such as machine break-downs, out of range conditions and material storage. It should be able to implement alternative actions in order to respond to unplanned resource interruptions.

The execute business process includes the following high-level functionality:

Transaction execution

Generates, assigns and maintains action items such as production targets, shift instructions (including preshift plant status and instructions) and safety information to operators. It captures and documents plant events such as equipment breakdowns, safety incidents, mode switches, etc. Transaction execution allocates resources and schedule stock transfers, product blends, sampling, inspections, etc. based on an optimized production plan. It also enables production mode switchover and routing WIP status, quantity and quality.

Distribute information

Comprises integration to production planning and scheduling, maintenance, safety and health, and in/outbound logistics scheduling and control systems.

Capture data

Captures all data related to plant, production and process, including process deviations, safety, health, risk, environment and quality-related data automatically/electronically or through manual intervention by the operator.

Inventory tracking

Inventory tracking comprises raw material, process material, WIP, off-spec, and final product storage location planning and management. This includes inventory overviews, consumption planning and take-off and stock reconciliation. Weighbridge business rules and instrument scheduling and management are some of the most important activities.

Process management

Process management includes SQC/SPC, calculation of performance indicators, real-time production costs, availability and utilization, SHERQ requirements, and management of abnormal conditions.

4.4.3   Process

Process includes all the features needed to physically produce products. It also provides the means to automate and control the process by either linking directly to the control system layer or through operator messages. The control system layers provide inputs and outputs, or status points of the process. It encompasses all processes the plant uses for producing many products, including the MMI to the process for operators and managers, process control and optimization and alarm management.

The *process monitor* function enables the setting up of methods and logical checks for starting and stopping production machinery, with the ability to establish parameters for complex system alarms, making the monitoring of parameters and system alarms possible.

This function passes requests for control actions to the machinery control function. It describes which control schemes are implemented within the process control function. The process monitor function should be integrated to the process control system, process alarm system, inventory tracking and management systems, planning system interface, work-order management, quality management system, maintenance management, etc. Data interchanged may be operator messages, product items, batch numbers, production lines, raw materials utilized, materials utilized and left, cycle times, changes, resource status, etc.

The *machinery control* function has the ability to directly (by interfacing with the control system layers) and indirectly (through operator messages) start and stop production machinery. It dictates which control actions are implemented within the machinery control function. The machinery control function should be integrated with process control system (SCADA, HMI, DCS, etc.). Data interchanged may be process status, process set-points, operator messages, etc.

4.4.4   Analyze

The analyze business process allows supervisory personnel to analyze meaningful data from all sources and therefore evaluate production performance, product quality, process capability and regulatory compliance. Data is also made available to ERP and SCM systems, as well as to the customer and suppliers for analysis purposes. Data may be real-time based or long-term historical data. Real-time data allows personnel to coordinate production actions in order to optimize process efficiency. Long-term analysis is applied for re-engineering actions in order to improve the overall process efficiency. The analyze business process supports management decision-making based on actual data/management information.

The *data validation* function is devoted to collecting raw data from external systems (HMI, Historian packages, etc.) and applying specific validation rules and congruence relation tests in order to obtain a set of recalculated and filtered data. This not only acts as an online check for the validity of data being used by the PROCESS business process, but also ensures that information made available to the planning system is an accurate representation of plant conditions.

Validation rules applied to process-monitored variables (energy consumption, temperature, pressure, speed, etc.) may be the average, standard/deviation, maximum, minimum, etc. Data validation may be integrated with HMIs, PLCs, DCSs, historian systems, batch management systems, etc. and the data interchanged may be order numbers, raw materials, cycle times or any of the other data captured by the other REPAC business processes.

The *product genealogy* function gives access to a full history of the manufacturing process with information stored on the components used, process start/stop times and process parameters for each unique product or lot. The most frequent use of this information is for warranty and product statistical information, as well as inventory information. It also provides quality assurance department with a vast statistical base for analysis and provides purchasing with better supplier product performance information. It describes specifically what and how data is to be collected i.e. manually or automatically.

The product genealogy function may be integrated with maintenance management, quality assurance, purchasing, inventory tracking, etc. and the data interchanged may be work-, item-, or lot-numbers, time and date of production, line number, raw material batch number, etc.

The *reporting* function collects detailed information related to production operations, material movements, process conditions and generates specific reports on demand. Process data, instructions and other production detail associated with the manufacture of a 'batch' or of a 'lot' are typically captured and stored in relational databases to enable their electronic retrieval.

Statistical process control is a quality control method that focuses on the continuous monitoring of a process rather than on the inspection of finished products. The SPC function is applied online to measure deviation from the set target for the process, to plot charts in real time to show the operator a graphical view of the process performance, to apply statistical tests to the data gathered and set off SPC alarms if any out-of-control condition are detected. It may be integrated with alarm system, quality management system, etc.

The *summarize data* function passes data downstream and upstream in order to update its suppliers and customers on material and product status. The system coordinates the lack of work orders on the shop floor, activating a request to an external scheduling system when a new production plan is required. It also ensures that materials for the works order actually exist in the warehouse before production is finally started.

The *regulatory compliance* function is applied anywhere compliance with government regulations and procedures is necessary. Design, implementation and integration of systems business processes have to satisfy the regulated functional requirements for the different business functions related to the manufacturing process. Electronic business records (EBR) is the efficient alternative to the vast amount of paperwork which used to be required to document a compliant manufacturing process.

This ensures that manufacturing operations are completed in accordance with approved specifications. All aspect of the operation should be recorded in a current good manufacturing practices (CGMP) regulated environment.

The *quality lab results* function collects detailed information related to tests done on the product and can compare the information with target values, or pass the information to the quality control function for SPC analysis, etc.

The *process capability and performance* function collects data regarding the actual performance of the process and measured values of the product from the process monitor function and reporting. It compares the data to the process and product specifications and thus supplies a decision support system. It may be integrated with reporting, ERP, routing, process Management, etc.

The analyze business process includes the following high-level functionality:

Production performance, product quality and regulatory compliance

The major functions are combining data from multiple components and assembling those for ERP, SCM, customers and suppliers. It includes data organization, analysis, reporting, archiving and performance measurement (refer Figure 4.4).

4.4.5   Coordinate

The coordinate business process coordinates plant operations with the enterprise and supply chain. It defines the optimized sequence of plant activities to meet the production requirements as defined by demand management and adjusts the behavior of the plant as financial drivers change. It optimizes the plant operations to meet the scheduled demands and converts sales requirements into operations plans. It ensures seamless delivery and procurement by coordinating with the delivery schedule of finished goods to meet scheduled dates. It also coordinates with maintenance for optimal availability of equipment, tools, etc.

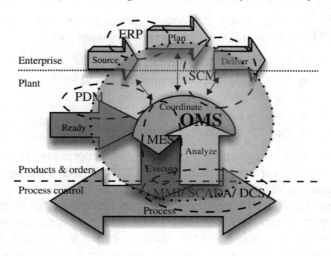

Figure 4.4
System view to REPAC

Required orders receives new order data from scheduling and passes information to the Execute-WOM module. This new information is then incorporated into the existing work order list.

Production delivery receives data from the execute-WIP function, when a work order is about to be completed and sends data to Execute-materials management function, to arrange for the completed work order to be transported to its destination.

Equipment maintenance receives data from the process-process monitor and machinery control modules concerning machine performance, running hours, etc. and transfers it to the maintenance system. Maintenance schedules from the maintenance system are passed to the scheduling system for inclusion in the planning cycle. The maintenance management system can be a very sophisticated integrated package, which provides historical, current and planned maintenance events.

The *process performance* function gathers data from the analyze-process capability and performance and from the execute-work order management, inventory tracking, WIP and Exception management and passes it on to the WOM system for processing to allow the fine-tuning of the current production plan.

Non-compliance correction defines the actions to be taken when analyze detects product manufactured outside production parameters.

Export production status and production performance function gathers data from the analyze business process and from the execute business process and transfers it to the scheduling system for processing.

4.5   Introduction to the IEC (6)1131-3 standard

Standards help in organizing specific industries. They help protect customers and suppliers by providing a solid base around which expectations, for users and developers, can be based. This ensures that a certain level of stability can be maintained, encouraging growth of the industry for both users and suppliers. In 1979 a working group was set up by the International Electro-technical Commission (IEC) to look at the complete standardization of PLCs.

The working group decided to develop the new standard (to be later called the IEC 1131 and then IEC 61131) in five separate parts:

- Part 1 – general information
- Part 2 – equipment requirements and tests
- Part 3 – programing languages
- Part 4 – user guidelines
- Part 5 – messaging service specification.

Part three, 'programming languages for programmable controllers', was issued in 1993 and provides a specification for PLC software. Part three covers PLC configuration, programing and data storage.

The IEC 61131-3 standard provides a framework for developing PLC programmes that are general and do not require manufacturer specific training. Most PLC and industrial control system manufacturers have adopted the standard. IEC 61131-3 is a world-wide standard that harmonizes the way people look to industrial control by standardizing the programing interface. It includes the definition of the sequential function charts (SFC) language, used to structure the internal organization of a program, and four interoperable languages: instruction list (IL), ladder diagram (LD), function block diagram (FBD), and structured text (ST). In addition, IEC 61131-3 structures the way a control system is configured.

4.5.1   Sequential function charts

The core language of the IEC 61131-3 standard, divides the process cycle into a number of well-defined steps, separated by transitions. The other languages are used to describe the actions performed within the steps and the logical conditions for the transitions. Parallel processes can easily be described using SFC.

4.5.2   Function block diagram

Function block diagram (FBD) is a graphical language that allows the user to build complex procedures by taking existing function blocks from the IEC 1131 compliance library, and wiring them together on screen. The diagrams are colorcoded and can be zoomed-in to view the whole diagram or specific areas in more detail.

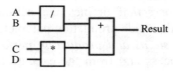

4.5.3   The ladder diagram

Ladder diagram (LD) is one of the most familiar methods of representing logical equations and simple actions. Contacts represent input arguments and coils represent output results.

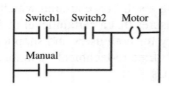

4.5.4   Structured text

Structured text (ST) is a high-level structured language with a syntax similar to Pascal but more intuitive to the automation engineer. This language is primarily used to implement complex procedures that cannot be easily expressed with graphical languages (e.g. IF/THEN/ELSE, FOR, WHILE . . .).

```
IF A > B THEN
  D := 1;
ELSEIF A = B + 2 THEN
  D := 2;
ELSE
  d := 3;
END_IF;
```

4.5.5   Instruction list

Instruction list (IL) is a low-level Boolean language similar to the simple textual PLC languages that are programed at the register level.

```
LD   A
ADD  B
MUL ( A
SUB  B
)
```

By implementing this standard on many programme development environments, users can move between different brands and types of control with very little training and exchange applications with a minimum of effort. To guarantee compliancy, certification by accredited institutes has been realized, increasing the common implementation of this standard.

Three levels of compliance are currently defined:

Base-level compliance

This level defines an essential core of the 61131 standard and the necessary features of each supported language. Base-level compliance criteria are defined for each of the 1131 languages. A product can be certified as base level compliant in one, several or all of the languages. Base-level compliance is restricted to include only a few and very basic data types, a restricted set of standard functions. Base level compliance indicates that the vendor is a serious member of the 61131 community and is committed to use of a standard syntax.

Portability level compliance

This level defines a much larger set of compulsory features where in addition, compliant products must incorporate an import/export tool that allow them to exchange 61131 software with other portability level compliant systems. The software exchange is based on an open file format that is over and above the IEC standard.

Full compliance

This level requires complete implementation of the 1131 standard.

4.6   S88 batch control standard

The S88 international standard on batch control provides standard models and terminology for defining the control requirements for batch manufacturing plants. It emphasizes good practices for the design and operations of a batch manufacturing plant. It is also used for better control of batch manufacturing plants regardless the degree of automation. S88 provides a standard terminology and a consistent set of concepts and models for batch manufacturing plants and batch control.

It improves communication between all parties involved, which will reduce the user's time to reach full production levels for new products. The S88 standard enable vendors to supply appropriate tools for implementing the batch control and also enables the user to identify their needs better. These standards make recipe development straightforward and contribute in the cost reduction of automated batch process.

In this standard, a batch process is a process that leads to production of finite quantities of material by subjecting quantities of input materials to an ordered set of processing activities over a finite period of time using one or more pieces of equipment. Batch processes are discontinuous processes and are neither discrete nor continuous.

A batch is:

- The material that is being produced or that has been produced by a single execution of a batch process
- An entity that represents the production of a material at any point in the process.

4.6.1   Process model

The sub-divisions of a batch process can be organized in a hierarchical fashion, called a process model, or entity-relationship diagram (refer Figure 4.5) with four entities.

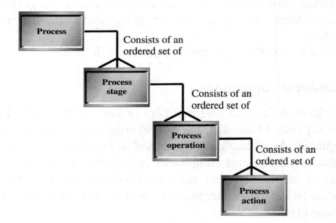

Figure 4.5
Process model

Process   Consisting of one or more process stages, which are organized as an ordered set, that can be serial, parallel or both.

Process stage   Usually results in a planned sequence of chemical or physical changes in the material being processed (process operations), and operates independently from other process stages. 'polymerize', 'dry', etc. for example.

Process operation   Each process stage consists of one or more process actions that carry out the processing required by the process operation. It describes minor processing activities that are combined to make up a process operation. e.g. 'addition of material', 'heating', 'maintaining fixed temperature'.

Process action Each process operation can be sub-divided into an ordered set of one or more process actions that carry out the processing required by the process operation. Process actions describe minor processing activities that are combined to make up a process operation. Adding catalyst in to a reactor is an example of a process action.

4.6.2   Physical model

The S88 standard describes a physical model which can be used to describe the physical assets of an enterprise. The physical assets of an enterprise are organized in a hierarchical fashion for batch manufacturing. The initial three levels are not in the scope of the S88 batch control standard (refer Figure 4.6).

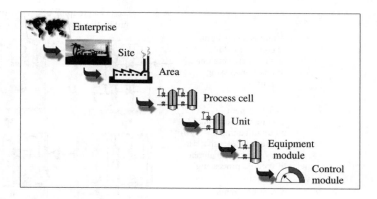

Figure 4.6
Physical model

Enterprise

An enterprise is responsible for determining the following:

- What products to be manufactured?
- In which site it will be produced?
- How it should be manufactured?

The boundaries of an enterprise are affected by many factors, which are not covered under S88 standard. An enterprise is a collection of one or more sites,

Site

A site is a physical, geographical or logical grouping determined by the enterprise.
The boundaries of a site are not dependent on technical criteria but are rather based on organizational or business criteria.

Area

An area is a physical, geographical or logical grouping determined by the site, which may contain process cells, units, equipment modules and control modules.

Process cell

All units, equipment modules and control modules required to make one or more batches are part of the process cell (refer Figure 4.7).

A process cell is a logical grouping of equipment that includes the equipment required for production. A process cell allows for production scheduling on a process cell basis and allows process cell-wide control strategies, which are very useful in emergency situations. By using various techniques and methods, process control activities respond to a combination of control requirements.

Train   Is a subdivision of a process cell and is composed of all units and other equipment required for a specific batch. All the equipments of a train are not always utilized by a batch. Multiple batches or products can use a train simultaneously.

Path   Is the systematic order in which a batch uses the equipment. A process cell may consist of multiple trains and all the equipment resides within the boundaries of the process cell.

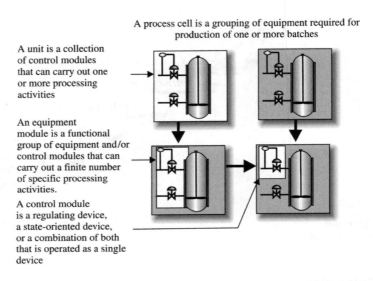

Figure 4.7
Equipment model

Unit

A unit consists of equipment modules and control modules. React, crystallize and make a solution, etc. are the processing activities which can be done in a unit. It is generally centered on a major piece of equipment (all necessary physical processing and control equipment behaves as independent equipment e.g. reactor). Units operate relatively independently from each other and are connected to other equipment, required to complete the major processing task. The S88 standard presumes that the unit does not operate on more than one batch at the same time. A unit can operate either on a complete batch or a portion of a batch.

Equipment module

An equipment module is part of a unit, or a stand-alone equipment group within a process cell. It is made of control modules and subordinate equipment modules. Stand-alone equipment groups can be used for an exclusive purpose or be available for shared use. The equipment module is described by the finite tasks (specific minor processing activities such as dosing and weighing, etc.) it is designed to carry out. It is generally centered around a piece of processing equipment, and combines all the necessary physical processing and control equipment to carry out specific processing activities.

Control module

A control module is a collection of sensors, actuators, other control modules and associated processing equipment, which is operated as a single entity. A control module can have other control modules within it. (A header includes several on/off automatic block valves, which coordinates and directs flow to one or several destinations based upon the set point directed to the header control module.) The equipment models are themselves made up of 'control modules'. The control modules implement any state-oriented, or algorithm-oriented, control. Control modules also provide the interface to the sensors and actuators within the unit. A control module is a collection of sensors, actuators and basic control logic that acts as either a regulating device, a state-oriented device or a combination that is operated as a single device. Examples of control modules are PID controllers and block valve controllers (refer Figure 4.8).

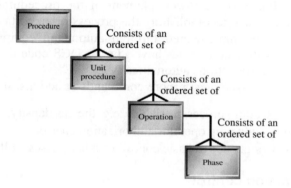

Figure 4.8
Procedural control

4.6.3 Batch control

Batch manufacturing consists of three types of control – basic control, procedural control, and coordination control.

Basic control

Basic control includes the control that maintains a specific state of equipment and process e.g. sequence control, regulatory control, etc. A basic control can be controlled by operator commands or by procedural or coordination control. It may respond to process conditions which influence the control output or trigger corrective action. Basic control is principally the same for batch and continuous processes.

Procedural control

It is a characteristic of batch processes that all equipment-oriented actions take place in a sequence to execute a specific process. This sequence of actions is ordered by procedural control and enables equipment to perform a batch process.

Procedural control consists of procedural elements – procedures, unit procedures, operations and phases.

Procedure The procedure is the highest level in the hierarchy and deals with the strategy related to batch-making through an ordered set of unit procedures. An example of a procedure is 'make product.'

Unit procedure An ordered set of operations producing continuously (in a sequence) within a unit. It is presumed that at any one time, one operation takes place in a unit. Generally, an operation is completed in a single unit but one procedure may also be carried out in multiple units.

Examples of unit procedures include the following:

- Polymerize substrate
- Recover residual substrate
- Dry product.

Operation An operation consists of an ordered set of phases. It defines a major processing sequence, which takes the material (to be processed) from one state to another and involves a chemical or physical change. It is preferable to have boundaries in order to suspend the operation safely. Preparation, charge, etc. are example of operations.

Phase Phase is the lowest element of the procedural control hierarchy, sub-divided into smaller parts accomplishing the process-oriented tasks in the phases. An equipment phase may be implemented through automated equipment, or it may be performed manually. Automated phases have PLC or DCS code that implements a series of control steps and control actions automatically.

A phase can issue one or more commands or actions, such as:

- Reading process variables, e.g. the gas density, gas temperature, etc.
- Conducting operator authorization checks
- Commands to basic control or other phases, collection of data, etc.

Coordination control

The execution of procedural control and utilization of equipment entities are directed, initiated and modified by coordination control. It is not structured along specific process-oriented tasks like procedural control but varies with time. Supervising the availability of capacity and allocating the equipment, requesting for allocation and coordinating with common resources for batches are examples of coordination control.

4.6.4   Batch recipe

A recipe is an entity that contains the minimum set of information that uniquely defines the manufacturing requirements for a specific product. Recipes provide a way to describe products and how those products are produced. Depending on the specific requirements of an enterprise, other recipe types may exist. However, this standard discusses only the four types of recipes typically found in an enterprise.

General recipe A general recipe is an enterprise-level recipe that is developed without any knowledge of the process-cell equipment that is to be used for manufacture of the product. It is not concerned with the process site, but identifies raw materials, their relative quantities and processing required. It communicates with multiple manufacturing locations and is used for enterprise-wide planning and investment purposes.

Site recipe The site recipe is specific to a particular site and is the combination of site-specific information and a general recipe. It is derived from the general recipe to meet the conditions of a specific manufacturing location and provides the level of detail necessary for site-level, long-term production scheduling.

Master recipe A master recipe can be derived from a general recipe or a site recipe. This type of recipe is targeted at a process-cell or a subset of the process-cell equipment, and can be created as a stand-alone entity if the recipe creator has the necessary process and product knowledge. This recipe not only contains material requirements but also process-cell equipment requirements to manufacture a product at a specific plant.

Control recipe The control recipe starts as a copy of the master recipe and is modified as required with scheduling and operational information to be specific to a single batch. It has product-specific process information required to manufacture a particular batch of product and also provides the level of detail necessary to initiate and monitor equipment procedural entities in a process cell. It can be modified to account for actual raw material qualities and actual equipment to be utilized. Modifications of a control recipe can be done several times over a period of time based on scheduling, equipment and operator information.

General/site recipes describe the technique i.e. how to do it in principle. Master and control recipes describe the task, that is, how to do it with actual resources. Recipes contain the following categories of information: header, formula, equipment requirements, procedure and other information (refer Figure 4.9).

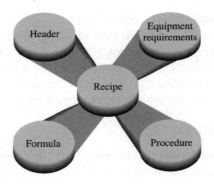

Figure 4.9
Recipe components

Recipe components

Header

The administrative information in the recipe is referred to as the header. Typical header information may include the recipe and product identification, the version number, the originator, the issue date, approvals, status and other administrative information. For example, a site recipe may contain the name and version of the general recipe from which it was created.

Formula

The formula is a category of recipe information that includes process inputs, process parameters and process outputs.

A process input is the identification and quantity of a raw material or other resource required to make the product. Process inputs include raw materials, energy and other resources such as manpower, which are consumed in the batch process during the manufacture of a product. Process inputs consist of both the name of the resource and the amount required to produce a specific quantity of finished product. Quantities may be specified as absolute values, or as equations based upon other formula parameters, or the batch or equipment size. The formula is not split between the material addition/activity sequence (procedure) and the quantities of added materials (ingredients), which may be a drawback in some industries. Process inputs may specify allowable substitutions, expressed in the same basic form.

A process parameter detail information such as temperature, pressure or time that is pertinent to the product but does not fall into the classification of input or output. Process parameters may be used as set points, comparison values or in conditional logic.

A process output is the identification and quantity of a material and/or energy expected to result from one execution of the recipe. This data may detail environmental impact and may also contain other information such as specification of the intended outputs in terms of quantity, labeling and yield.

The types of formula data are distinguished to provide information to different parts of an enterprise and need to be available without the clutter of processing detail. For example, the list of process inputs may be presented as a condensed list of ingredients for the recipe, or as a set of individual ingredients for each appropriate procedural element in a recipe.

Equipment requirements

Equipment requirements constrain the choice of the equipment that will eventually be used to implement a specific part of the procedure. In the general and site recipes, the equipment requirements are described in general terms, such as allowable materials and required processing characteristics. It is the guidance from and constraints imposed by equipment requirements that will allow the general or site recipes to eventually be used to create a master recipe, which targets appropriate equipment.

Recipe procedure

The recipe procedure defines the strategy for carrying out a process. The general and site recipe procedures are structured using the levels described in the process model since these levels allow the process to be described in non-equipment specific terms. The master and control recipe procedures are structured using the procedural elements of the procedural control model, since these procedural elements have a relationship to equipment.

The S88 model also provides a convenient means to integrate manufacturing operations with corporate enterprise resource planning (ERP) systems. Most systems on the factory floor, such as DCS or SCADA systems, have an equipment centric view of the factory. This is absolutely necessary for the correct operation of the factory, but does not match the product centric view of an ERP system. However, batch automation systems know the association of product to equipment, and as such they know what materials went into a specific batch, what materials were produced by the batch, what equipment was used to make the batch and what personnel worked on the batch. Batch automation systems provide a natural point of interface to business systems, and they complement the functionality of the business system through bridging the gap between costing and scheduling activities, and the factory floor equipment.

4.7   S95 Enterprise-Control System Integration Standard

The S88.01 model for batch manufacturing stop at the process cell but real batch manufacturing facilities also deal with issues of area management, site management and the integration of manufacturing facility systems with other business systems. The

International Standard Association (ISA) S95 committee have addressed these areas through the definition of a new standard, S95.01 Enterprise-Control System Integration, which contains a model for the functions of area and site management, and a model of the interfaces between an enterprise's business systems and its manufacturing control systems. The ISA S95 standard provides standard models and terminology for defining the interfaces and interactions between an enterprise's business systems (ERP) and their manufacturing control systems (MES).

4.7.1 Equipment hierarchy

The physical assets of an enterprise involved in manufacturing are usually organized in a hierarchical fashion. This figure (refer Figure 4.10) is an extension of the S88 Physical model and it includes the definition of assets for discrete and continuous manufacturing. The equipment hierarchy model additionally defines some of the objects utilized in information exchange between functions.

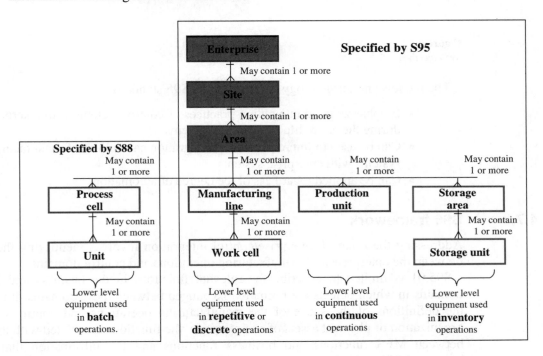

Figure 4.10
S88 and S95 physical models

4.7.2 Objective of S95

Integration of manufacturing control systems with the rest of the business is one of the most difficult problems, which is not related to technology but more to people and organizational problems. The viewpoint of IT organizations that typically develop, purchase and maintain business systems (ERP) are entirely different than that of the engineers of manufacturing systems (MES). It is mostly a cultural problem: the two groups do not share a common terminology, their viewpoints of what is important differ and their critical success factors are different.

The two groups often use the same terms for different things, or different terms are used for the same things. The S95.01 standard provides a solution to this situation by defining a common set of terms and definitions of the information and activities associated with logistics and manufacturing integration. The terms include definitions of the activities of business logistics systems, the activities of manufacturing control and coordination systems and reduce the need for custom-integration solutions, simplify multi-vendor integration, and improve reusability and transportability of functions across the enterprise (refer Figure 4.11).

Figure 4.11
S95 objectives

The models and terminology in part 1 of the S95 standard

- Emphasize good integration practices of control systems with enterprise systems during the entire life cycle of the systems
- Can be used to improve existing integration capabilities of manufacturing control systems with enterprise systems
- Can be applied regardless of the degree of automation.

4.7.3   S95 framework

Addressing the issue of enterprise/control integration involves identifying the boundary between the enterprise and manufacturing operations and control domains.

S95.01 is limited to describe the relevant functions in the business and MES level domains in which objects are normally exchanged between these domains. It is limited to the definitions of the scope of the manufacturing operations and control domain, the organization of physical assets of an enterprise, the functions associated with the interface between MES functions and business functions and the information that is shared between MES and business functions (refer Figure 4.12).

4.7.4   S95 model

The S95.01 standard uses multiple models to explain the elements of enterprise-control system integration. The standard provides models and information in various levels of detail and abstraction. These levels are illustrated in Figure 4.13, which serves as a map to the S95.01 standard. Each model and diagram increases the level of detail defined in the previous model. Multiple models are defined in the standard.

- Functional hierarchy model
- Equipment hierarchy model
- Functional data flow model
- Object model.

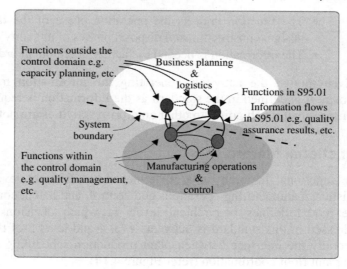

Figure 4.12
S95.01 framework

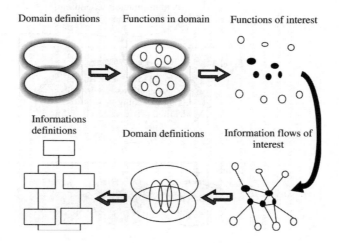

Figure 4.13
S95 models

Hierarchy models describe the levels of functions and domains of control associated within manufacturing organizations. A data flow model describes the functional and data flows within manufacturing organizations. An object model describes the information that may cross the enterprise-control system boundary.

4.7.5   Criteria for inclusion in manufacturing operations and control domain

The criteria for deciding what activities are in the control domain are:

- The function is critical to maintaining regulatory compliance and that includes safety, environmental and CGMP compliance.
- The function is critical to plant reliability.

- The function impacts the operation phase of the facility's life, as opposed to design, construction and disposal phases of a facility's life.
- The information is needed by facility operators in order to perform their jobs.

Even though the activities generating the information may be outside the control domain, it is defined in the standard as the information is required in the control domain. It ensures that touch points to other enterprise activities are not forgotten.

4.7.6   Functional hierarchy

The functional hierarchy model defines three different levels: business planning and logistics, manufacturing operations and control, and batch, continuous or discrete control. The model defines hierarchical levels at which decisions are made. The interface addressed in this standard is between level 4 and level 3 of the hierarchy model. This is generally the interface between plant production scheduling and operation management and plant floor coordination (refer Figure 4.14).

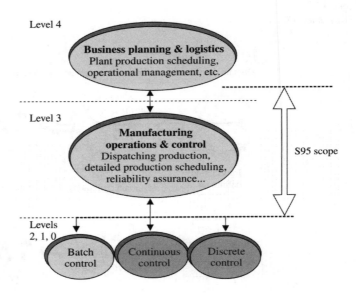

Figure 4.14
Functional hierarchy model

Levels 2, 1 and 0 define the cell or line supervision functions, operations functions and process control functions, and are not defined in the S95.01 standard. The S88 standard already discussed defines the batch control functions.

Level 3 activities

Level 3 activities include:

- Reporting on area production including variable manufacturing costs.
- Collecting and maintaining area data on production, inventory, manpower, raw materials, spare parts and energy usage.
- Performing data collection and off-line analysis as required by engineering functions. This may include statistical quality analysis and related control functions.

- Carrying out needed personnel functions such as work period statistics (time, task, etc.), vacation schedule, work force schedules, union line of progression, in-house training and personnel qualification.
- Establishing the immediate detailed production schedule for its own area including maintenance, transportation and other production-related needs.
- Locally optimizing the costs for its individual production area while carrying out the production schedule established by the level 4 functions.
- Modifying production schedules to compensate for plant production interruptions that may occur in its area of responsibility.

The additional descriptions of the activities contained in level 3 are based on the MESA model of activities. These activities are in the following general areas:

- Resource allocation and control
- Dispatching production
- Data collection and acquisition
- Quality management
- Process management
- Production planning and tracking
- Performance analysis
- Operations and detail scheduling.

The section also defines activities where the enterprise and the control systems generally share responsibility. These areas include: document control, labor management and maintenance management.

Level 4 activities

The level 4 activities include:

- Collecting and maintaining raw material and spare parts usage and available inventory, thereby providing data for purchasing of raw material and spare parts.
- Collecting and maintaining overall energy use and available inventory data, and providing data for the purchasing of energy sources.
- Collecting and maintaining overall goods in process and production inventory information.
- Collecting and maintaining quality control information as they relate to customer requirements.
- Collecting and maintaining machinery- and equipment-use and life history data necessary for preventive and predictive maintenance planning.
- Collecting and maintaining manpower use data used by personnel and accounting.
- Establishing the basic plant production schedule.
- Modifying the basic plant production schedule for orders received, based on resource availability changes, energy sources available, power demand levels, and maintenance requirements.
- Developing optimum preventive maintenance and equipment renovation schedules in coordination with the basic production schedule.
- Determining the optimum inventory levels of raw materials, energy sources, spare parts and of goods in process at each storage point. These functions

also include material requirements planning (MRP) and spare parts procurement.
- Modifying the basic production schedule as necessary whenever major production interruptions occur.
- Capacity planning based on all of the above activities.

4.7.7   Functional data flow model

The functional data flow model presents

- The functions of an enterprise involved in manufacturing.
- The information flows between the functions that cross the enterprise-control interface.

The enterprise-control interface is described using a data flow model. The model is defined using the Yourdon notational methodology.

Symbol	Definition
Function (labeled ellipse)	A function is represented as a labeled ellipse. A function is a group of tasks that can be classified as having a common objective. Functions are organized in a hierarchical manner and are identified with a name and a number. The number represents an identification of the data model hierarchy level
External entity (labeled rectangle)	An external entity is represented as a labeled rectangle. An external entity is a component outside the model boundaries that sends data to and/or receives data from the functions
Data flow name →	A solid line with an arrow represents a grouping of data that flows between functions, data stores, or external entities, which is defined in the enterprise/control integration model. All solid lines have a name for the data flows. Data flows at one level of the functional hierarchy may be represented by one or more flows at the lower level of the hierarchy
- - - - - - →	A dashed line with an arrow represents a grouping of data that flows between functions, data stores or external entities, which is not pertinent to the enterprise/control integration model, but is shown to illustrate the context of functions. Dashed line data flows without names are not identified in this model

The wide dotted line (------) illustrates the boundary of the enterprise-control interface. The manufacturing control side of the interface includes most of the functions in production control and some of the activities in the other major functions. The lines indicate information flows of importance to manufacturing control. The wide dotted line intersects functions that have subfunctions that may fall into the control domain, or fall into the enterprise domain depending on organizational policies. The model structure does not reflect an organizational structure within a company, but an organizational structure of functions (refer Figure 4.15).

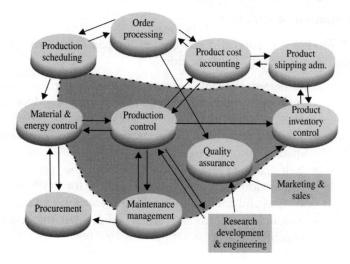

Figure 4.15
Functional data flow model

The data flow model gives the activities further definition and models the relationships between the activities. This model identifies specific enterprise functions and defines how these functions interact with the process control functions of manufacturing.

Functions

Order Processing (1.0)

There is generally no direct interface between the function of order processing and the manufacturing control function.

Functions of order processing

Customer order handling, acceptance and confirmation of orders, sales forecasting, waiver and reservation handling, gross margin reporting, converting customer orders to production orders.

Production scheduling (2.0)

The production scheduling functions interface to the manufacturing control system functions through a production schedule, actual production information and production capability information. Detailed scheduling, within an area, is defined as a control function.

Functions of production scheduling

Determines the production schedule, identifies long-term raw material requirements, determines the pack-out schedule for end products and determines product available for sale. The information generated or modified by the production scheduling function includes:

- The production schedule
- The actual production vs the planned production
- The production capacity and resource availability
- Current order status.

Production control (3.0)

The production control functions encompass most of the functions associated with manufacturing control and include:

- Issue requirements for raw materials, control of transformation of raw materials into end product in accordance with production schedule and production standards, plant engineering and updating of process plans.
- Produce reports of performance and costs; evaluate constraints to capacity and quality.
- Self-test and diagnostics of production and control equipment, creation of production standards and instructions for SOPs (standard operating procedures), recipes and equipment handling for specific processing equipment.
- Production control is further subdivided into operations planning, operations control and process support engineering.

Material and energy management (4.0)

This manages the inventory, transfers, and quality of material and energy. It generates requests for purchasing of materials and energy based on short-and long-term requirements. It calculates and reports inventory balances and losses of raw material and energy utilization. Receives incoming material and energy supplies, requests quality assurance tests and notifies purchasing of accepted material and energy supplies (after receipt and testing).

Procurement (5.0)

This manages the placing of orders for raw materials, supplies, spare parts, tools, equipment and other required materials with approved suppliers. Monitors progress of purchases and reports to requisitioners, releasing incoming invoices for payment after arrival and approval of goods. Collects and processes unit requests for raw materials, spare parts, etc., for order placement to vendors

Quality assurance (6.0)

This manages the testing and classification of materials and sets standards for material quality. Issues standards to manufacturing and testing laboratories in accordance with requirements obtained from technology, marketing and customer services. Collects and maintains material quality data and releases materials, for further processing or final delivery, after each quality step. It certifies that the product was produced according to standard process conditions, and compares product data and statistical quality control routine results against customer requirements to assure adequate quality before shipment. It relays material deviations to process engineering for re-evaluation and processes upgrades.

Product inventory control (7.0)

This function manages inventory of finished products, making reservations for specific product in accordance with product selling directives and packs end product in accordance with the delivery schedule. It reports on inventory to production scheduling and on balance and losses to product cost accounting. It arranges physical loading/shipment of goods in coordination with product shipping administration.

Product cost accounting (8.0)

This function calculates and reports on total product cost, provides cost results to production for adjustment and sets cost objectives for production. It collects raw material, labor, energy and other costs for transmission to accounting.

Product shipping administration (9.0)

This function organizes transport for product shipment in accordance with accepted order requirements, and negotiates and place orders with transport companies. It accepts freight items on site and release material for shipment whilst preparing accompanying documents for the shipment (bill of Lading, customs clearance). It confirms shipments and release for invoicing to general accounting, whilst it reports shipping costs to product cost accounting.

Maintenance management (10.0)

This function manages the provision of maintenance to existing installations. It provides preventative maintenance programmes and manages equipment monitoring to anticipate failure. Allows for the placing of purchase requests for materials and spare parts, develops maintenance cost reports and coordinates outside contract work effort. It provides status and technical feedback information on performance and reliability to process support engineering.

Research development and engineering

The general functions of Research, Development and Engineering include:

- Development of new products
- Definition of process and product requirements related to the production of the products.

Marketing and sales

The general functions of marketing and sales include:

- Generating sales and marketing plans
- Determination of requirements and standards for products
- Interaction with customers and their requirements, etc.

Data flow

The information flows of interest between these functions are:

The schedule information flows from the production scheduling (2.0) function to the production control (3.0) function.

The production-from-plan information flows from the production control (3.0) function to the production scheduling (2.0) function.

The production capability information flows from the production control (3.0) function to the production scheduling (2.0) function.

The incoming order confirmation information flows from the material and energy control (4.0) function to the procurement (5.0) function.

The material and energy order requirement information flows from the material and energy control (4.0) function to the procurement (5.0) function.

The long-term material and energy requirements information flows from the production scheduling (2.0) function to the material and energy control (4.0) function.

The short-term material and energy requirements information flows from the production control (3.0) function to the material and energy control (4.0) function.

The material and energy inventory information flows from the material and energy control (4.0) function to the production control (3.0) function.

The production cost objectives information flows from the product cost accounting (8.0) function to the production control (3.0) function.

The production performance and costs information flows from the production control (3.0) function to the product cost accounting (8.0) function.

The material and energy receipt information flows from the material and energy control (4.0) function to the product cost accounting (8.0) function.

The quality assurance (QA) results information flows from the quality assurance (6.0) function to the product inventory control (7.0) function and the production control, operations control (3.2) function.

The standards and customer requirements information flows from the marketing and sales function to the quality assurance (6.0) function, and from quality assurance to production control (3.0).

In-process waiver request information flows from production control (3.0) to the quality assurance (6.0) function.

Finished good waiver information flows from the order processing (1.0) function to the quality assurance (6.0) function.

The finished goods inventory information flows from the product inventory control (7.0) function to the production scheduling (2.0) function.

The process data information flows from the production control (3.0) function to the product inventory control (7.0) function and the quality assurance (6.0) function.

The pack out schedule information flows from the production scheduling (2.0) function to the product inventory control (7.0) function.

The product and process requirement information flows from the research development and engineering (RD&E) function to the quality assurance (6.0) function.

The product and process information request flows from the production control (3.0) function to the RD&E function.

The maintenance request information flows from the production control (3.0) function to the maintenance (10.0) function.

The maintenance response information flows from the maintenance (10.0) function to the production control (3.0) function.

Maintenance standards and methods information flows from the production control (3.0) function to the maintenance (10.0) function.

Maintenance technical feedback information flows from the maintenance (10.0) function to the production control (3.0) function.

Product and process technical feedback information flows from the production control (3.0) function to the RD&E function.

Maintenance purchase order requirements information flows from the maintenance management (10.0) function to the procurement (5.0) function.

Production Order information flows from order processing (1.0) function to production scheduling (2.0) function.

Availability information flows from the production scheduling (2.0) function to the order processing (1.0) function.

Release to ship information flows from the product shipping administration (9.0) function to the product inventory control (10.0) function.

Confirm to ship information flows from the product inventory control (7.0) function to the product shipping administration (9.0).

4.7.8   S95 information exchange categories

The S95 information exchange categories (between the enterprise control domain and the operations control domain) are defined, and the object models are specified in Part 1 of the S95 standard. Part 2 of the standard defines the object model attributes. Part 3 defines the production operations model, and the areas of information overlap and exchange (refer Figure 4.16).

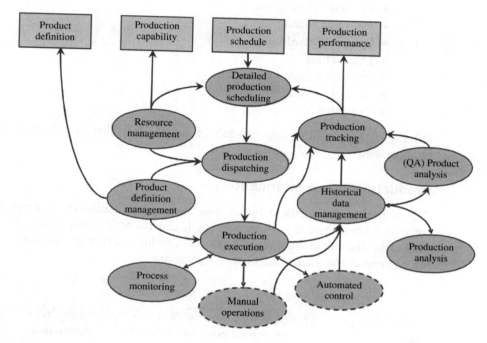

Figure 4.16
Production operations model

The S95 information exchange categories are the following:

Product definition information

Product life cycle management (PLM) information exchanged on how to make a product.

Production capability information

Information exchanged on required and available production resource capability and capacity.

Production schedule information

Information exchanged on what to make when and where to make it and resources to use.

Production performance information

Information exchanged on what was made and what resources were used. This includes feedback information needed to respond to the business system request to make product.

4.7.9   Activity models

Part 3 of the standard defines eight activities (as oppopsed to the 11 MES modules in Figure 4.17) and the high level data flows between the activities (refer Figure 4.18). These activities are:

- Definition Management
- Resource Management
- Detailed Scheduling
- Dispatching
- Execution Data Collection
- Analyze
- Tracking.

All the eight activities are generic in terms of quality, production and maintenance requirements.

Product definition information

S95 definition management defines the product information that the enterprise level keeps for recipe management such as S88 batch general recipes and site recipes. It also defines the batch, discreet and continuous process production rule information exchange between the enterprise and the manufacturing operations.

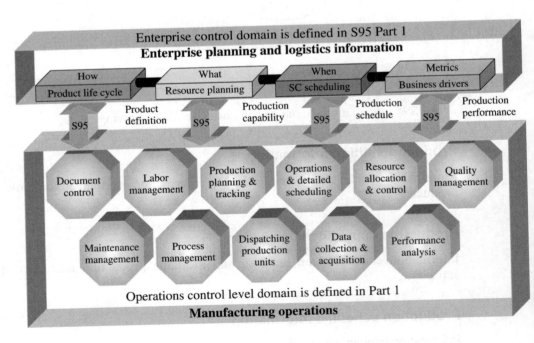

Figure 4.17
S95 Part 1 and 2 Definitions

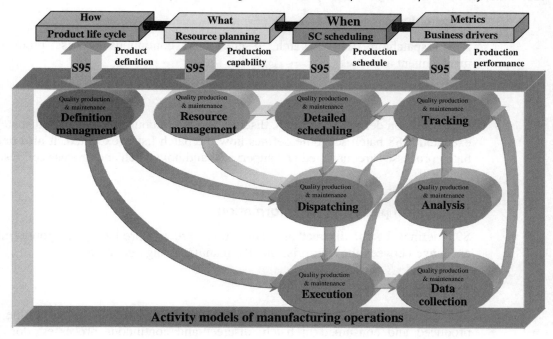

Figure 4.18
S95 Part 3 activities

Definition management

The S88 batch standard defines manufacturing operations information such as recipe management activities, master recipes and control recipes.

Production capability information

S95 defines batch, discreet and continuous process resource capability information exchange between the enterprise and the manufacturing operations.

Resource management

Resource management defines how capacity is managed in terms of available, committed and attainable capacity for batch, discreet and continuous manufacturing operations. S88 defines how batch materials are stored, transformed and moved between manufacturing resources.

Production schedule information

S95 defines batch, discreet and continuous process production schedule information exchange between the enterprise and the manufacturing operations.

Detailed scheduling

Detailed scheduling defines the site and area level finite capacity schedule activities for batch, discreet and continuous processes.

Dispatching

S95 dispatching defines batch, discreet and continuous process production flow management activities.

Execution

S95 execution defines how batch, discreet and continuous process work dispatch lists are executed. S88 batch schedule defines how the batch list is executed. It also defines how batch processes are managed and supervised and how batch and process cell resources are managed.

Production performance information

S95 defines batch, discreet and continuous process production response information exchange between the enterprise and the manufacturing operations.

Tracking

S95 tracking defines product and resource tracking activities. It tracks the resources produced and consumed in batch, discreet and continuous processes, does cost and performance analysis and reports the results. It follows material movements and determines product genealogy.

Analysis

S95 analysis defines product and process analysis activities for defines batch, discreet and continuous processes. It enables constraint analysis, SPC, SQC and quality test analysis.

Data Collection

S95 data collection collects, retrieves and archives data related to execution of batch, discrete and continuous process production requests. S88 keeps batch history through the recording of batch events.

4.7.10   Object models

This section of the standard is used to identify where in any particular company the functions are performed, what their local names are, and identifies the local names for the information exchange.

Some of the information in each of the areas must be shared between the manufacturing control systems and the other business systems. The standard is concerned with defining a model and common terminology for that information. Nine object models are defined in S95 part 2 (refer Figure 4.19).

The nine objects are divided into four categories:

Resource information, that includes personnel, equipment, material and process segments.
Capability information, that includes production capability and process segment capability.
Product definition information, and
Product scheduling information, that includes production schedule and production performance.

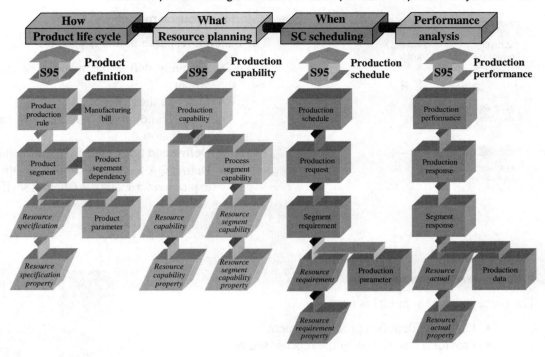

Figure 4.19
S95 object models

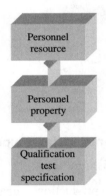

Resource Information

Personnel resource model

The personnel resource model is used to:

- Uniquely identify the person or personnel class
- Define skills and training for individuals or groups
- Define qualification tests, results and results expiration for individuals (such as licensing and certification).

Equipment resource model

The equipment resource model is used to:

- Uniquely identify the equipment or equipment class
- Define a description of the equipment
- Define the capability of the equipment
- Define capability tests, results and results expiration for equipment (such as safety inspections)
- Define and track maintenance requests.

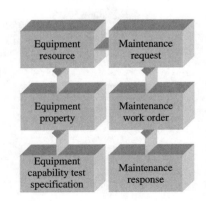

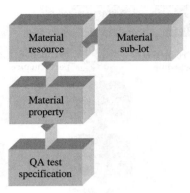

Material resource model

The material resource model is used to:

- Uniquely define the material or material class property
- Describe the material
- Define and track material lot and sub-lot information
- Define and track material location information
- Define QA test specifications, results and results expiration for materials (such as shelf-life).

Process segment model

The process segment model is used to:

- Uniquely identify a process segment
- Provide a description of the process segment
- Define resources used by the segment (personnel, equipment and materials)
- Define capabilities of the process segment
- Define process segment execution order (procedure).

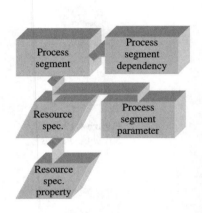

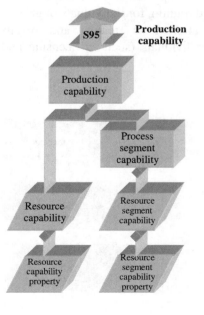

Capability information

Production and process segment capability model

The production capability model is used to:

- Uniquely define the production capabilityfor a specific element of the equipment model
- Provide a description of or other information about the production capability
- Provide current state of the capability (available, committed or attainable)
- Define a location for the capability
- Define the physical level of the capability (enterprise, site, area, Process cell, etc.)
- Define a start time and end time that defines the life cycle time for the capability
- Document the publish date for when the capability as published or generated.

Product definition information

The product definition model is used to:

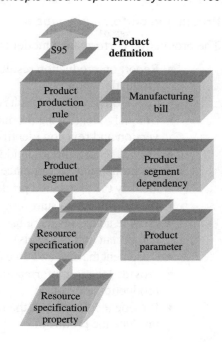

- Uniquely specify the product production rule
- (Recipe, Work Instructions, etc.)
- Provide a publish date and version for the product production rule
- Provide a description of or other information about the production rule
- Specify the bill of materials and material routing to be used
- Specify Product Segment requirements for the production rule (personnel, equipment and materials)
- Specify product segment execution order.

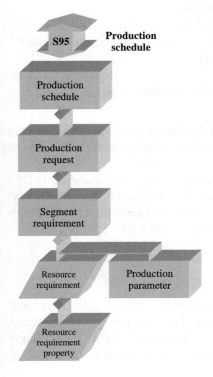

Product scheduling information

Production schedule model

The production schedule model is used to:

- Request production to make a specific product
- Provide a unique identifier for the request
- Provide a description of or other information about the schedule
- Provide a start time and end time for the schedule request
- Document the date and time for the published schedule
- Identify the location and equipment type for the schedule request (site, area, process cell, production line, etc.).

Production performance model

The production performance model is used to:

- Report on production results from the execution of scheduled production requests
- Report on production results due to an event during production
- Uniquely identify the production performance including version and revision identification
- Provide a description or additional information about the production performance
- Identify the associated production schedule
- Provide an actual start time and end time for production
- Provide actual resource use information (personnel, equipment and materials)
- Document the date and time that the performance was published
- Provide location information that identifies where the production took place
- Provide a definition of the physical equipment used to produce the product.

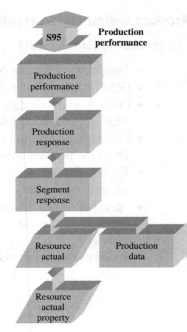

4.8   Code of Federal Regulation (CFR) Title 21 part 11

21 CFR part 11 is a USA federal regulation that came into effect on 20 August 1997. The regulation addresses the security and integrity of electronically stored data. Title 21 defines ALL aspects of food and drug research, clinical trials, manufacturing and distribution. Title 21 for food and drugs consist of 1404 parts.

Examples of some of the parts are:

Part 11   –   electronic records; electronic signatures
Part 120   –   hazard analysis and critical control point systems
Part 210   –   cGMP in manufacturing, processing, packing or holding of drugs
Part 211   –   cGMP for finished pharmaceuticals
Part 803   –   medical devices reporting
Part 806   –   medical devices; reports of corrections or removals
Part 820   –   cGMP for medical devices.

Part 11 adds no new requirements for additional record-keeping or signatures other than those already required, but other parts of Title 21 define those requirements, referred to as the predicate rule.

If electronic records are generated to satisfy any food and drug administration (FDA) requirement or submission, 21 CFR Part 11 compliance is not optional, but mandatory. Bio/pharmaceutical manufacturers selling product (bulk or finished) in United States must validate their compliance with Title 21, including 21 CFR Part 11 if they want to continue doing business in the USA.

4.8.1   Objective of 21 CFR part 11

The objective is to establish criteria under which electronic records = hardcopy records and electronic signatures = Handwritten signatures. This is to ensure the authenticity, reliability and trustworthiness of electronic records with the same integrity as hardcopy records. The regulation specifies how legally valid electronic records can be created so as to reduce the paperwork load on the industry.

4.8.2   Impact on software systems

The major areas of impact on systems are security and audit trails. This means that system access needs to be limited and controlled, much in the same way as hardcopy documents. This includes approval levels, authorizations and levels of access.

Systems that require compliance with this regulation need to automatically generate time-stamped audit trails for electronic records, so that it can be determined that the correct processes were followed and that only people with the correct level of authority approved product status-change.

This requires that systems have built-in operational system checks that enforce permitted sequencing steps/events. The system should therefore not allow materials to move to unauthorized areas or subsequent processes without the approval/authorization of someone with the correct level of authority.

Systems should also be able to keep track and determine the existence of altered records, a copy of the record before alteration and the identity of the person who affected the change. To enable the above, system security measures need to be implemented to verify individual identities of personnel accessing the system.

Obviously, systems must also have the ability to generate reports on all of the above to indicate compliance with this and other regulations.

4.9   Continuous improvement (Kaizen)

Kaizen means continuous stepwise improvement. It can be done as a sequence of small steps or as innovation using a new technology. The improvement starts with the problem identification and ends with the solution of the problem. The improvement increases with every solution of the problem.

Kaizen focuses on process-oriented quality control rather than result-oriented control and requires a continuous effort to improve the processes. Kaizen uses a set of methods and tools e.g. QC, SQC, QC-circle and TQC. The success of an innovation step is guaranteed with the application of Kaizen after each innovation step. An incremental development strategy is achieved by continuous improvement.

Maintenance is considered as a specialization of system development, a system develops through change. Thus system development and maintenance is a process of progressive change. Therefore, Kaizen can be considered as an engineering process for system development and maintenance (refer Figure 4.20).

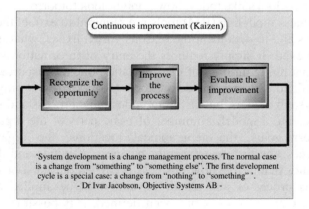

Figure 4.20
Continuous improvement (Kaizen)

5

Business process and system modeling tools and packages

Learning objectives

- To become familiar with different system and process modeling tools.
- To become familiar with various modeling packages that can be used for business process and system modeling.
- To understand where and how the various tools and packages can be applied.

5.1   Introduction

In today's business environment, change is constant and companies must respond quickly to take advantage of the changes, or cease to exist. Key decisions have the power to create a tremendous ripple effect – not only throughout the internal organization, but also across your entire supply chain. As soon as a process has been updated or a new system implemented, a rapid cycle of obsolescence begins. In an environment like this, organizations need the flexibility to change as new opportunities arise. Only, quick decisions are not enough, smarter decisions are needed, decisions that support business objectives and reduce the risks inherent in change. To improve business, a multi-dimensional, integrated business blueprint is required. A virtual schematic, that shows what a business looks like today, how it might look tomorrow and what it will take to get there.

Business modeling (BM) tools are required to extract the basic definitions of a business process and to document the reasons why it exists, what is expected from it and how this is achieved. It simply requires the entry of information starting with an idea or concept and continuing until all of the rules and policies, resource requirements and relationships are defined. Business modeling tools (BMTs) are not only used to model current process and system models, but also future, desirable process and system models.

To create a model, a number of basic parts of the organization will be documented – business goals, the geographical locations, the organizational structure, available resources, the different types of business activities and the relationships between them. The rules that define how business is carried out must be defined and associated with the various processes and resources, and must be applied consistently across the process model. Once a process has been defined, it is possible to derive object definitions. An object will represent a single component or element of a business process and will be associated with all of the methods and data that define its actions.

The relationship between objects can be shown through workflow diagrams that incorporate links into IT-based applications. Various companies in various industries are involved in software development projects, and as such there are wide differences in the business environment from project to project. Business modeling enhances the understanding of the business processes and helps companies answer key questions such as: what are we doing, why are we doing it, where is it being done, by whom, when and in what way?

On the BM level, many of the conflicts that exist within an organization can be corrected. Business modeling is a top-down exercise that starts at the top of the business and goes down to the lowest levels as more detail is added. A successful BM exercise should provide a complete understanding of the business processes, let everyone know where their activity fits into a bigger picture, and obtain agreement regarding the whole picture at all levels of the organization.

Once the business model has resolved the what and why issues, the process models can start to address the how, who, where and when issues. Each business process needs to be broken down into the specific tasks and activities required to complete it. The next step down the model hierarchy from a process model is an object model, and is used to define individual objects or classes.

A BMT is used during business and process analysis and can form the basis for software logic. The focus of BMT's decisions should be to suit the requirements of the industry, environment and level of detail in which it operates.

5.2 Generic BMTs

Companies can benefit from the use of generic modeling tools that can be used irrespective of the development environment the client provides i.e. Oracle or Microsoft. This will reduce the duplication of work during software development, starting from the initial analysis process to testing and maintenance of the software. The BMT should be generic enough to use in different project phases and in different environment. A BMTs output of a phase should be the input to the next phase. Therefore, a project supported by a BMT should ensure continuity of the project as it passes through various phases and personnel involved.

5.2.1 The need for BMT

Many of the activities performed during the development of a business model can be time-consuming and error-prone. For this reason, many business modelers use a variety of general-purpose and specialized tools to assist in developing a business model. Computer-based (packaged) tools can improve the productivity and quality of a modeler's work.

Businesses have multiple functions and the lack of a standard BMT causes inefficiency and repetitions in software development. A generic BMT helps to ensure:

- Minimum or no duplication of work, if the techniques are correctly applied and standards are followed
- The maintenance of a central repository of definitions associated with business modeling products
- Consistency-checking of diagrams against a repository
- Higher system integrity
- Consistency between different facets of a business model
- Consistency and re-use of developed system modules

- Shorter development time, and responding to rapid changes
- Cross-functional application of resources
- Make it easy to present different views of a single model.

A suitable BMT will help deliver a higher quality, lower cost system.

Business process modeling is used to analyze processes and set-ups, and further suggests different ways to improve the business. Business process modeling is a technique, which refers to the process in which a view of a business enterprise is taken and represented in an easily understandable format.

5.2.2   BMT capabilities

A BMT should be capable of the following:

- Be able to model people, tasks, IT, future plans such as:
    - The functions of each process in the business and their performance within their constraints, sequence, conditions, etc.
    - The components (or objects) used in the processes
    - Who does what? – the key role-players in the business.

- Generate one general model including all features of a business with the ability to model both people and the IT architecture.
- Be flexible enough so that components can easily be removed/altered.
- Provide a diagrammatic and straightforward model easy for everybody to understand.
- Be able to display various subparts of the system at a time.

Business processes, by their nature, are complex and a business model should be able to divide this complexity into smaller 'easy-to-manage' chunks. The model should be precise and clear in its definitions. Model components should be well defined with well-defined, simple interfaces.

The final model should be analyzed to identify any inconsistencies e.g. bottlenecks, deadlock, etc. The business model should be easy to interpret.

5.2.3   Business modeling techniques

There are many different BMTs. Some techniques have several variations on the same theme. A few basic techniques work well in most situations. Often the techniques focus on different aspects of the enterprise, one of which is modeling. When applied together, the techniques reinforce and complement each other (refer Figure 5.1).

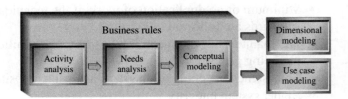

Figure 5.1
Business modeling tools

Even when each of the business modeling techniques above and their products are well understood, there are a number of problems that can hamper the development of a successful model.

- How to apply the technique, which is appropriate to the problems being addressed.
- Different people often apply the same technique differently.
- The same person may apply the same technique inconsistently.
- It is often difficult for a business modeler to decide which technique is appropriate at a particular phase of a project.
- People working on different projects may apply the techniques in different ways.

Activity models

Models the work that needs to be done in order to support an enterprise's mission independently from who performs the work, the forms used, or databases and information systems that may be used:

- Defines the business scope of a system
- Aligns the requirement of a system to the business activities performed
- Re-engineers the way the business currently works
- Helps to understand the context of the business
- Provides a framework for a structuring need analysis.

Information requirement

Identifies and focuses the business information requirements to plan, perform and monitor the activities:

- Identifies information required to perform an activity
- Focuses on business needs rather than software requirements
- Is structured around activities and technology
- Maintains a high level of redundancy.

Entity relationship models

Models the concepts (people, places, things, etc.) of interest to an enterprise and helps to identify the data that must be captured and stored to satisfy information requirements. The entity relationship (ER) modeling technique is a discipline used to illuminate the microscopic relationships among data elements. The highest art form of ER modeling is to remove all redundancy in the data. This is immensely beneficial to transaction processing because transactions are made very simple and deterministic. The transaction of updating a customer's address may devolve to a single record lookup in a customer address master table. This lookup is controlled by a customer address key, which defines uniqueness of the customer address record and allows an indexed lookup that is extremely fast. It is safe to say that the success of transaction processing in relational databases is mostly due to the discipline of ER modeling (refer Figure 5.2).

- Identifies the concept of relevance
- Concept can be abstract or concrete
- Structured around objects
- Optimized for updates
- Reusable and generic, no redundancy
- Relevant to current and future activities.

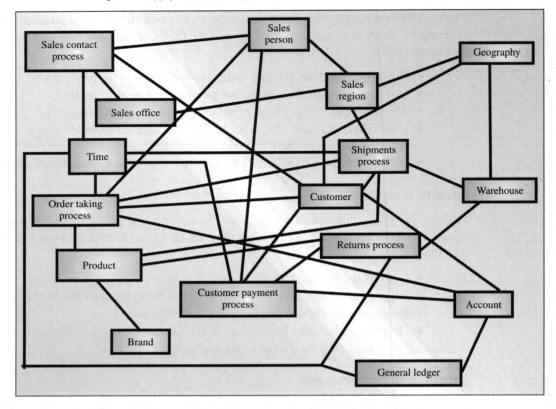

Figure 5.2
ER model

Dimensional models

Dimensional models (DMs) are intuitive and identify the data required for business analysis and decision support. The DM is a logical design technique often used for data warehouses. It is the only viable technique for databases that are designed to support end-user queries in a data warehouse. Every dimensional model is composed of one table with a multi-part key, called the fact table, and a set of smaller tables called dimension tables. Each dimension table has a single-part primary key that corresponds exactly to one of the components of the multi-part key in the fact table (refer Figure 5.3).

- Structured to assist with measurement, analysis and discovery
- Optimized for query.

Use case models

Identifies roles of users who will interact with software and the way a user will use a software system to support the business activities. It also emphasizes not on the internal, but on the external perspective of software. Use case modeling is a critical technique in developing an application. Within UML, Use cases are used primarily to capture the high-level user-functional requirements of a system. This long-winded description is important as use cases can neither be used to capture non-functional requirements nor can they be used to capture 'internal' functional requirements. The use case is not only important as a unit of requirement definition, but also as a unit of estimation and a unit of work.

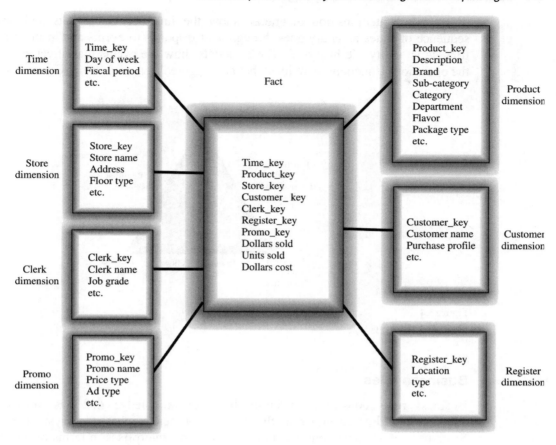

Figure 5.3
Dimensional models

- Actors (who is going to be using the system directly) and use cases define system boundaries and scope.
- Use cases support activity execution.
- Is structured around application use.
- The data elements map to information needs.
- The attributes map to the conceptual model.

Entity life cycle models

Identifies the life cycles of those things of interest to the organization and emphasizes the internal perspective of software rather than the external. There are three aspects of an organization that are the subject of most modeling techniques in use today. The three aspects are:

1. The objects that the business manipulates: the things of significance about which the organization must hold and maintain information
2. The functions or activities performed by the business
3. The events in the world that cause those functions to be performed.

The relationships between functions and objects are usually represented in function/entity matrices, showing which functions create, retrieve, update and delete which entities. Relationships between events and functions are shown effectively in 'essential data flow diagrams'; 'state/transition diagrams' show the link between events and objects. The sequence of states an entity goes through as it responds to events and in the middle of the triangle is 'entity life histories'. These models show the events that affect each entity and the operations (functions, activities) that are triggered as a result (refer Figure 5.4).

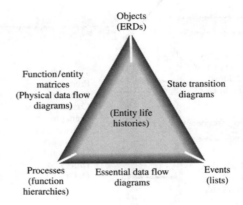

Figure 5.4
Entity life cycle model

Business rules

Business rules constitute an entire body of knowledge that has mostly not been adequately addressed either in the analysis or design phases of system development. Systems analysts have long been able to describe enterprises in terms of the structure of the data those enterprises use, and the organization of the functions they perform, but have neglected the constraints under which the enterprise operates. A business rule is important in its own right. It is different from the definition of data structure in a data model and from the definition of processes.

Business rules identify the global policy and rules that ensure integrity of a business model and are used with all of the above techniques:

- Defines and constrains some aspect of other models.
- Action rules define the dynamic aspects of a model.
- Structural rules define the static aspects of a model.
- Derivation rules define the way in which information can be derived from one of the models.

Matrices

Matrices can be used to cross-reference the products of all of the above techniques and provide a systematic approach to analysis of business models, which highlights omissions. Matrices are very useful for making systematic comparisons between two different types of things. Individual instances of the first type of thing are arranged as row headings while members of the second type are arranged as column headings. Each cell of the matrix can then show a comparison of the things in the intersecting row and column (refer Figure 5.5).

For example, activities and the people performing the activities can be compared by arranging the activities as the row headings of a matrix and the people performing the activities as the column headings. Each cell where an activity and person intersect can be used to indicate if the person performs the activity.

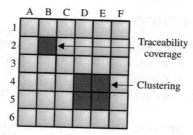

Figure 5.5
Matrices

5.3 IDEF0

IDEF0 is a method designed to model the decisions, actions, and activities of an organization, process or system. IDEF0 was derived from a well-established graphical language, the structured analysis and design technique (SADT). Effective IDEF0 models help to organize the analysis of a system and to promote good communication between the analyst and the customer. IDEF0 is useful in establishing the scope of an analysis, especially for a functional analysis.

It is important to represent the process flow or concepts of a model visually to make it easy for the user to interpret. Visual diagrams avoid misinterpretation and provide a standard for different users. As a communication tool, IDEF0 enhances domain expert involvement and consensus decision-making through simplified graphical devices. As an analysis tool, IDEF0 assists the modeler in identifying what functions (an activity, process, or transformation identified by a verb or verb phrase that describes what must be accomplished) are performed, what is needed to perform those functions, what the current system does right, and what the current system does wrong. Thus, IDEF0 models are often created as one of the first tasks of a system development effort. It is necessary that the diagram should be as simple and user-friendly as possible.

Different tools are used to generate visual representations.

5.3.1 Characteristics

- It is comprehensive and expressive, capable of graphically representing a wide variety of business, manufacturing and other types of enterprise operations to any level of detail.
- It is a coherent and simple BMT, providing for rigorous and precise expression, and promoting consistency of usage and interpretation.
- It enhances communication between systems analysts, developers and users through ease of learning and its emphasis on hierarchical exposition of detail.

5.3.2 Information and process flows

IDEF0 diagrams consist of at least one activity (represented as a box) with one input and one output arrow (refer Figure 5.6a). IDEF0 diagrams are effectively used to represent information and process flows. Decomposition (the partitioning of a modeled function into its component functions) is used to describe the process in further detail, which enables the moving from general to more specific descriptions.

5.3.3   General conventions

Diagrams are based on simple box and arrow graphics.

Text labels are used to describe boxes and arrows, and glossary with text defines the precise meanings of diagram elements.

Boxes indicate specific actions or processes, and are labeled in text starting with a verb or verb phrase.

A hierarchical structure with the major functions at the top gradually exposes detail within successive levels of subfunctions, revealing well-bounded detail breakout.

A 'node chart' provides a quick index for locating detail within the hierarchical structure of diagrams.

A limitation of detail to no more than six subfunctions on each successive function is recommended as this decreases diagram complexity.

5.3.4   Numbering

The simplest IDEF0 diagram is called context diagram and is numbered A, which describes the context and borders of the business process being modeled and consist only one box. There is only one context diagram per model.

The top-level decomposition diagram is called A0 and each of the boxes in the A0 level is numbered as A1, A2, etc. The boxes in the A0 diagram is then decomposed into a more detailed diagram/s called child activities where it is labeled as A11, A12, etc. as children of A1 and A21, A22, etc. as children of A2. These decomposition diagrams (child activities) can be decomposed further if required (refer Figure 5.6a).

5.3.5   Diagrams and activity boxes

One IDEF0 model should have at least one context diagram and one decomposition diagram. The 'box and arrow' graphics of an IDEF0 diagram show the function as a box and the interfaces to or from the function as arrows entering or leaving the box. An activity is represented by a box, which conforms to the rules of naming and numbering. Boxes should have a number at the right lower inside corner with the numbering sequence starting at the upper leftmost box and continuing to the lower right corner of the diagram.

Decomposition diagrams contain between 3 and 6 activities and represent the detail of a specific parent activity (refer Figure 5.6b).

5.3.6   Arrows

An IDEF0 activity box has four sides with standard and specific meanings (refer Figure 5.7a):

- The left side of the box represents INPUT and is reserved for input arrows.
- The right side is reserved for OUTPUT arrows (at least one per box) indicate the process/activity output.
- The top represents CONTROL inputs and there should be at least one arrow per activity box.
- The bottom arrow represents a MECHANISM and shows what is used to perform the task (manual, system X, batch-card, weighbridge, etc.).

Boxes represent the activities that transform inputs into outputs, control arrows constrain or determine under what conditions transformation occur, and mechanisms describe how the function is accomplished. All arrows should be attached to the side of a box (but not at the corner) and must have a name with a noun or noun phrase.

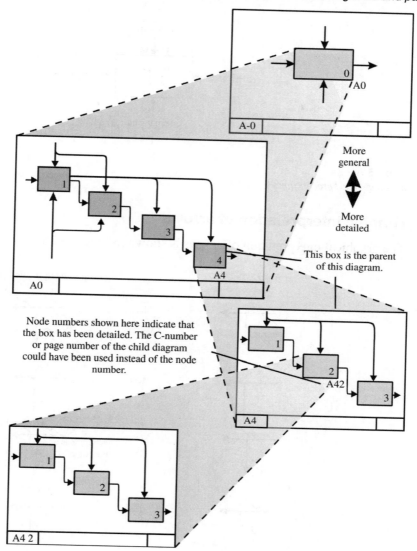

Figure 5.6a
IDEF0 hierarchy of diagrams

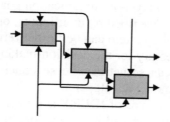

Figure 5.6b
Decomposition diagram

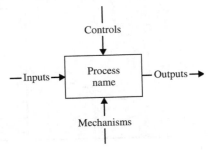

Figure 5.7a
Information and process flows

5.3.7   Graphic interpretation of arrows

The graphical interpretation of arrows is shown in Figure 5.7b.

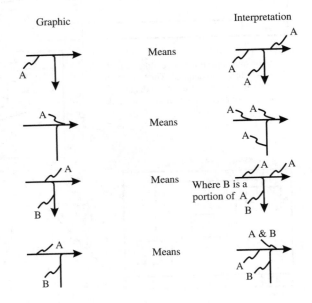

Figure 5.7b
Graphic interpretation of arrows

5.3.8   Strengths and weaknesses of IDEF0

IDEF0 has proven effective in detailing system activities for function modeling, the original structured analysis communication goal for IDEF0.

Activities can be described by their inputs, outputs, controls, and mechanisms (ICOMs) and the activities of a system can be further refined into minute detail until the model is as descriptive as necessary for the decision-making task.

IDEF0 models can sometimes be so concise that they are understandable only by a domain expert or model developer.

The hierarchical nature of IDEF0 facilitates the ability to construct models that have a top-down representation and interpretation, but which are based on a bottom-up analysis process.

There is a tendency that IDEF0 models are interpreted as a sequence of activities though that is not the original intent. It is natural to order the activities left to right because, if one activity outputs a concept that is used as input by another activity, drawing the activity boxes and concept connections is clearer. Thus, without intent, activity sequencing can be imbedded in the IDEF0 model. In cases where activity sequences are not included in the model, readers of the model may be tempted to add such an interpretation. This anomalous situation could be considered a weakness of IDEF0.

The concept away from timing, sequencing and decision logic allows concision in an IDEF0 model, but such abstraction also creates confusion among readers outside the domain. This particular problem has been addressed by the IDEF3 method.

5.3.9 Example

An example of IDEF0 is shown in the Figure 5.7c.

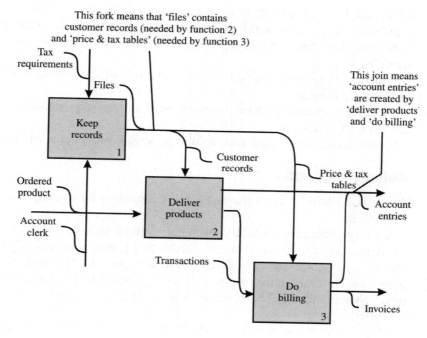

Figure 5.7c
Example of IDEF0

5.4 Unified modeling language

Unified modeling language (UML) is the standard language for specifying, visualizing, constructing and documenting all the artifacts of a software system as well as for business modeling and other non-software systems. The UML represents a collection of best engineering practices that have proven successful in the modeling of large and complex systems.

Using UML, programmers and application architects can make a blueprint of a project, which, in turn, makes the actual software development process easier.

5.4.1 The purpose of UML

The UML is an open system that supports the entire software development life cycle. It is based on the experience and needs of the user community and is supported by multiple tools. It is also applicable to diverse applications areas.

Good models are essential for communication among project teams and to assure architectural soundness and understanding of the requirements. As the complexity of systems increase, so does the importance of good modeling techniques. There are many additional factors of a project's success, but having a rigorous modeling language standard is one essential factor. UML was designed with following primary purposes in mind:

- Business process modeling with use cases
- Class and object modeling
- Component modeling
- Distribution and deployment modeling.

5.4.2   Diagrams

UML defines twelve types of diagrams, divided into three categories: four diagram types represent static application structures, five represent different aspects of dynamic behavior, and three represent ways in which to organize and manage application modules.

Structural diagrams include the class diagram, object diagram, component diagram, and deployment diagram.

Behavior diagrams include the use case diagram, sequence diagram, activity diagram and state chart diagram.

Model management diagrams include packages, subsystems and models.

5.4.3   Activity diagrams

The logical place to start exploring UML diagrams is by looking at activity diagrams (refer Figure 5.8).

Activity diagrams show the flow of control and activities as rounded rectangles. Activities are typical action states – states that transit automatically to the next state after the action is complete. The filled-in circle represents the start of the activity diagram where the flow of control starts. Arrows represent the transition from one activity to the next. Synchronization bars show how activities happen in parallel and guards the transition.

UML is versatile in nature; so activity diagrams may be used at the beginning of the life cycle or in different phases entirely. Many people use activity diagrams to show the flow between the methods of a class. They can be used as swim lanes which clearly exhibit to all parties the individual or group responsible for a specific activity.

5.4.4   Use case diagrams

The next diagram to look at is the use case diagram. Use case diagrams are created to visualize the relationships between actors and use cases. An actor is someone or something external to the system, but that interacts with the system. Actors are represented as stick figures. An actor does not have to be a person – it can be anything that interacts with the system from outside the system. A use case is a sequence of related transactions performed by an actor in the system. A use case is a group of functionality that provides value to the actor.

Use cases are shown as ovals, and the easiest way to find them is to look at each of the actors and identify why they want to use the system. Once it is identified, each use case needs to be documented with the flow of events, from the actor's point of view. It should detail what the system must provide to the actor and indicate how the use case starts, finishes and what the use case have to accomplish? (refer Figure 5.9).

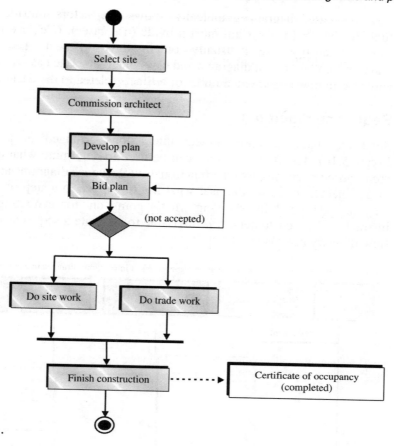

Figure 5.8
Activity diagrams show flow of control

Use case diagrams are created to visualize the
relationships between actors and use cases.

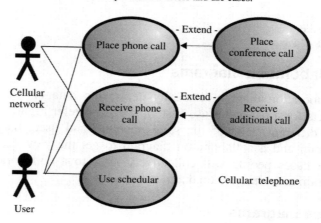

Figure 5.9
Use case diagram

A use case diagram graphically shows the actors outside the system and the functionality that the system must provide (use cases). Use case realization explains the 'how' of the use case. It usually contains three different types of diagrams: sequence diagrams, collaboration diagrams and class diagrams. Use case realizations are basically a way of grouping together a number of artifacts related to the design of a use case.

5.4.5   Sequence diagrams

Sequence diagrams show object interactions arranged in a time sequence (refer Figure 5.10). The flow of events can be used to determine what objects and interactions are required to accomplish the functionality. Sequence diagrams are great tools at the start of a project as it shows the user what has to happen in a step-by-step fashion. Sequence diagrams are good for showing what's going on, for extracting requirements and for interacting with customers. It is advisable to generate a sequence diagram for every basic flow of every use case.

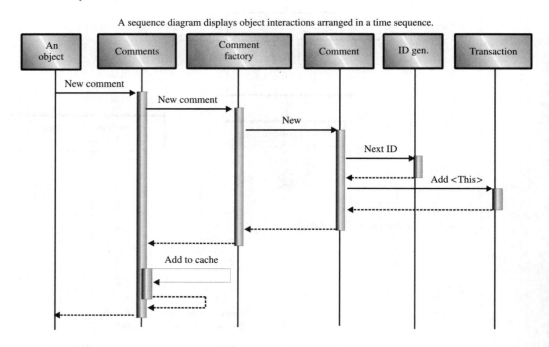

Figure 5.10
Sequence Diagram

5.4.6   Collaboration diagrams

Collaboration diagrams visually depict the data or messages that pass between two objects for a particular use case or scenario (refer Figure 5.11). A collaboration diagram is a different view of the same scenario and users may switch between a sequence diagram and a collaboration diagram to get the view that best illustrates the point. In some cases people will collectively refer to a collaboration diagram and a sequence diagram as an interaction diagram.

5.4.7   Class diagrams

A class is a collection of objects with a common structure, common behavior, common relationships and common semantics. They are identified by examining the objects in sequence and collaboration diagrams, and are represented in the UML as a rectangle with three compartments. The first compartment shows the class name, the second shows its structure (attributes) and the third shows its behavior (operations). The structure of a class is represented by its attributes (refer Figure 5.12).

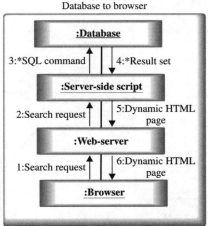

A collaboration diagram displays object interactions organized around objects & their links to one another
Database to browser

Figure 5.11
Collaboration diagram

The UML modeling elements found in class diagrams include:

- Classes and their structure and behavior
- Association, aggregation, dependency, and inheritance relationships
- Multiplicity and navigation indicators
- Role names.

5.4.8   Multiplicity and navigation

Multiplicity defines how many objects participate in a relationship and it is the number of instances of one class related to one instance of the other class. For each association and aggregation, there are two multiplicity decisions to make, one for each end of the relationship. Multiplicity is represented as a number and a * is used to represent a multiplicity of many.

An arrow shows navigation, and although associations and aggregations are bidirectional by default, it is often desirable to restrict navigation to one direction. In that case, an arrowhead is added to indicate the navigational direction.

5.4.9   Relationship diagrams

Relationships represent a communication path between objects. There are three types of UML relationships (refer Figure 5.13).

Association   A bi-directional connection between classes and represented in the UML as a line connecting the related classes.

Aggregation   A stronger form of connection where the relationship is between the whole and its parts, represented in the UML as a line connecting the related classes with a diamond next to the class representing the whole.

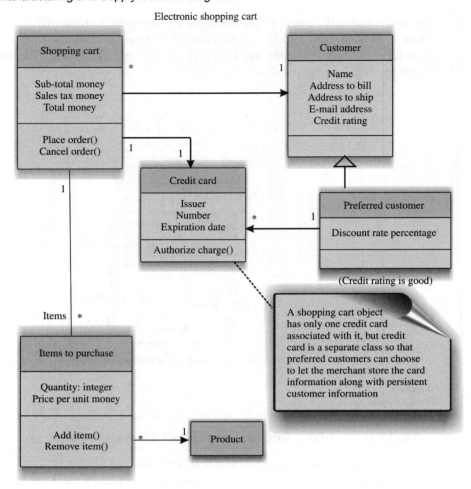

Figure 5.12
Class diagram

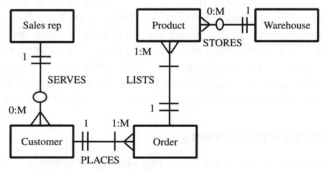

* 1 instance of a sales rep serves 1 to many customers
* 1 instance of a customer places 1 to many orders
* 1 instance of an order lists 1 to many products
* 1 instance of a warehouse stores 0 to many products

Figure 5.13
Relationship diagram

Dependency A weaker form of connection showing the relationship between a client and a supplier where the client does not have semantic knowledge of the supplier, represented in the UML as a dashed line pointing from the client to the supplier.

As mentioned in the sequence diagram, if two objects need to communicate, there must be a relationship between their classes.

5.4.10   Inheritance

Inheritance is the relationship between a superclass and a subclass and shows uniqueness and commonality, and adds new behavior without changing the super-class. Inheritance is represented with a triangle and should only be used when there is an inheritance situation.

5.4.11   State transition diagrams

A state transition diagram shows the life history of a given class. It shows the events that cause a transition to change from one state to another and the actions that result from the state change.

5.4.12   Object diagram

Object diagrams are built during the analysis and design phase and illustrate data/object structures.

5.4.13   Component diagrams

These diagrams are used to illustrate the organization and dependencies of software components, including source code components, run time components, and executable components (refer Figure 5.14). Components are shown as a large rectangle with two smaller rectangles on the side. This diagram shows what interfaces are used by executables to identify where the impacts may occur when the interface changes.

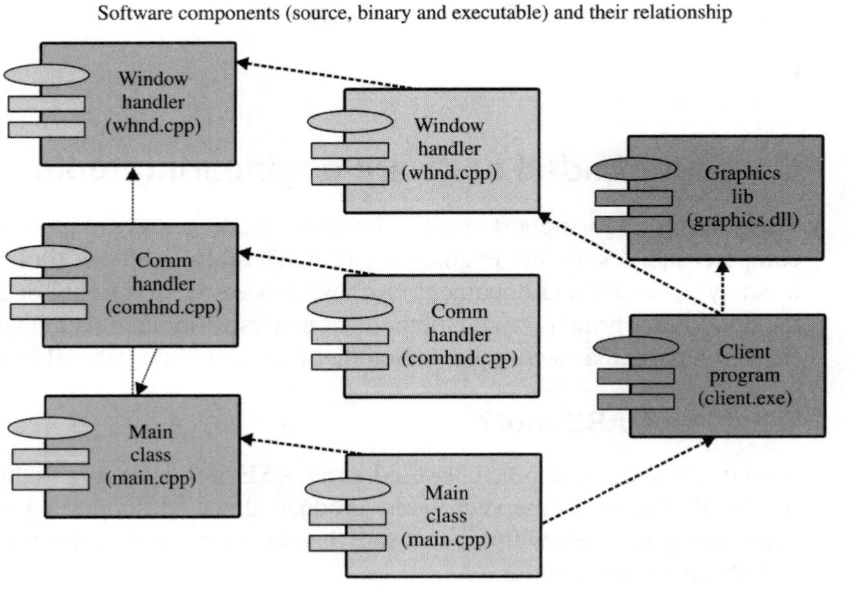

Software components (source, binary and executable) and their relationship

Figure 5.14
Component diagram

5.4.14 Deployment Diagrams

Deployment diagrams (deploying the system) are crucial as they show the processors on the system and the connections between them (refer Figure 5.15). They are also used to visualize the distribution of components across the enterprise and indicate what executables are running on the processors.

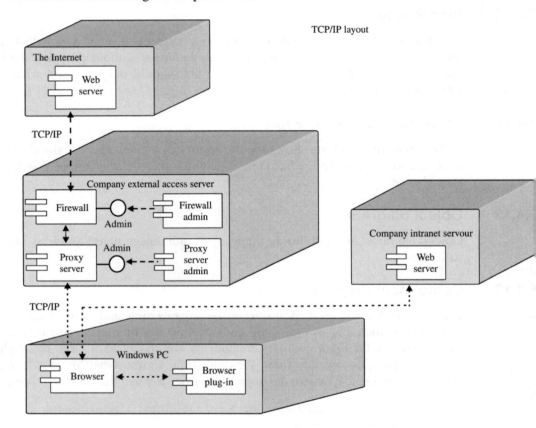

Figure 5.15
Deployment diagram

5.5 Computer-aided software engineering tools

Tools designed to support specific business modeling techniques are often referred to as computer aided software engineering (CASE) tools. A CASE tool is a product that helps to analyze, model and document business processes. It is a tool or a toolset that supports the underlying principles and methods of analysis. Some tools are specifically designed to support a particular technique while other tools are more general in nature.

5.5.1 What is a CASE tool?

A software package can be classified as a CASE tool according to the following definition. A CASE tool is a computer-based product aimed at supporting one or more software engineering activities within a software development environment (refer Figure 5.16).

CASE tools are used for:

- Diagramming
- Computer display and report generation

- Analysis tools
- Central repository
- Document generation
- Code generation.

Computer-aided software engineering is used especially to organize and control the development of software on large complex projects, involving many software components and people. It allows designers, code writers, testers, planners and managers to share a common view about the project position (stage wise). It helps to ensure a disciplined, check-pointed process. It may also serve as a repository for or be linked to document and programme libraries containing the project's business plans, design requirements, design specifications, detailed code specifications, the coded units, test cases and results, and marketing and service plans. It encourages code and design re-use, reducing time and cost and improving quality.

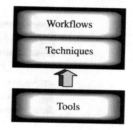

Figure 5.16
Business modeling tool

Many CASE tools are based on UML, the industry-standard language for specifying, visualizing, constructing and documenting the components of software systems. It simplifies the complex process of software design, generating a 'blueprint' for development.

Before buying a CASE tool, one has to look objectively at such products and evaluate the benefits. The following questions should be answered before any decision:

- Is there parity between advantages offered and cost?
- Will a CASE tool stop the interactive nature of the current development life cycle and the continuous development associated with work duplication?
- Will the CASE tool improve the overall development process when choosing a certain tool?

5.5.2   Products

There are several hundred business- and process modeling tools and toolsets available. These range from integrated development tools to specific, single function tools. The tools do not all provide the same functionality and some are based on different technologies and techniques/methodologies.

Some of the main classifications used to evaluate the tools are the ability to handle:

Workflow

After the process models and the object models have been developed, workflow is the basis for an understanding of what an organization is doing and why, along with a shell that can be used to implement the processes as an IT solution. Traditional workflow is the answer to the controls required that ensures that the process is executed correctly with a mixture of manual and automated tasks. Development of a workflow helps to identify the obvious flaws in the automation that may exist such as, where data is not managed properly and where there are unnecessary waits for parallel steps to complete. The workflow should define and describe (refer Figure 5.17):

- The activities which are to be performed at each stage of a project
- The correct sequence of applying the techniques and any dependencies which exist between them
- Information which ensures that the application of a technique is repeatable, measurable and can be improved over time.

We have defined two different types of workflow:

1. Core workflows which guide the application of a specific technique
2. Product workflows, which define the activities required to produce a particular business-modeling end product.

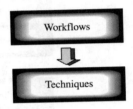

Figure 5.17
Workflow

Simulation

Simulation is a key part of any business modeling solution and offers an organization the opportunity to optimize processes in a variety of ways. The most obvious requirement is to minimize the elapsed time required for the process to complete, but it is also common for optimization to be based on cost or resource usage. For process simulation all the key information is included in each step of the process – how long it takes, the costs involved and the resources used. Then a workload is applied to the model to see how smoothly the operation works. Steps that take too long will result in a backlog and inefficiencies further down the chain. It should be possible to make modifications to the model quickly so that fine-tuning can be carried out online. In complex environments with multiple geographic locations and distributed processes, the whole process of modeling comes to life within simulation. There is often a balancing act to be achieved between the cost of the process and the time it takes to complete.

Activity-based costing

Many business modeling products have implemented accounting methods for analyzing the value of business processes, and the most popular of these is activity-based costing (ABC). It assigns costs to each step in the process according to the amount of resource consumed. These costs are then associated with the end product of the process along with any overhead costs. The purpose of ABC is to arrive at the true cost of a product and each of the steps within the process that creates it. It assists in calculating competitive pricing for any product based on accurate information, for developing budgets and also just to measure the overall performance of the business processes. It identifies the origin of the costs, how effectively resources are being used and where resources add the most value. The ABC fits very well into the business modeling environment as it requires the cooperation of all levels of a business to understand how things are done and why.

Object-orientated methodologies

Object-oriented analysis and design methodologies take full advantage of the object approach when it comes to modeling the objects in a system. However, system behavior continues to be modeled using essentially the same tools as in traditional systems analysis such as state diagrams and dataflow diagrams (DFDs).

State diagrams

State diagrams are often used to represent the dynamic behavior of systems. The circles in a state diagram correspond to states of the system being modeled, and the arcs connecting those circles correspond to the events, which result in transitions between those states. The state diagram thus defines a set of possible sequences of events and states. Each state diagram must include at least one initial state and one final state (double circle). All sequences must begin with an initial state and continue until they terminate with a final state.

The set of states included in a state diagram can be thought of as a one-dimensional attribute space where the single attribute has values, which correspond to the possible states. A system behavior corresponds to a sequence of these states and each state diagram defines a process, that is, a set of such behaviors. For example, the state diagram permits the event sequences ac, abac, ababac, abababac, and so on. This entire set of sequences can be described by the regular expression a(ba)*c (refer Figure 5.18).

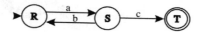

Figure 5.18
State diagram as a class of possible event sequences

Dataflow diagrams

Dataflow diagrams are intended to show the functionality of a system: the various processes, and the flows of information and material which link them to each other, to inventories (data stores) and to various agents external to the system. A DFD consists of a collection of processes, stores and terminators linked by flows (refer Figure 5.19).

Processes, shown as circles in the DFD, are the component actions or sub processes which together constitute the overall process or system being represented in the diagram. Each process transforms a set of inputs into outputs or outgoing flows e.g. receive order as input (a flow of orders) and produces (as outputs) order detail, billing information, and notification of invalid orders.

Stores, represented by pairs of parallel lines in the DFD, are repositories of the data or material carried in the flows. An incoming flow represents an update to a store (additional data or material), a modification to existing data or the deletion of data. An outgoing flow represents access to the data or material in a store. In the case of physical material, an outgoing flow represents an actual transfer of material from the store. In the case of information, an outgoing flow is a non-destructive read of the data in the store. The invoices store is updated by a flow of billing information from the receive order process and is a source of customer and invoice information for the collect payments process.

Terminators, shown as rectangles in the DFD, represent the actors, which interact with the various system processes. Terminators may be sources of information and material which flow into the system, and may also receive information and material, which flow out of the system. For example, customers are a source of orders, payments and payment inquiries. Customers also receive invoices, statements and notification of invalid orders.

Flows, shown as arrows in the DFD, represent the movement of information or material between processes, terminators and stores. A flow can diverge into several sub-flows (duplicate data/material is being carried in each sub-flow, or the data/material has been split into separate components each of which is carried in a sub-flow) and several flows can converge into a single super-flow (carries an aggregate of the data/material contained in the individual flows). A typical flow is 'orders' which carries orders from customers to the receive order process.

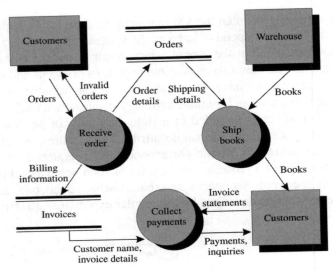

Figure 5.19
Dataflow diagram (order processing)

Based on specific requirements one can choose suitable products. When multiple functionalities are required, products must be chosen carefully so that it will provide the benefit of an integrated CASE tool.

Some of the well-known business process modeling tool products includes:

Product	Supplier
ARIS	IDS Prof Scheer
BPWin	Computer associates
ProVision	Winsoft
Visio2000	Microsoft
Designer 2000	Oracle

5.6   ARIS

5.6.1   Background

ARIS means architecture of integrated information systems. It is a framework or concept used to describe companies and business application systems. The ARIS concept and ARIS model types can help to represent the business structure of a company, application software or procedures. This gives the opportunity to comprehensively document the organizational and procedural structure of company and/or individual areas of it.

ARIS is a collection of tools for developing and optimizing business processes. It is targeted at the development of an IT solution rather than simply documenting and understanding the business, which will result in the implementation of new business processes rather than the traditional business models. As some companies have a requirement to document knowledge within an organization, tools are made available to all types of users – especially those without any specialist knowledge of business models or IT products. This is achieved using simple user interfaces and by providing as much flexibility as required.

5.6.2 Product

ARIS offers management of the complete life cycle of a business process from its initial conception through to the final implementation. The ARIS Framework enables companies to describe every major component of their organization accurately – people, data, activities, products, processes and objectives – including all of their inter-relationships, and stores the information to make smarter business decisions. There are also tools for process design, optimization, analysis and execution, in a single department or across the enterprise.

ARIS simplifies the complexity of business so that business can be run more efficiently by:

- Doing a better job of analysing the underlying business processes.
- Determining an organization's future ('should-be') state before implementing any change saves time and money.
- Significantly reducing the risk of unsuccessful or suboptimized system or process implementations.
- Creating unique, valuable assets (e.g. knowledge transfer and change management activities) and after the implementation, users have the knowledge and tools necessary to take ownership of their new, optimized business processes. Users can enjoy the benefits of their IT implementation investment on an ongoing basis if the framework is maintained and kept up-to-date continuously.

The ARIS product suite consists of the following integrated components:

- The major offering is the ARIS toolset, which offers a variety of functions and supports a large number of modeling methods – including UML. Enterprise modeling, model management and process analysis helps to manage and improve a company's business processes.
- An integrated, scalable, business process management system that enables companies to capture, communicate, evaluate and improve their business processes on a continuous basis.
- The toolset consists of a number of graphical tools built on a shared database that holds all of the definitions. The tools allow definition of the events and functions that make up a process, along with the time-line and relationships that make up the flow.
- In addition to this, there is organizational and static business data that provide additional background documentation for multimedia objects – videos, graphics, sound and so on.

The ARIS toolset supports a wide range of process-related projects including:

- Definition, analysis and optimization of business processes
- Knowledge management
- Supply-chain management and e-business modeling
- Change management
- Organizational handbooks
- Benchmarking

- Development and implementation of standard company processes
- Simulation and management of personnel requirements
- Process cost management
- Software selection and implementation
- Workflow specifications
- Certifications
- Business process-oriented training.

ARIS shows how the various components are related, how different projects are related and how everything relates to business objectives. The ARIS toolset, with the e-business process method, can determine what processes to move to the Internet, and how best to configure, support and improve them over time.

The balanced scorecard is one of the most influential management trends to emerge in recent times and gives a holistic, birds-eye view of business processes and their relationship to business objectives. The information in the scorecard can be set at essentially any level to define these relationships at the project level, the divisional level or enterprise wide.

One of the most common causes of failure is a lack of understanding among employees of how the improved process works and what it's supposed to achieve. ARIS easy design provides a quick route through the definition process at the cost of little customization and flexibility. It allows distributing process design activities for improved roles and responsibilities. Employees are also able to see how their job fits into the company's overall processes and objectives.

ARIS for mySAP.com can link into the process-related information held within the ERP environment.

In order to analyze operational processes, ARIS provides its process performance manager (PPM). The PPM takes a set of performance indices and carries out regular measurements to see how the various parts of a business process are performing. This can include end-to-end views of processes that cross multiple systems. All of the data can then be viewed graphically and used to implement a policy of continuous process improvement.

5.6.3   Function allocation diagram

Function allocation diagrams (FAD) (refer Figure 5.20) can be used to display object allocations to object types. The event control as well as the transformation of input data into output data and the representation of data flows between functions represent links between the data view and the function view of the ARIS concept.

The transformation of input into output data is represented in function allocation diagrams (I/O), which are very similar to input/output diagrams in other methods. Parts data, inventory data, BOM data and shipping data are the input data of the determine delivery data function. Inquiry data are input data as well as output data. Thus, an FAD consists of functions of the function view as well as information objects of the data view.

An information object represents an input data element; an output data element or both elements can be determined by arrow. The information objects can be data clusters, entity types, relationship types or attributes of the data view, and the actual purpose of the FAD is to represent the input and output data of a function.

5.6.4   Objective diagram

Objective diagram defines (business) objectives and creates objective hierarchies. An objective is the definition of future company goals, which is supposed to be reached by supporting the critical success factors and realizing new business processes. All possible critical factors can be arranged in a hierarchy and allocated to the objectives which they support. Critical factors are the aspects which need to be considered to reach a particular company objective and are allocated to the company objectives in the objective diagram. It is linked to the other diagram types by means of the function object type.

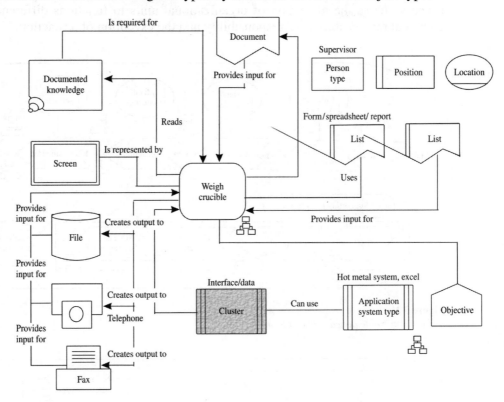

Figure 5.20
Function allocation diagram

5.6.5   Organizational chart

The organizational chart represents the organizational structures in which the formed organizational units and their inter-relationships are reflected according to the selected structuring criteria. Organizational units are the tasks performers, carrying out the tasks in order to reach the business objectives. The relationships are the links between the organizational units. The organizational chart illustrates the distribution of the business tasks, where the functional responsibilities are shown in boxes and for individual positions, an independent object type position is available. Several positions can be assigned to one organizational unit. The positions and organizational units can be assigned to persons holding these positions. Separate objects are also available for persons in ARIS. Organizational units and people can be grouped to form a type. Thus it can be defined whether an organizational unit is a department, a main department or a group. In process chains, only certain person types can execute a function or only certain person types can have access to an information object.

5.6.6   Value-added chain diagram

The value-added chain diagram (VACD) specifies the functions in a company which directly influence the real added value of the company. These functions are linked to one another in a sequence of functions and thus form a value-added chain. The functions are arranged in a hierarchy similar to a function tree where process-orientated sub-ordination or superiority is mentioned. A VACD represents (refer Figures 5.21 and 5.22) the links between functions and organizational units and functions and information objects. In process chains the allocation of organizational units to functions differentiates between technical responsibility, IT responsibility and the execution of a function.

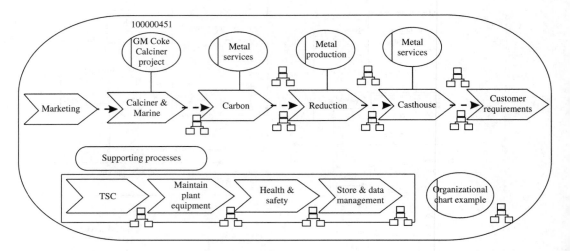

Figure 5.21
Value chain: value-added chain diagram

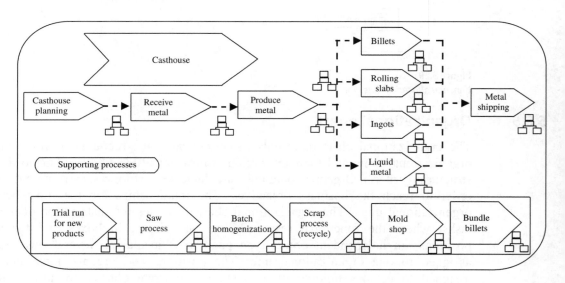

Figure 5.22
Process level: value-added chain diagram

5.6.7   Extended event-driven process diagram

Event-driven process diagram (EPC) defines the procedural organization of the company as well as the links between the objects of the data, function and organizational view. A procedural sequence of functions is illustrated by means of process chains. In the diagram, the start and end events of every function are specified.

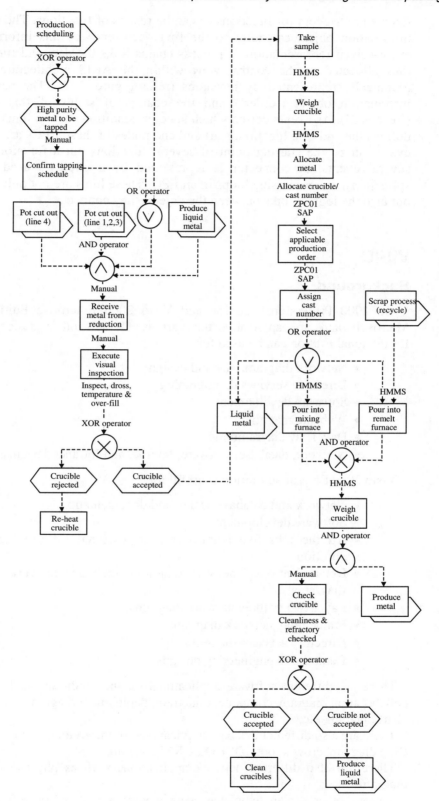

Figure 5.23
Extended event driven process diagram

Events are triggers for functions or can be results of functions. The changed state of an information object can refer to the first occurrence of this information object. The event-driven representation of process chains links the data and function views and is thus allocated to the control view within the ARIS architecture where events are graphically represented by hexagons (refer Figure 5.23). The name consists of the information object ('order') and the change of status of this information object ('received'). As events define which state or condition starts a function and which state defines the end of a function, start and end nodes of these EPCs are always events. One event can be the starting point of several functions but a function can also result in several events. A connector is represented by a circle is used to illustrate these splits/joins and processing loops in an EPC. These links are not only graphic connectors but also the logical links between the objects they connect.

5.7   VISIO

5.7.1   Background

Visio 2000 Professional Edition and Visio 2000 Enterprise Edition are products of Microsoft and are applicable for various IT and other design applications. The Professional Edition can be used for:

- Network diagramming and design
- Directory services diagramming
- Software development
- Web site mapping
- Data flow diagramming
- Network, database, software, Internet and Intranet design and documentation.

Visio 2000 Enterprise Edition is useful for:

- Network and database design and documentation
- Software development
- All the jobs listed above for professional Edition but it requires an IT solution
- Detailed physical network diagrams, including network rack or server room diagrams
- Layer two or three network diagrams
- Frame relay network diagrams
- Directory services diagrams
- Database re-engineering designs.

There are different software applications for these editions such as activity diagram, collaboration diagram, component diagram, deployment diagram, sequence diagram, state chart diagram, etc.

There are also different database applications for these editions such as entity relationship (ER) diagram, crow's foot, IDEF1X, ORM diagram, etc.

There are also different network applications such as physical and logical network diagrams.

Various other design templates using a wide variety of predefined symbols such as equipment drawings, electrical diagrams, IDEF0, building architecture, process flow, electronic diagrams, etc. are also available.

5.7.2 Products

Professional Edition	Enterprise Edition
It gives a clear picture of the networks, databases, software, processes and Web sites, and even helps to start planning for the new ones. Visualize legacy systems. Document development projects. Prototype architectures	It is an automated IT design and documentation tool. Enterprise Edition files can be viewed and annotated in Professional Edition, ensuring that as projects move through the organization, the associated documentation can move with them
Professional edition delivers industry-standard tools diagramming networks, databases and software applications	Enterprise edition offers all of the features in Professional Edition, plus automated tools for IT design and documentation

Both can help streamline the planning, implementation and management of IT systems.

5.7.3 Database solutions

Visio database solutions include visualization tools, which helps define, communicate and implement robust database designs. Both Visio 2000 editions provide database documentation and design tools. In Professional Edition, new and existing databases are visualized with reverse engineering capabilities whereas database re-engineering can be automated with Enterprise Edition.

Both the editions are having following features:

Reverse engineering Reverse engineer leading desktop and client/server databases, including code.

DBMS support Support for client/server and desktop databases from IBM, Informix, Microsoft, Oracle, Sybase, Corel and others.

Notation support Design databases using relational, IDEF1X or crow's Foot notation.

Integration with other tools Import Erwin, ERX files or legacy InfoModeler models.

Object role modeling stencil Create conceptual models using object role modeling (ORM) notation.

Visio Enterprise has some special features apart from the above:

Error checking Dynamically verify DBMS-specific logical and physical validity of database models.

Model-database synchronization Synchronize the database with changes made to the model or update the model to reflect database changes.

Database schema generation Automatically generate a DDL script and/or database schema from a model.

User-defined types Define custom data types for frequently used data types in models.

Workgroup collaboration tools Use collaborative database design tools to merge individual database diagrams into a single model.

Database reports Create customizable reports on conceptual, logical, and physical database models.

ERwin export Export database models to ERwin ERX format.

Microsoft repository integration Import models from and export them to Microsoft Repository 2.0.

Natural language business rules   Create conceptual data models by working with natural language business rules.

Model and schema generation   Map conceptual data diagrams to normalized ER models and automatically generate database schemas.

5.7.4   Software solutions

Visio 2000 software solutions are used to document existing architectures, map out new structures, and detail process and data flows for development projects.

Software diagrams can be created with Professional Edition using a variety of methodologies or it can be done with the automated design and modeling tools of Enterprise Edition. Some of the important features are:

Multiple software notations   Visio create models using Booch (object-oriented design), Rumbaugh (object modeling technique), Schlaer-Mellor (object-oriented systems analysis), Express-G, Jacobson (object-oriented software engineering) and other notations.

UML notation   It creates all UML 1.2 diagrams i.e. activity, component, collaboration, deployment, sequence, state chart, static structure and use case.

Reverse engineering   Extracting application code directly from Microsoft Visual Basic, Visual C++, and Visual J++ to generate UML class diagrams.

Visio enterprise edition   Automatically diagnose semantic errors, generate customizable code from UML class diagrams, generate UML report and Import models from and export them to Microsoft Repository 2.0.

5.7.5   Features and diagrams

Business diagrams in Microsoft Visio® Standard and Visio Professional help to visualize, document, and share ideas with attention-grabbing flowcharts (refer Figure 5.24), organization charts, office layouts and more. With the familiar Office interface and tools, Visio makes it possible for anyone to visualize and share ideas with maximum impact and results.

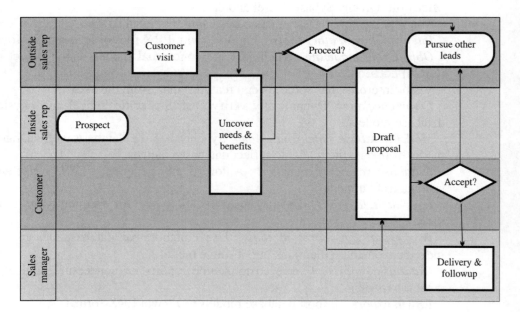

Figure 5.24
Flowchart

Diagrams can be built/assembled quickly by dragging task-specific Microsoft SmartShapes® onto user pages. Unlike clip art, Visio shapes behave like the real-world objects they represent. Text boxes stay with their shapes; connectors stretch and change angles while moving the things around (refer Figure 5.25).

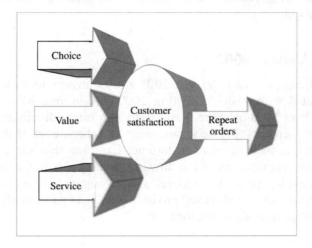

Figure 5.25
Design

The perfect shape required for the job can be located quickly, whether it's stored on computer or online. Style and impact can be added to diagrams with vibrant backgrounds, titles, borders and color schemes.

Visio diagrams can be transformed to/published as Web pages, complete with online zoom capabilities, a custom property viewer and built-in navigation for multi-page diagrams.

A broad range of diagrams can be generated, assisting in the documentation and sharing of ideas. Data of every-day work can be visualized and organization charts, timelines and Gantt charts can be generated from existing information automatically. Visio can store business data in custom properties and then use it to generate reports or export it for analysis to other programmes, such as Microsoft Excel. Visio features the same interface, tools, and functionality as other Office programs (refer Figure 5.26).

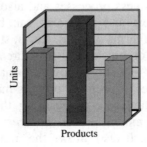

Figure 5.26
Column chart

Visio features Office XP user-interface enhancements such as personalized menus, customizable toolbars, and task panes, plus Office classics: autocorrect, background spelling checker and keyboard shortcuts. Visio diagrams can be inserted into Office documents or e-mail messages by simply copying and pasting. Online diagrams can be enriched by choosing images, sounds and videos from among the thousands of items in the Design Gallery Live.

CAD drawings can be converted into Visio objects on a layer-by-layer basis for easy CAD drawing editing within Visio. CAD drawings can be annotated with Visio shapes and drawings.

5.7.6   Visio Viewer 2002

The Microsoft Visio® Viewer 2002 allows anyone to view Visio drawings and diagrams (if created with Visio 5, 2000 or 2002) inside their Microsoft Internet Explorer version 5.0 or later Web browser. Visio users can freely distribute Visio drawings and diagrams to team members, partners, customers or others, even if the recipients do not have Visio installed on their computers. Internet Explorer also allows for printing, although this is limited to the portions of the drawings that appear in the current view. The Visio Viewer 2002 enables users to discover the advantages of using Visio drawings, charts and illustrations in a Web-based environment. It is particularly useful for teams that need to collaborate from different locations.

5.8   Oracle Designer

5.8.1   Background

Oracle Designer was created to enable the development of database-oriented applications built upon the Oracle relational database management system (RDBMS). The three major functions that it supports are:

1. The capture of requirements and design – including reverse engineering legacy applications
2. Modeling the design of databases and applications
3. Generating the complete applications as a result.

5.8.2   Product

Process modeler is one of the tools within the designer toolset, which offers a graphical representation of business processes and also enables the addition of key metrics so that analysis of time, cost and resources can be carried out. Process modeler describes business processes in terms of steps, flows, stores, events and organization. It shows how different parts of the organizational structure are involved with the process over time, and becomes a workflow that shows how information is passed between each step. All of the information that is held within the process modeler is stored in an underlying repository. This has the important role of ensuring that there is a single source for all of the components of the model, which ensures consistency and accuracy throughout the development process. It also allows the implementation of team development facilities and full integrity checking.

Oracle Designer also has an ER Diagrammer for managing the data-centric components and database design and a Function Hierarchy Diagrammer that offers a higher-level view of the business processes. It also provides function point analysis and links through to the final application design. Matrix diagrams and data flow diagrams are all available to support the ongoing development process.

5.8.3 Disadvantages

Oracle Designer is a powerful product with excellent graphics and is totally integrated with its repository to provide a strong modeling foundation, but chooses only to support the Oracle database. It does not model or support business processes that execute manually. Designer has to be considered when an organization seeks to develop an IT automation solution to support business processes using an Oracle database. Organizations that have business modeling needs beyond the IT department have to incorporate other CASE tools along with Designer.

5.9 Bpwin

5.9.1 Background

Bpwin, a product of computer associates, provide users with the ability to define and optimize business processes. This is done through the creation of a top-down activity model that allows business processes to be seen in terms of a number of functions.

5.9.2 Product

BPwin/Professional is a modeling tool for analyzing, documenting and improving complex business processes. BPwin automates the capture, validation, analysis and optimization of business or IT processes. A BPwin model documents important factors e.g. what activities are needed, how they are performed and what resources are needed. It supports function, dataflow and workflow modeling in one tool, integrating three key business perspectives. It meets the basic needs of any business – speed to market, improved competitiveness, enhanced product quality, more innovation and better responsiveness to customers. It simulates the flow of business processes and apply Activity-based costing techniques to process optimization.

5.9.3 Disadvantages

BPwin emphasizes too much on functional flow, data flow and workflow and not enough on answering the 'why' questions. It should be seen as a tool for developing business process knowledge and constructing a model for IT development, rather than a vehicle for documenting the thoughts and reasoning behind why an organization has these processes in the first place. BPwin is as yet not a complete business modeling life cycle tool.

6

Enterprise planning and supply-chain interaction

Learning objectives

- To learn about the different e-manufacturing modules that operate on an enterprise level.
- To become familiar with the functions performed by the different enterprise modules.
- To understand how the enterprise modules interact with the external supply chain.

6.1 Introduction

Continuous reconfiguration of the supply chain is important and depends on a combination of internal and external factors. A company's life cycle influences the supply chain considerably. Historically, the focus was on creating the chain, with a lot more emphasis on assured availability of inputs, which resulted in relationships between established players. During the next phase, cost were not as much of an issue as the ability of the vendors to ramp up their supplies when the need arose, which required a different kind of supplier. Next, companies will invest in creating suppliers and lift their technological and process capabilities, and work more closely with them in meeting customer needs.

Changing company strategies are influencing the supply chain e.g. as customization is becoming a differentiator, suppliers are being chosen for their versatility. In an environment where customer tastes are changing fast, the supply chain needs to be fast and flexible. Companies in mature markets are building long supply chains and product brands, trying to optimize supply chain effectiveness and speed. Companies competing on cost in the market place pick suppliers on the basis of price. The most successful company will be the one that meets the specific needs of different customer segments, and the supply chain can play a crucial part in assisting companies to achieve this.

Supply chain management (SCM) is about optimizing business processes and the visibility of the entire supply chain. The management of an effective supply chain depends upon the flow of relevant information and communication from the lower to the upper levels and vice versa. This should be two-way communications, with a certain source of information, such as customer demand, that is the driving force behind the flow of information. The 'chain' needs every link to be functional and communication forms these links.

Companies are not looking at SCM purely from an efficiency or cost reduction perspective anymore, as superior customer service, growth and revenue enhancement, getting the competitive edge over their competitors are common objectives of SCM these days. The focus has shifted toward customer, value/benefits instead of process/efficiency (an internal focus). Business success depends on the ability to react to the changing needs of customers. Global supply and sourcing, unpredictable demand, fluctuating pricing strategies, shorter product life cycles and decreasing brand loyalty create uncertainty.

Demand forecast tools are used to micro-tailor orders for suppliers, and inventory levels are continuing to reduce as manufacturers source materials only when they are needed for firm orders. Manufacturers are starting to set target costs for their suppliers instead of asking for quoted prices and are helping their vendors in the cost reduction process.

6.1.1   Basis of relationships

The following table indicates how the relationship between companies and suppliers changes, depending on the relational strength of companies and suppliers and on what basis they compete in the market.

Supplier/Company	Cost	Quality	Speed	Flexibility	Relationship
Multiple Suppliers, Single Company	*	$	*	$	*Transaction based
Multiple suppliers, Multiple Companies	*	-	*	*	$ Company Dictated
Single Supplier, Multiple Companies	@	@	@	@	@Supplier Dictated
Company Supplier Joint Venture	$	@	@	@	

Both supplier and company operate in a dynamic market environment where the crucial issue will be determining the parameters for choosing between competing vendors. Information originating with the customer regarding a specific product need must be communicated to the various companies throughout the value chain and translated into product specifications, which should in turn be adhered to by the various suppliers.

A study conducted by AT Kearney shows that the lack of integration within the supply chain, arising from inaccurate forecasting of customer needs, can bloat inventories by as much as twenty-seven times. Unless customer information is captured accurately and disseminated into the supply chain, the distortions keep multiplying. The Sloan Management Review (1997) demonstrates how poor supply-chain coordination amplifies variability from customer to the manufacturer, and this is evidenced by high inventories. To optimize business operations, it is essential that the information flows in real time, using various communication techniques.

6.1.2   E-business and supply-chain integration

E-business is the use of Internet technology to improve and transform key business processes. The use of Internet-based computing and communications to execute both front-end and back-end business processes has emerged as a key enabler to drive supply-chain integration. E-business has gone from concept to reality in the past few years.

E-business impacts four critical dimensions of the supply chain:

1. Information integration
2. Synchronized planning
3. Coordinated workflow
4. New business models.

E-business and supply-chain integration ensures reduced costs, increased flexibility and faster response times. Effective e-business and supply-chain initiatives need the following:

Information integration

Information needs to be shared among members of the supply chain. This includes any type of data that could influence the actions and performance of other members of the supply chain e.g. demand data, inventory status, capacity plans, production schedules, promotion plans and dispatch schedules. This information should be easily accessible by the respective parties on a real-time, online basis.

Planning synchronization

Planning synchronization defines what is to be done with the information shared in the supply chain. Supply-chain members need to jointly design and execute plans for product introduction, forecasting and replenishment. Supply-chain members coordinate the order fulfillment plans to meet the ultimate customer demands.

Workflow coordination

Defines how the integration takes place between the supply-chain members. It enables streamlined and technology-based automated workflow activities between the partners e.g. procurement activities from a manufacturer to a supplier can be tightly coupled so that efficiencies in terms of accuracy, time and cost can be achieved.

New business models

Any e-business approach changes the roles and responsibilities of members in order to improve overall supply-chain efficiency. A supply-chain network may jointly create new products, pursue mass customization, and penetrate new markets and customer segments as a result of integration.

Integration cannot be complete without well-defined and maintained channels of communication, performance measures and incentives for members of the supply chain. The success of supply-chain integration depends on close cooperation and mutual benefit that fosters trust and commitment within the channel partners. The impact of e-business on supply chain integration with respect to the various business processes involved is tabulated below.

Dimension of SC Integration	Procurement	Order Fulfillment	Product Design	Post-Sales support
Information integration	Supplier information sharing	Information sharing across the supply chain	Design data sharing, product change plan sharing	Customer usage data linkages
Planning synchronization	Coordinated replenishment	Collaborative planning and coordination, demand and supply management	Synchronized new product introduction and rollover plans	Service supply chain planning coordination
Workflow coordination	Paperless procurement, auctions, auto-replenishment auto-payment	Workflow automation with contract manufacturers or logistics providers, replenishment services	Product change management automation, collaborative design	Auto-replenishment of consumables
New business models	Market exchanges, auctions, secondary markets	Click-and-mortar models, supply chain restructuring, market intelligence and demand management	Mass customization new service offerings	Remote sensing and diagnosis, auto-test, downloadable upgrades
Monitoring and measurement	Contract agreement compliance monitoring	Logistics tracking, order monitoring	Project monitoring	Performance measurement and tracking

6.2 Logistics planning and optimization

Logistics essentially refers to the management of goods and materials and incorporates the movement and storage of goods and their associated information flows from source to customer both internal and external to the company. It is the crucial link of the supply chains through which materials flow from the supplier to manufacturer to customer.

6.2.1 What is logistics?

Logistics is that part of the supply chain process that plans, implements and controls the efficient, effective forward and reverse flow and storage of goods, services, and related information between the point of origin and the point of consumption in order to meet customers' requirements.

'The process of planning, implementing, and controlling the efficient, effective flow and storage of goods, services, and related information from point of origin to point of consumption for the purpose of conforming to customer requirements.' This definition includes inbound, outbound, internal and external movements, and return of materials for environmental purposes (Council of Logistics Management).

6.2.2   Inbound and outbound logistics

Competitive pressures of global markets have placed heavy demands in terms of speed and reliability on the logistics operations of both the inbound and outbound sides. This has led to the growth of specialized third party logistics providers (3PL) for total integrated door-to-door logistics solutions.

In today's fast-changing environment, the whole logistics mechanism is customer-service driven. The timely, reliable and accurate delivery of products free from damage are thus the major elements of customer service. Competing in the global arena requires efficient supply chains where the cumulative efficiency of all the partners matter. The timely movement of freight helps to cut costs by cutting down inventories and reducing cycle-times in supply-chains, requiring a high degree of logistics skills to manage. It involves a specialized approach to order processing, billing, documentation for statutory clearances, warehousing, payment collection at the delivery points and replenishment mechanisms.

The material flow from supplier to the manufacturer is referred to as inbound logistics and from manufacturer to customer as outbound logistics (refer Figure 6.1).

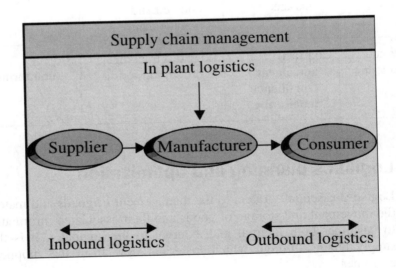

Figure 6.1
Logistics

Logistical chains are full of connecting linkages between various parties in the chain known in logistics as potential 'handoffs.' These are all points of potential delays and increase the inventory in the chain, tying up capital.

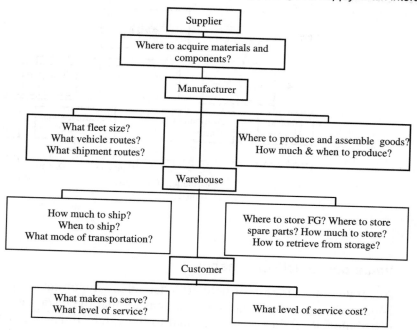

6.2.3 Time and space concept

The traditional model of stock keeping at depots and warehouses are slowly being replaced by manufacturers. With good planning, some companies immediately transfer incoming shipments into outgoing shipments without even taking it into the warehouse. Stocks at depots and warehouses are increasingly being replaced by virtual inventory (inventory in transit). The concepts of time and space are key aspects of the logistics system. The integration of logistics emphasizes the need for companies to remove functional barriers and focus on streamlined processes.

It requires a high degree of synchronicity between the production systems and the related processes of supply chain partners to maintain the time commitments on delivery and simultaneously, inventories are to be kept low across the chain. The quality of information flow across the chain participants, concerning inputs and finished goods movement, have enormous impact on response times across the chain and on inventory build-ups. Integrated information management has become a key feature in all logistics management systems.

Logistics plays a vital role in determining the competitive advantage of the supply chain in terms of cost, customer satisfaction and market share.

6.2.4 Fundamentals of logistics

The functional aspect of logistics should be more efficient than that of the competitors. Transportation facilities should be configured in such a way that raw materials, components and finished goods are delivered at right time, the right place and at the right quantity/quality with the least cost and time.

Logistics as operations

Logistics operations include the following components that need to be considered and coordinated to ensure efficient operation.

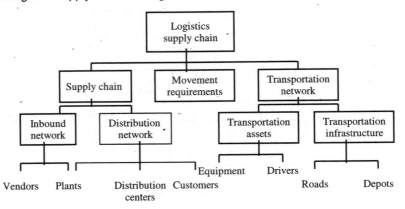

Value-based delivery channel

Accurate configuration of delivery channels is important for outbound logistics. This is necessary in order to optimize and ensure that the resources allocated to servicing a distributor or retailer is proportional to the value (financial, tactical or strategic) they add to the company. A company should ensure that value is generated from each selling point.

End-user service

Customer satisfaction as the objective of logistics operations can be translated into goals like zero defects, on time delivery, etc. that contributes toward building a brand image. Satisfying 'local' customers within the global arena is a major focus for logistics concepts such as the following:

Lean logistics is generally used in an inbound logistics system, ensuring quicker flow of goods into and out of the plant. Lean logistics is about doing more with less, and is used to describe manufacturing practices that operates effective pull systems. It works well for high volume, low-variety and predictable markets.

Agile logistics is generally used as an outbound logistic system, which primarily addresses the issue of customer satisfaction by quick, efficient deliveries regardless of the demand uncertainties and geography. Agile logistics is useful where demand variety is high and in less predictable environments. Agility recognizes that centralization brings benefits, but satisfying local customers or markets is a pre-requisite.

Reverse logistics is a logistics a planning system needs to include reverse logistics processes for the return of goods, containers and packaging to the source of supply or distribution. Reverse logistics is considered an opportunity to improve logistics systems.

Reduce cycle time

Integrated and effective logistics management can dramatically reduce turn-around times, lower working costs, make product life cycles shorter and delight the end user. It enables the tailoring of inventory exactly to market demand.

	Supply-Chain Planning	**Transportation Planning**	**Shipment Planning**	**Vehicle Routing**	**Warehousing**
Strategic	Site location Capacity sizing Sourcing	Site location Fleet Sizing	Outsourcing Bid analysis Fleet sizing	Fleet sizing Service-day balancing Frequency analysis	Warehouse Layout Material handling Design
Tactical	Production Planning Sourcing	Routing strategy Network alignment	Consolidation strategy Mode strategy	Routing strategy Zone alignment	Storage allocation Order picking strategy
Oper-ational	MRP, DRP, ERP	Load matching	Shipment dispatching	Vehicle dispatching	Order picking

6.3 E-fulfillment

The advent of e-commerce has added a new dimension to logistics and given rise to a term e-fulfillment. The fulfillment strategy is an essential component in the development of any e-commerce strategy.

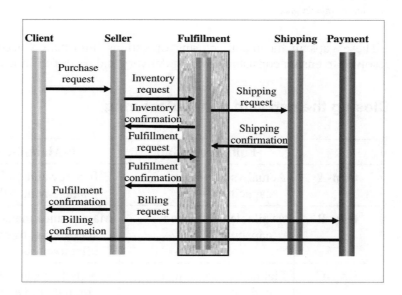

Figure 6.2
Typical fulfillment activities

6.3.1 Fulfillment activities

The current trend is to outsource non-core capabilities such as transport, distribution and logistics activities resulting in virtual supply chains (refer Figure 6.2).

To ensure effective and efficient operation of outsourcing activities, effective communication between the various parties involved in the supply chain is required. This is made more difficult by the unknowns presented by manufacturing facilities. Manufacturing is by nature a chaotic system, which leads to gaps in fulfillment activities (refer Figure 6.3).

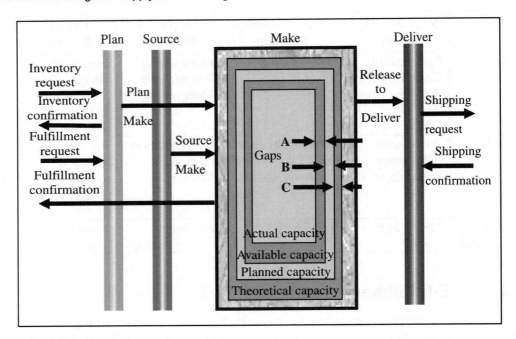

Figure 6.3
Fulfillment capacity gaps

These gaps in the manufacturing operations information needs to be identified an resolved to ensure consistent product delivery and order fulfillment.

6.3.2   Closing the gap with e-manufacturing

	Fulfillment Gap	**E-Manufacturing Enablers**
Gap-A	Actual vs available capacity	Efficiency and yield optimization, Batch management, SPC
Gap-B	Available vs planned capacity	Maintenance management finite capacity scheduling, overall equipment effectiveness
Gap-C	Planned vs theoretical capacity	Key performance indicator, supply chain simulation, Demand and supply planning, forecast collaboration

Integrated fulfillment is an e-fulfillment approach with a broad end-to-end supply-chai performance objective. It combines:

- New warehousing, transport and customer service solutions that delive operational excellence in fulfillment.
- New relationships between supply-chain partners create integrated an responsive supply-chain networks.
- New technologies and services to support these networks.

6.3.3 B2B e-fulfillment channels

This includes activities such as order management, customer service, procurement and inventory management, warehousing and shipping, processing and disposition of returns. B2B business models evolved to satisfy the explosive growth of demands from consumer as a result of e-commerce/e-procurement. B2B e-fulfillment migrates from long-term one-to-one relationships to fluid many-to-many relationships, depending on the specific organizational need (refer Figure 6.4).

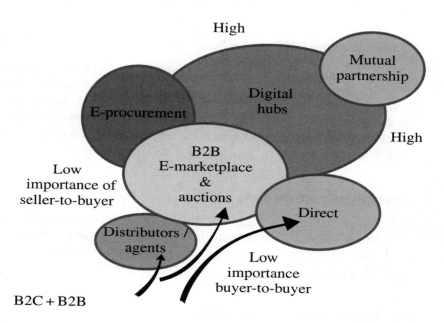

Figure 6.4
E-commerce channels

Mutual partnership (B2B)

The suppliers and customers are strategic partners to each other's business and often emphasize high logistics availability to support fast replenishment.

This relationship makes use of electronic data interchange or the Internet, as well as vendor-managed inventory (VMI) services. In case of VMI, the supplier takes the responsibility to monitor material levels at the customer's plant, also known as consignment stock.

E-procurement (primarily B2B)

In this channel products are supplied from multiple supplier locations to a single customer location and several possible product characteristics influence the fulfillment solution, which involves consolidating items from multiple sources into a single delivery or a repetitive set of deliveries.

Agents and distributors (B2C and B2B)

Two categories of fulfillment models exist in this environment – stock and stockless. The fulfillment solution to support a stockless environment is similar to the e-procurement environment. Agents and distributors maintain a buffer stock to meet the delivery lead-time. Agents and distributors considered following fulfillment strategies:

- Segregating large and small items into different picking facilities.
- Using trans-shipment locations to transfer stock into smaller vehicles for delivery.
- Holding fast moving items at local replenishment facilities and slow moving items centrally.
- Different lead-times and availability policies for different products.

Direct channel (B2B and B2C)

Direct channels range from 'low-cost direct' to 'high-cost interactive' transactions. In low-cost transactions, the emphasis is on reducing the transaction cost of customer and supplier. High-cost interactive transactions focus on increasing breadth of service provided to the customer. In both solutions, customers are able to track their orders placed on a supplier. Direct channel is applied to bypass existing distributor, wholesaler and retailer channels.

Exchanges and auctions (B2B)

Exchanges and auctions are best suited to very low transaction volumes where these fulfillment channels can support the anticipated transaction volumes. In these solutions, real time is the key in two areas: confirming the availability of fulfillment capacity and providing a price for delivery.

Digital transaction hubs (B2B)

Digital transaction hubs focus on reducing the cost of integration between buyers and sellers. It allows member companies to collectively outsource fulfillment activities that are not their core activity or a competitive differentiator. By doing this, the member companies achieve enormous economies of scale, which is impossible for individual companies. Fulfillment service providers can integrate with the hub only, instead of with individual companies.

E-fulfilling factors

While dealing with e-fulfilling solutions companies should consider the following factors:

- Speed of implementation
- Degree of customer contact and operational control
- Reliability and flexibility of service
- Initial and subsequent costs.

6.3.4   Keys to integrated fulfillment

The variety of distribution channels available to an enterprise creates complexity in the enterprise's supply chain, which emphasizes the importance of aligning the three key aspects of the enterprise and fulfillment providers.

Merging operational efficiency with e-commerce opportunities

In order to maximize e-commerce opportunities, the following fundamental but important areas need to be considered:

- Segmentation of customer according to needs
- Customized logistics network
- Demand and supply planning integration
- Product, information and financial flows to be integrated through supply chain
- Strategic sourcing
- Supply-chain performance metrics.

Realizing new relationships and services

New relationships are emerging as a result of e-commerce solution providers, stretching the traditional relationships between fulfillment service providers and users (refer Figure 6.5). Technology is no longer the primary competitive advantage to realizing integrated fulfillment, but the key differentiators are rather the ability to change, developing new kinds of relationships and implementing new solutions.

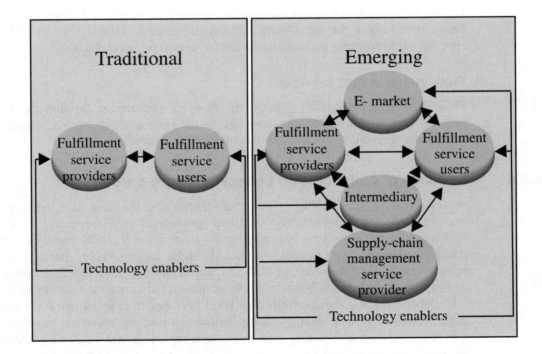

Figure 6.5
The changing fulfillment solution landscape

The characteristics of these new relationships are:

- More collaboration between the different role-players
- Win–win commercial arrangements

- Reassessment of activities to understand the core competencies and what is to be outsourced
- Arrangements that provide services to networks and individual companies.

6.3.5 Emerging capabilities

In the past, companies provided fulfillment services to individual companies such as 3PL, freight forwarding, parcel carrier, etc. E-commerce demands require new fulfillment capabilities to support B2B2C. These capabilities are classified into three broad types of services.

Fulfillment e-marketplace

Fulfillment e-marketplaces match buyers and sellers of a product or service or they provide a mechanism (exchange, auctions and reverse auctions) for setting a price or bidding on a commodity or both.

Public marketplace   These are open to any carrier or shipper that wishes to participate.

Private marketplace   These are restricted to members only, and typically constructed around a specific opportunity/requirement and tends to focus on specific geographies.

Informediaries

Informediaries provide information throughout a network that supports synchronized decision-making at the operational, tactical and strategic levels. They provide users with specific track and trace information, industry trends, prices and news.

Flow management services

Flow management services manage the flow of transactions through the network and provide supply-chain planning services to network participants i.e. end-to-end order management visibility throughout the entire order life cycle.

6.3.6 Essence of collaboration between trading-partners

A collaborative relationship is one that defines the business as a team of suppliers, manufacturers, distributors and retailers providing goods/services to a consumer. Collaboration needs to be a two-way beneficial relationship with participation by all the trading partners. Organizations should only aggressively pursue collaborative relationships when the barriers between businesses are minimized, surprises are few, activity costs are low and business plans are integrated to meet the consumer's needs.

Collaboration at a business process level between trading partners has demonstrated high returns. The best example of a well-developed collaborative business process is CPFR (collaborative planning forecasting and replenishment). Advanced Manufacturing Research Inc. completed a review of several test cases showing the following benefits.

Retail Retailer Benefits	Manufacturer Benefits
Better in-stock 2–8%	Lower inventory level 10–40%
Lower inventory 10–40%	Faster replenishment 12–30%
Higher sales 5–20%	Higher sales 2–10%
Lower logistics costs 3–4%	Better customer service 5–10%

6.4   Business process optimization

Electronic business process optimization is a new class of decision intelligence software that integrates and optimizes all business processes at the planning and execution levels. It allows enterprises to achieve velocity in their operations, that is, to intelligently execute at maximum speed. eBPO provides:

- Forward visibility and responsiveness
- Flexibility to cope with changes in organizational structure
- Support of strategic business goals
- Global integration and optimization
- Performance under extreme conditions.

Processes have two important characteristics: they have customers (internal or external), and they occur across or between organizational subunits. Processes are generally identified in terms of beginning and end points, interfaces and organization units involved.

6.4.1   Characteristics of business process optimization

Multi-enterprise

Today, businesses focus on core competencies, and emphasize partnering as the route toward delivering value to the consumer. The shift toward partnering creates an imperative for companies to be more closely integrated with their trading partners.

Advances in object technology reduced the technology gaps and made it possible to integrate within and between companies using various enterprise systems, including enterprise resource planning (ERP), sales force automation (SFA), advanced planning and scheduling (APS) and others.

Multi-functional

The advent of ERP systems allowed a more realistic workflow in a company's information systems. For example, upon the entry of an order, the production schedule is modified to reflect that order, and the company's accounts are modified to show the inflow of cash and the outflow of inventory. This means that all functions of the organization is affected and influenced by changes in other functions.

Complexity

Business processes are remarkably complex, with multiple courses of action available at any given point. For example, the order processing process needs to take cognizance of the following:

- The profitability of various product sources
- The competing status of various customer orders
- The management of incomplete orders
- Credit-management issues
- The feasibility of orders
- Available-to-promise issues

It is the rich complexity of business processes that poses a severe challenge to ERP systems, in that their role is to execute a transaction, rather than to generate the complex analysis necessary to compare and evaluate the multiple avenues of opportunity that arise at almost every step in the process.

Necessity of speed

Due to changes in business and economic conditions, the speed of execution of business processes is critical. For example, a company faces time pressures due to:

Product life cycle   Product shelf-life issues.

Globalization   Competitive pressures from all over the globe affect previously insulated companies and industries.

Rising customer expectations   Customers have become more educated, are more demanding and expect service at 'Internet speed', partially because of the increased competition caused by globalization, as mentioned above.

Policy changes   Quick, unpredictable and continuous policy changes characterize numerous markets.

6.4.2   Enterprise solutions

Both major classes of enterprise solutions (ERP and APS) approach business processes in their own way:

Advanced planning and scheduling

Advanced planning and scheduling (APS) systems provide modeling and analysis capabilities for intelligent decision support but focus on a single enterprise 'supplychain,' i.e. manufacturing, distribution and transportation functions.

Enterprise resource planning

Enterprise resource planning (ERP) applications have succeeded in integrating data from multiple functional areas, but their scope is limited to a single enterprise, and are typically slow, transaction based and unable to encapsulate the complexity required in cases where multiple courses of action are available.

6.4.3   Electronic business process optimization

Electronic business process optimization (eBPO) initiatives and product offerings leverage the strengths of APS and ERP systems, while mitigating their weaknesses. eBPO offers enterprises the ability to closely couple multiple functions across multiple enterprises and quickly react to micro and macro changes in demand, supply, competition and partner relationships. It allows enterprises to achieve velocity in their operations; that is, it maximizes an enterprise's ability to execute its business processes more quickly and with intelligence (the ability to compare and evaluate multiple courses of action).

Figure 6.6 provides a graphical representation of the movement of enterprise information systems toward velocity. As enterprise software systems have evolved, incremental advances in speed of execution and intelligence have been made. In most cases, however, while the gains in the speed of execution have been significant, the gains in the intelligence of systems (their ability to support decision-making) have been marginal. The exception to this has been the movement from ERP to APS, where the increase in intelligence was quite large.

Business decisions are complex in nature and are implemented in a step-wise, risk-reduced manner.

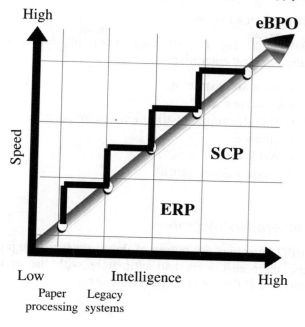

Figure 6.6
Movement toward velocity

Forward visibility and responsiveness

Business process workflow focuses on transaction processing in order to improve customer service levels and lower cost. Business process optimization focuses on the optimization and execution of processes through the use of intelligent decision support algorithms. This forward visibility yields significantly higher net value for companies than traditional ERP.

Flexibility to cope with changes in organizational structure

Mergers, acquisitions and downsizing require flexible enterprise information systems. BPO's object-oriented architecture allows such flexibility and supports multi-enterprise activities.

Support of strategic business goals

Business process optimization strongly supports companies' efforts to achieve leadership positions in one or more of the areas through industry-specific templates and solutions for operational efficiency (SCM), product leadership (product life cycle management) and customer service excellence (customer management).

6.4.4   Global integration and optimization

Integration as it applies to information systems can be thought of in three distinct dimensions:

Functional integration

The eBPO links together the major operating functions of manufacturing-centric and distribution-centric enterprises:

- Sales and marketing
- Customer service
- Manufacturing, distribution and logistics
- Finance
- Workforce
- Product development.

Multi-enterprise integration

The eBPO leverages the power of the Internet to integrate companies with their trading partners, their customers and ultimately with the end-consumer. This integration is accomplished at the data and workflow level.

Integration with enterprise systems

No one vendor can meet all of the technology needs for everyone without leveraging other commonly used products. The eBPO leverages existing standards wherever possible for:

- Messaging
- Object Standards
- Data access Middleware
- Security
- Software.

In order to integrate single or multiple enterprises at the planning and execution level, numerous data sources and applications must communicate with one another, ideally through the adoption of a common data model.

6.4.5   eBPO processes

eBPO enables maximum velocity by optimizing the core business processes of an enterprise.

- Product life cycle management
- Supply-chain management
- Customer management
- Inter-process planning
- Strategic planning.

The five business processes span a continuum of organizational impact, from strategic planning, to tactical planning, to operational planning and execution. The operational processes allow a company to create competitive advantage by focusing on product superiority, operational superiority or customer service superiority. eBPO optimizes and integrates the workflows of all major business processes (refer Figure 6.7).

Figure 6.7
The E-BPO business processes

Supply-chain management

Supply-chain management involves the effective anticipation of market demand, the optimal positioning of enterprise resources to meet demand, and the efficient fulfillment of demand as it is realized. It is the comprehensive combination of all business sub-processes that enables velocity, i.e. the fast and intelligent exchange of information and the movement of goods and capital between end-consumer and component (raw-material) suppliers. The SCM includes the following sub-business processes:

Demand fulfillment The focus of the demand fulfillment process is to provide fast, accurate and reliable delivery date responses to customer orders. This is mainly an execution level subprocess that includes demand creation, order capturing (including configuration), customer verification, order promising, backlog management and order fulfillment.

Demand planning The objective of the demand planning process is to understand customers' buying patterns and develop aggregate and collaborative forecasts. Demand planning is by definition a planning process and feeds into the demand fulfillment and supply planning processes. Demand planning spans over long-term, intermediate and short-term time horizons.

Supply (or Master) planning This planning level subprocess spans across strategic and tactical supply-planning processes. Strategic planning, inventory planning, distribution planning, collaborative procurement, transportation planning and supply allocation are all part of this subprocess.

Business benefits

The major benefits from high-velocity customer-to-supplier business processes include:
Inventory-lean supply chain: With customer information propagating through to the raw-materials level, there is need for limited finished goods inventories and safety stocks at any of the intermediate manufacturing, storage or transportation points.

Reduced end-to-end supply chain planning and execution cycle-time   Finely scheduled supply chains minimize the 'touch time' of the product and minimize the time products remain in inventory. Optimized scheduling assures that no product is started if it is going to end up waiting for a key part at a point in this process.

Reduced depreciation exposure   Products that are already ordered at a fixed price face limited depreciation risk en route to the customer.

Reduced exposure to demand volatility   As products proliferate, the reliability of forecasting at the finished good level declines rapidly, and the total level of inventory required for maintaining service levels start to increase exponentially. Avoiding finished goods inventories allow manufacturers to maintain service reliability without prohibitive inventory levels and the risks associated with them.

Improved forecasting   Manufacturers can collaborate with their customers, suppliers and trading partners to create optimal plans, based upon realistic forecast.

This process attempts to ensure:

- Reduced assembly-to-customer cycle-time
- Collaborate with suppliers and customers
- Optimization of the total physical transportation to maximize reliability and minimize costs
- Optimization of product mixes decisions while respecting customer priorities.

Customer management

Customer management (CM) is centered on 'customer intimacy'. It focuses the effort of companies on serving the needs of carefully targeted market and customer segments. Depending on the industry and the nature of products or services sold, businesses may need to emphasize different aspects of customer management such as:

Sales and marketing activities   Defining markets, acquiring customers, managing brands, categories and accounts.

Service and support management   Maintaining customer relationships, supporting customers, providing spare parts, field service and scheduled maintenance.

The CM solution encompasses   Sales and marketing planning maximize the effectiveness of demand-creation for a company's products and services through activities directed both at consumers (end-users) and customers (channel intermediaries). It includes brand management, category management, channel and account management and customer commitment management.

Service planning defines the set of decision support processes and solutions that manage the after-sales processes including service, support and additional sales to existing customers. The need for such a solution is driven by two related business drivers:

1. The increasing importance of customer service in retaining customers.
2. The revenue opportunity in selling additional services and products to existing customers.

Business benefits

- Increased profit margins
    - Decrease marketing and promotion expenses
    - Improve channel efficiency
    - Increase parts and service revenues
    - Improved demand-creating collaboration.

- Increased return on service/support investments

    - Reduce service parts inventory
    - Reduce physical storage space of service parts
    - Improve service personnel productivity
    - Increase equipment uptime, first-time fix rates
    - Improved inventory forecasting with service parts suppliers
    - Increased ability to integrate product configuration and order pricing with available-to-promise and capable-to-promise capabilities.
    - Multiple modes – client/server, detached, web based
    - Present exhaustive product information electronically
    - Dynamic sales document creation – creation of quotations/proposals in a matter of minutes with specific solution configuration and pricing, instead of taking several days.
    - Profitability/margin analysis by quotation/order
    - Increased customer satisfaction and responsiveness
    - Reduce lost sales and stock outs due to poor planning
    - Increase critical parts availability
    - Increase service response time.

This process attempts to ensure:

- The most remunerative mix of maintenance agreements that should be sold to customers.
- The most profitable location of service centers that maintains reliable, efficient service to customers.
- How can field service engineers handle more calls on a daily basis?
- Impact on the growth of primary products effects the service support operations.
- Real-time or near real-time price and availability quoting for products offered as build-to-order.
- Volumes of transactions.

Inter-process planning

Inter-process planning (IPP) focuses on the tactical decision-making process at the enterprise level, and balances the competing goals of various business processes and functions, generating an enterprise-wide optimal use of resources: capital, people, physical assets. Inter-process planning is a tactical planning tool that integrates across multiple eBPO processes and resources within an organization to achieve an enterprise-wide optimal solution.

Business benefits

- Achieve better utilization of human, financial and physical resources.
- Obtain frequent, detailed updates and validation of strategic plan.
- Detect problems or react to changes in the marketplace in days, not months, even for large complex enterprises.

This process attempts to ensure:

- The alignment of strategic goals for day-to-day operational activities.
- Resolve conflicting resource requirements amongst the various processes i enterprise.
- The reconciliation between long-term product objectives with short-term revenue and profit goals.

Strategic planning

Strategic planning (SP) is the process where senior executives set company performanc objectives and make longer-term decisions of revenue planning, product portfoli management and supply-chain design. It represents a coherent, unifying and integrativ pattern of decisions a firm makes.

Business benefits

- Feasible aggregate financial plans (AFPs) replace infeasible annual budgets.
- Critical strategic decisions are optimized with planning engines rather tha executive assertion.
- Strategic planning cycle time is shortened.
- With a shortened cycle, AFPs are based on recent information and therefor more accurate.
- Better-managed information flows to the investment community.
- Accurate enterprise model allows for effective cash flow management.

This process attempts to ensure:

- Optimum supply-chain design
- Optimum capacity design
- Most beneficial location of distribution centers
- Optimized product portfolio management
- Introduction design of new products
- Optimum revenue planning
- Appropriate relative budget allocations for various business units.

6.5   Procurement management

Purchasing managers operating in complex situations are rethinking strategies an procurement policies in light of new economic marketplace and budgeting restrictions Listening to the needs of both internal and external customers has forced them t undertake measures to improve the cost-effectiveness of the acquisition and suppl processes.

6.5.1   What is procurement?

The main outlay of a company's cash goes toward material purchases. The efficiency c any business activity depends upon having materials, supplies and equipment available i proper quantity with the proper utility at the proper place, time and price. Purchasing i its narrow sense refers merely to the act of buying an item at a price. Procurement refer the managerial activity that goes beyond the simple act of buying and includes th planning and policy activities, research and development, service selection, etc. Thus procurement is buying material/service of the right quality, the right quantity, at the right time, at the right price from the right source with delivery at the right place (refer Figure 6.8).

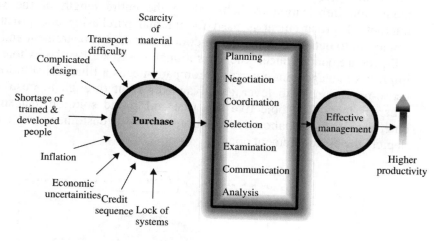

Figure 6.8
Elements of material procurement

Transformation of purchase departments into procurement functions provides global information and enables continuous process improvement. It enables better decision-making by centralizing strategic decisions and decentralizing low value-add tasks.

Supplier performance analysis, item/commodity analysis and forecasting are part of the activities of a procurement function. Some of the other activities are tabulated below.

Backend Buyer System	Backend Supplier System
Purchase requisition/order	Sales order
Contract price	Availability check
Goods receipt	Expected date of delivery
Supplier evaluation	Alternative product
Quantity	Back order quantity

For effective procurement, the various procurement systems across the enterprise need to be connected to suppliers to provide real-time information and visibility i.e. providing visibility to suppliers for new purchase orders and purchase order changes as they are generated, and linking supplier purchase order delivery details with the company's planning operations.

Organizations attempting to be world-class in the field of strategic procurement and e-commerce can enhance their competitiveness by re-engineering their procurement, payment and reporting processes. This can be achieved by significantly reducing the cost of the acquisition of goods by:

- Reducing the resources expended such as, staff, time, paperwork, etc.
- Reducing turnaround time, between requisition and receipt of goods or services.
- Improving in the quality and timeliness of financial reporting.

6.5.2   E-procurement

E-procurement simply means buying products and services over the Internet. In this sense, most companies are already using e-procurement. However, the term is fast becoming: 'the automation of the whole purchasing process and making order and requisition information available along the entire length of the value chain via the Internet'. E-procurement is used for internal purchasing and procurement and is also connected to outside systems of vendors to automate procurement and payment.

E-procurement reduces the operational costs across the whole supply chain and improves management control. A company needs a financial solution first so that it can use e-procurement to leverage advantages right through to customers, suppliers and other partners. The objective is a true end-to-end solution that enables companies to enter into truly collaborative relationships with customers and suppliers. Figure 6.9 depicts the e-procurement concept.

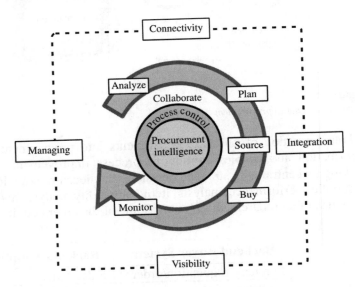

Figure 6.9
E-procurement concept

Analyze   Conduct company-wide spending analyses to identify and prioritize savings opportunities, including supply base rationalization and purchase aggregation (buying in bulk).

Plan   Develop optional sourcing and procurement strategies based on existing and future purchase requirements across the enterprise.

Source   Identify, evaluate, negotiate and configure optional trading relationships.

Buy   Communicate, execute and settle payment against negotiated trading agreement and contracts.

Monitor   Track and enforce internal content compliance and external supplier performance.

Collaboration   Engage in intra- and inter-enterprise collaboration across all procurement processes.

Process Control   Standardize and enforce common processes across the enterprise and supply chain.

Procurement intelligence   Provide a single point of truth of all procurement-related information.

6.5.3   E-procurement software

There are a lot of e-procurement packages available which simply automate the ordering process i.e. to place orders over the Internet. This saves time and cuts postage and paper costs, but without integration to back-office finance systems it is impossible to achieve cost savings. To take full advantage of e-procurement, companies should tie procurement systems not only into financial systems, but also into the systems of customers and suppliers to create a virtually paperless system from one end of the value chain to the other. Integration is very important, as e-procurement can become an isolated step, without moving automatically throughout the value train, and the purchasing information stops in procurement.

6.5.4   Procurement cycle

The diagram below illustrates the different steps associated with the overall procurement cycle, from the order placement to the accounts payable, in steps through a web based e-commerce network (refer Figure 6.10). It includes customers' inventory and supplier information and is readily accessible at any time and from everywhere.

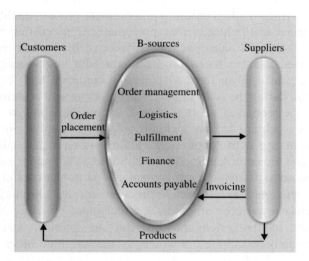

Figure 6.10
Procurement cycle

E-procurement allows two-way communication of real-time financial and purchasing information. Management at both the buying and selling end can see what's happening at any given moment and identify trends and problems. A shorter purchasing cycle enables companies to maintain lower inventory levels and respond more quickly to stock-outs.

6.5.5   Procurement execution

Procurement execution helps companies reduce transaction costs and streamline the purchasing process by allowing disparate procurement organizations to generate purchase orders and collaborate purchase order details with the entire supply base. Transactions that were once labor intensive and handled manually can now be electronically communicated as they are generated. Providing this level of visibility to new purchase orders and purchase order changes allows procurement professionals to focus more of their time on strategic issues and less-time fire-fighting.

- Reduce transaction costs and streamline the procurement process
- Improve visibility and tracking of all procurement transactions
- Decrease maverick buys while improving contract compliance
- Improve ability to meet customer demand
- Reduce inventory-carrying costs
- Improve supply chain efficiencies
- Link supplier's deliveries with company's operations
- Focus buyer's attention on strategic issues.

Advantage of e-procurement execution

E-procurement brings a set of immediate advantages. Firstly, all the resources that were previously tied up in administrative functions related to procurement become available for core functions. Secondly, as a result of the procurement automation and safeguards, the number of mistakes related to procurement functions is decreased drastically. Thirdly, it manages all logistics associated with fulfillment and deliveries. E-procurement delivers instant savings while increasing procurement process quality and efficiency.

E-procurement reduce written procedures offers costing options that auto-generate order plans and monitor sales against the forecast that, if exceeded, auto-generates production plans based on sales. It can also be interfaced to any forecasting module. Capacity planning functions are used in finished goods procurement and 'capacity exceeded' warnings are automatic during the issue of purchase orders as well as domestic work orders.

Once the processes are streamlined, the business has greater control over its purchasing, and excess administration is eliminated. There is substantial savings from the selection of and partnering with suppliers, including supplier prequalification, request for proposal, tender and the contracting processes. Strategic sourcing and partnering arrangements have contributed to dramatic improvements, in particular capital project processes.

Suppliers gain a lot from e-procurement by eliminating paper-based catalogs and suppliers can process orders as a direct input by using XML. Suppliers can notify the customer of deliveries and transmit an invoice, and the customer's system can automatically match that invoice to the original order and post information to the financial back office. Buyers also can notify suppliers automatically of payment, based upon a match with the delivery notice, relieving the suppliers of generating an invoice.

6.5.6   Future trends

Companies are automating the purchasing process from requisition to payment all along the value chain creating a paperless, fast and secure procurement process (refer Figure 6.11). It permits all users to share the information on a real-time basis where managers readily match expenditures against budgets, and employees can check the status of orders at any moment. Nowadays large companies are moving toward commitment accounting, where managers set a budget against an item, and the average spend is automatically reduced once that item is purchased.

Some of the latest software has built-in rules on employee purchasing authorization and single-click access to trading partners. For example, if an employee is authorized to purchase a copier, the e-procurement system should automatically direct the purchase to the vendor with the lowest price. It also makes sure that the employee should not exceed his or her level of authorization.

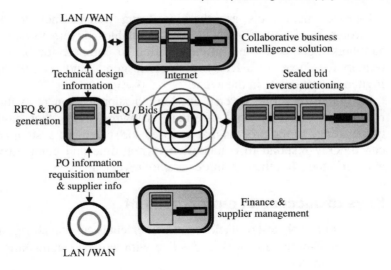

Figure 6.11
Paperless e-procurement with collaborative business intelligence

6.5.7 Manufacturing and e-procurement

Plant and process industries have achieved economies of scale with increased efficiencies using plant design. Now there is a need for strategies that will deliver maximum efficiency and the lowest cost. Procurement has been the most important area to reduce cost, using efficient procurement systems. Online procurement solutions primarily serving manufacturing and process plants have request for quote (RFQ) mechanisms used for highly specialized engineered and fabricated goods in manufacturing plants. It provides maintenance and procurement support for the complete life cycle of a facility.

Most online procurement solutions focus on the purchase of maintenance, repair and operations (MRO) items, from a catalog. This web-based process helps reduce time and costs associated with design and procurement processes, and leverages historical information to improve maintenance and customer response time. With RFQ, process manufacturers can begin to realize quick benefits as they reduce the need to send documentation back and forth between vendors, suppliers and business partners. All facets of a manufacturing enterprise benefit from the data that is integrated because it helps extract, build and manage knowledge from historical business practices to streamline future operations.

A procurement manager can collect and aggregate relevant RFQ data online, including financial and technical information. Dynamic 2D and 3D views of complete manufacturing plants, including technical specifications, rotation and zooming of objects assist users with the visualization of the required items. The financial data, such as the purchase requisition and the authority to engage in the transaction, comes from the ERP system. In the RFQ process the procurement manager can invite appropriate suppliers, attach information and then send the entire package to the online trading engine.

Suppliers are notified and then can log onto the auction, consult all necessary RFQ information and make bids. The solution supports an interactive communication path between the procurement manager and prospective bidders. The procurement manager can then evaluate the bids, make a decision (often not only based on price), generate and send a purchase order online to the selected supplier.

E-procurement tools are there to bring transparency to the competitive bidding environment. Standard reverse auction enables buyers to save money by aggregating suppliers together in one online location to compete for business. A buyer submits a request for quote (or a RFQ) for specific goods and services and several suppliers are invited to participate in the event.

In online, real-time reverse auctions, a supplier can re-evaluate and adjust a bid in response to other bidders' offerings. Knowing what other bidders are proposing adds transparency to the process that a closed- or sealed-bid system cannot achieve. Suppliers can quickly respond to competition in real-time and competitive bidders may see one another's bids, but their identities are not revealed.

6.5.8   Keys to success of e-procurement

- The biggest obstacle to e-procurement is cultural change and not technical change.
- E-procurement is ineffective without first establishing a strategic supply-chain system.
- Part number standardization/rationalization is required to realize the full effect of strategic sourcing and e-procurement.
- Procurement and contracting fundamentals must be addressed with preferred suppliers before attempting to facilitate the transactions with electronic tools.
- Purchasing skill-sets are considerably different in the e-procurement environment and requires training or replacement.
- Involvement of preferred suppliers in the development (planning) process is a must for the success of e-procurement programs.
- The implementation of strategic sourcing and e-procurement cannot be relegated to a secondary function, low priority or part-time job for key personnel.
- Executive sponsorship of the teams implementing e-procurement projects is of paramount importance.

E-procurement is not just about automating the procurement process and strategic sourcing is not simply about consolidating purchases.

6.5.9   Threats and security of e-procurement

E-procurement threats

E-procurement involves a lot of data transfers related to business transactions of an organization and are exposed to multiple threats related to online business.

- Masquerading
- Unauthorized disclosure
- Unauthorized access
- Loss of information integrity
- Denial of service
- Theft of service and resources.

Security

Security issues are complicated and not easily implemented in a short time. Various security layers protect data that is visible on the Internet via Web pages and corporate communications. If breached, corporate image and/or communications can be compromised (refer Figure 6.12).

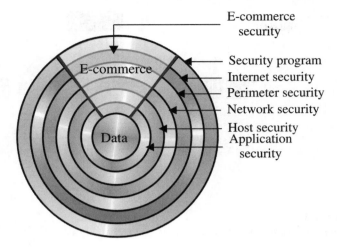

Figure 6.12
Security layers

E-commerce security Protects the data while communicating it across the organization and outside the organization. If breached, all corporate layers of security can be compromised.

Security program Is the overall foundation to protect IT systems and to set policies for other security layers. It includes monitoring, detection and response.

Perimeter security Is the first layer of logical protection through the implementation of perimeter devices.

Network security Is the first internal layer of protection. If breached, loss of control of data movement and/or data modification is possible.

Host security Protects computer applications and data. If breached, data could be altered and/or deleted.

Application security Protects application and data. If breached, data could be altered, copied and/or deleted.

6.6   Supplier relationship management

In today's accelerating world economy, manufacturing companies are faced with the market realities of ever more demanding customers, shrinking product life cycles and steep price erosion. The drive to continually cut costs and focus on core competencies has driven many to outsource some or all of their production. With increasing reliance on outsourcing and strategic procurement, managing close relationships with the large supply base on a real-time basis becomes crucial.

Supplier relationship management (SRM) focuses on maximizing the value of a manufacturer's supply base by providing an integrated and holistic set of management tools focused on the interaction of manufacturer with its suppliers.

6.6.1   Competitive advantage

The SRM becomes an important technology investment as companies recognize the value of managing their supply base as a competitive weapon (refer Figure 6.13).

Figure 6.13
Enterprise applications architecture

The SRM solutions provide competitive advantage in three important areas:

1. Dramatic cost savings. The SRM drives cost reductions by enabling streamlined business processes and improved information flow.
2. Tighter replenishment execution, reduction of excess and scrap inventory, elimination of non-value-added tasks and leveraging scale with strategic suppliers provides substantial savings to the manufacturers.
3. Increased flexibility and responsiveness to customer requirements. Visibility, communication and collaborative planning with suppliers give flexibility and responsiveness.

Faster cycle times are achieved by collapsing the latency of information flow through faster communication, automated replenishment and streamlined process management to ensure real-time information for buyers and suppliers.

6.6.2   Mechanisms

Supplier relationship management contributes to competitive advantage through three primary mechanisms:

Improved business processes across the supply chain

The SRM focuses on cross-functional and inter-enterprise processes and participates in these processes. It provides seamless integration between these processes and non-SRM processes. A well-designed SRM system has the ability to improve business processes in the following areas: strategic supply management, supply-chain collaboration and procurement execution.

The relationship between these three is critical as they form the foundation of strong supplier relationships (refer Figure 6.14).

Strategic supply management   Aligning supply management goals with the manufacture's business goals. The benefits include an efficient, consolidated supply base, stronger company-wide business relationships and contracts.

Supply chain collaboration   Provides efficient exchange of planning information between trading partners, which ensures optimized flow of parts and materials into plants.

The types of exchanged information includes material plans, inventory availability, performance scorecards, etc. The main benefits are increased customer responsiveness and reduced inventory costs.

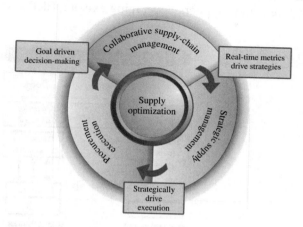

Figure 6.14
Integrated business processes enhance SRM

Procurement execution Focuses on tactical manufacturing execution and ensures the most efficient and cost-effective delivery of parts and materials into the plant. The benefits are reduced cycle times, lower shortages and minimum expediting.

Empowering inter-enterprise workflow

SRM takes a role-based business process-centric approach to ensure streamlined business processes between manufacturers and suppliers. The benefits are:

- Ability to rapidly adapt business processes
- Information supports efficient decisions within the business process.

Abstracting the ERP system

SRM system abstracts the ERP system, leveraging its capabilities, while hiding its details from the user who is instead provided a highly tuned role-based process information-set aimed at efficiently supporting the user's business objectives (refer Figure 6.15).

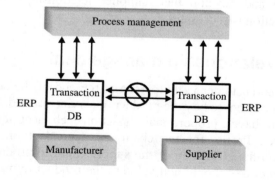

Figure 6.15
Process management

6.6.3   Next generation architecture

New generation Internet-based solution architectures is deployed on top of existing manufacturing systems (refer Figure 6.16). It supports business processes between manufacturers and suppliers, leveraging existing ERP systems.

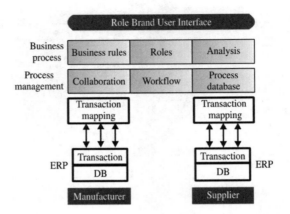

Figure 6.16
SRM architecture

The Internet makes real-time interaction possible and these SRM architectures include:

- A many to many data model
- A process-focused workflow
- Strong transaction integration
- Rich role support.

6.6.4   Rapid product cycle and new product introduction

Strategic advantage is measured in terms of rapid time to market, market share and profitability. The ability to achieve this depends on manufacturer's responsiveness toward market requirements, developing new products and synchronizing new product introduction with market growth. Collaborative product introduction for competitive advantage requires strong relationship with suppliers. Collaborative product introduction is done with strategic trading partners. The SRM is the foundation for manufacturers to assess, manage and develop their supplier relationships to ensure that collaborative product introduction is successful.

6.7   Customer relationship management

The end- or final-customer is the person at the end of the supply chain who makes the decision whether or not to buy the product or service offered. The purchasing decisions of the end customer have a proportional impact on each facet of the chain contribution to the finished product. Profits for the chain are dependent on the customer's purchasing decisions and on the efficiency of the supply chain to produce the product. All internal and external processes should add value to the product to maximize the profit. Processes that do not add value diminish the levels of profit that can be achieved. The level of profits that can be achieved and the continuation of these relationships form the basis for managing and integrating the supply chain. The importance of the end customer helps to make the chain more efficient and responsive to the needs of the customer.

To understand the customer there must be some direct link with the customer and it is essential that information channels speak the language of the customer. Information about customers' preferences, buying habits, attitudes toward particular products and service satisfaction can be collected to form scientific customer profiles which are not based on assumptions and perceptions, but on collected and transformed information. Getting closer to the customer, compiling information about the customer's first enquiry, repeat ordering patterns, service satisfaction and customer attitudes provide inputs for designing products and value packages that matter to the customer. Relationships with customers go a long way to determine future profitability, and companies are making bigger and bigger investments to do just that. Customers are more knowledgeable about the service they should be getting and are spending money based on the experience they receive.

6.7.1   What is CRM?

Customer relationship management (CRM) requires a customer-centric business philosophy and culture to support effective marketing, sales and service processes. It is a business strategy that selects and manages the most valuable customer relationships. The CRM applications can enable effective CRM, provided that an enterprise has the right leadership, strategy and culture. It is designed to reduce costs and increase profitability by solidifying customer loyalty.

True CRM brings together information from all data sources within an organization (and where appropriate, from outside the organization) to give one, holistic view of each customer in real time.

This allows customer-facing employees in such areas as sales, customer support and marketing to make quick yet informed decisions on everything from cross-selling and up-selling opportunities, to target-market strategies, to competitive positioning tactics.

CRM has become a strategic success factor for companies in all types of industries, as companies are continually faced with significant challenges in acquiring, engaging and retaining profitable customer relationships. With rising customer expectations, these objectives are becoming more and more difficult to achieve.

The following table lists differentiating attributes. The left-hand column contains the internal issues (product related) and other three columns list a variety of issues related to external performance.

Physical Products	Pre-Sales Support	Availability	After-Sales Support
Features	Ease of ordering	In stock	Field service
Performance	Technical support	Fast delivery	Technical support
Reliability		On-time delivery	Warranty policy
Quality		Complete delivery	Service parts
			Accurate billing
			Payment terms

CRM should always start with a business strategy which drives changes in the organization and work processes, which are in turn enabled by information technology.

6.7.2   The end customer

CRM technology enables a systematic way of managing customer relationships on a large scale. Companies have always understood the importance of focusing on customers with the best potential for sales and profits, and provide good service so that they come back again and again. Research has shown that retained customers are more profitable than new customers, for the following reasons:

- The cost of acquiring new customers can be substantial. A higher retention rate implies that fewer customers need be acquired and these can be acquired more cheaply.
- Established customers tend to buy more.
- Regular customers place frequent, consistent orders and therefore usually cost less to service.
- Satisfied customers often refer new customers to the supplier at virtually no cost.
- Satisfied customers are often willing to pay premium prices for a supplier they know and trust.
- Retaining customers makes market entry or share gain difficult for competitors.

It is an important fact that the features that delight the customer today will gradually move toward spoken needs and eventually become basic expectations:

- Basic spoken needs are the needs the customer regards as so elemental that it does not need mentioning. Customer will be dissatisfied if these are not met.
- Spoken needs are the expressed features that the customer would like as part of the total value package and are options to satisfy the customer.
- Unspoken needs are the unexpected delights that the customer had not anticipated as part of the value package.

Transaction Focus	Relationships Focus
Orientation to single sales	Orientation to customer retention
Discontinuous customer contact	Continuous customer contact
Focus on product features	Focus on product benefits
Short timescale	Long timescale
Little emphasis on customer service	High customer service emphasis
Limited commitment to meeting Customer expectations	High commitment to meeting customer expectations
Quality is the concern of production staff	Quality is the concern of everybody

Retention of customers is the focus for 'relationship' marketing and the principle is that the total package offered to the customer comprises a core product and a service package that surround it.

Successful CRM initiatives start with a business philosophy that aligns company activities around customer needs, only then can CRM technology be used as it should be used – as a critical enabling tool of the processes required to turn strategy into business results.

Customer transaction data alone is not enough to succeed in today's fierce competitive environment. Organizations should not only gather and store transaction data on individual customers, including purchase patterns, current status, contact history, demographic information, sales results and service trends, but also data that is 'actionable', so that managers and employees can use it for decision support.

6.7.3   e-CRM

Electronic customer relationship management is the new face of CRM and presents a company to the world. It provides a company with the technology to capture customer information into database and uses the Internet to enhance customer relationships and get more intimate with them. It provides a global real-time view of customers that is acquired and retained and helps build customer loyalty.

The e-CRM is customer-centric and makes relationships profitable, sharing information across platforms for a seamless customer experience. e-CRM applications allow enterprises to interact directly with customers via corporate Websites, e-commerce storefronts and self-service applications. Customer relationship management covers a broad range of applications that are designed to help companies maximize the customers' experience at each stage of interaction.

The objective of CRM is to enable companies to build deeper, more profitable long-term relationships, by reaching customers with the right message at the right time and by providing superior customer service. The CRM applications include sales-force automation, marketing automation, customer service and support technologies, and customer interaction management.

6.7.4   Key elements of CRM

A CRM strategy consists of clear and well-constructed people, process and technology strategies – each one of which is carefully constructed as both an individual strategy as well as part of the larger holistic strategy. The people throughout a company, from the CEO to each and every customer service representative need to buy into and support CRM to ensure success.

A company's business processes should be re-engineered to bolster its CRM initiative. Firms should select the right technology to drive these improved processes, provide the best information to the employees and be easy enough to operate so that users will feel comfortable using it. If any of these three foundations is not sound, the entire CRM structure will crumble. It is a process that brings together lots of pieces of information about customers, sales, marketing effectiveness, responsiveness and market trends.

For CRM to be truly effective, an organization must obtain buy-in from its staff and ensure them that CRM will benefit them. Then it should analyze its business processes to decide which need to be re-engineered and how best to go about it.

Next is to decide what kind of customer information is relevant and how it will be used. Finally, a team of carefully selected executives must choose the right technology to automate the required processes and data acquisition.

6.7.5   Implementation of CRM

The most consistent thing about CRM is the inconsistency that surrounds it. Companies spend millions of dollars on software purchases, integration and implementation, but success is more often than not directly related to their ability to focus on the basics of CRM.

The CRM is more about managing data than customers. Customers do not want to be controlled. They do not want to be told what to do or how to buy. Customers want to manage themselves, and companies should focus more on relationship aspects of CRM to succeed. Analysis is important for CRM success, which is done by slicing and dicing information for analysis purposes. It gives unique pictures of customer purchase habits, contact behaviors and company involvement, but it cannot provide the needs of customer. The customer wants true and human personalization.

Technology does not ensure relationships with customers. The CRM success does not come from full-scale implementations, and CRM experts normally suggest phased implementation. Organizations can make sure each preceding phase is working and creating its own return on investment before moving on to the next phase of CRM implementation. Full-scale implementations demand corporate culture change. It is preferably done in stages to generate internal acceptance from employees and external acceptance from customers due to the new ways of doing business. Employees do not always adapt to and adopt new technology. A change in mindset is required before companies can expect their people to embrace CRM technology.

The high failure rates attributed to CRM implementations are due to three factors:

1. Early adoption of a technology platform before making certain the company had a strategic CRM plan in place along with the appropriate processes and the people to implement CRM (The software cannot make CRM work – the software is only an enabler).
2. Asking the employees to adopt CRM, instead of taking them through a culture change and preparing them for their new way to work.
3. Focusing on the internal workings of the new CRM implementation instead of getting customers involved and creating a CRM process that works the way the customer wants to do business with the company.

6.7.6   CRM and organizations

There are a lot of software companies and large consulting companies that are regularly engaged to build and deploy CRM solutions. Very few companies really have a CRM plan that emanates from their business plan, and many outside vendors do their best to work with companies and give them what they ask for. Where this is the case, it is vital that the software vendor has an intimate understanding of the business into which a CRM solution is being deployed. The solution must be aligned with the CRM process, which emanates from the business plan – not the other way around. It is necessary for the organization to have a clear-cut plan about CRM and its implementation.

Managing change

Companies should consider what they want to achieve and select the right people and tools to help them realize those goals from the very beginning. Customer relationship management implementation should be lead by the top management. If a company forces employees to change the way they work, it will create resentment within the user community. Projects that do not focus on winning the support of users at every stage in the process will usually end in total, abject failure. If no one is using the system, it will slowly be abandoned. Research indicates that 75–85% of CRM implementations fail as result of inadequate change management.

ROI

Companies should not forget that CRM is all about making sales and increasing profit margins. Distribution, billing and customer care are important considerations; customers insist rightly that these issues are given top priority. However, for the company, increasing customer satisfaction almost certainly leads to increased revenues.

Integration

Organizations should choose systems that integrate well with existing infrastructure, while providing powerful built-in customization tools on a standard operating platform. This approach helps to minimize the risk of installing CRM and reduces the disruption to the organization as the systems are deployed or upgraded.

Right vendor

Companies should also choose CRM partners that have proven experience of the CRM solution they plan to implement. A small highly specialized team of skilled people, such as project managers, system integrators and implementers can help the company meet CRM objectives faster and assist in building a proper strategy.

6.7.7 CRM service providers

The CRM solutions allow the organization to manage customer relationships throughout the selling side of the value-added chain. They are based on the approach of managing customer relationships for competitive advantage and differentiation, and assists in the devising of strategies for the acquisition of new customers and enhancing customer loyalty. It includes tapping all the communication resources and tools available to enhance customer relationships.

The scope of services includes:

- Installation and implementation support
- Data mapping and system conversion
- Interface design and development
- Report design and development
- GUI design and customization
- System testing
- Operations and support
- Upgrading
- Training, knowledge transfer and post-implementation support.

Consolidation among CRM vendors is sparking technology improvements that let non-technical people view and act on real-time customer data. All CRM vendors are emphasizing data accessibility and collaboration in their latest marketing-automation packages. Companies have to make a choice between stand-alone, specialized marketing-automation software and full-function CRM suites. The suites handle everything from outbound campaign management to inventory management and fulfillment. Specialized packages are generally better at handling Web marketing campaigns, while the bigger suites boast better integration with sales and customer service.

6.8 Material returns management

Replacement or return of an item or product is resorted to whenever the item or product deteriorates in its function or fails to work. Most companies handle material returns poorly. And when the returns are associated with Internet purchasing, the process becomes even more complex. Fortunately, reverse logistics, which encompasses product returns, service contract returns and product recalls, is an area that has been helped by new Internet-based tools and services. These offerings can streamline both traditional and e-business-based returns. The Center for Logistics Management at the University of Nevada estimates that typical return rates associated with e-business is 6–8% of all goods.

6.8.1   What is reverse logistics?

Reverse logistics is the process of moving goods from their typical final destination to another point, for the purpose of capturing value otherwise unavailable, or for the proper disposal of the products.

6.8.2   Objectives

Reverse logistics is an integral component of the supply chain and manufacturers have come to realize that an effective reverse logistics program is an important and strategic part of their business profitability and competitive position. In attempting to better manage the constant flow of returned goods, manufacturers have identified target areas key to streamlining the reverse logistics process, and have concentrated their efforts in the following primary areas:

- The process by which products come back through the pipeline
- The technology available that enables products to move expeditiously while providing causal information as to why they were returned
- The disposition options available for products, once they are returned to the retailer or the manufacturers.

6.8.3   Functions

Product returns are the most common aspect of reverse logistics and goods are assessed once received in the return centers. The items are automatically routed to their final destination based on the condition of the returned merchandise and the manufacturer's choice for disposition. Disposition options related to asset recovery include repair, upgrade, refurbishing (including repackaging), remanufacture, parts reclamation and recycling (particularly of pallets and containers).

Disposition logic also includes channel or routing logic, which means the returned items and components can be sent back to the customer, routed to a warehouse or sold in secondary markets. Reverse logistics activities include:

- Processing returned merchandise for reasons such as damage, seasonal, restock, salvage, recall or excess inventory
- Recycling packaging materials and reusing containers
- Reconditioning, remanufacturing and refurbishing products
- Obsolete equipment disposition
- Hazardous material programs
- Asset recovery.

6.8.4   Reverse logistics providers solutions

Most companies prefer to work with a third-party logistics (3PL) providers, as logistics is not their core competence and with the proliferation of e-commerce, new companies have emerged to focus on various aspects of the Internet order return process. Traditionally manufacturers used to run a distribution center and material returns center out of the same facility. Returned materials were not a priority for distribution center managers. Today, e-commerce has made the materials returned function one of the key determinant factors in achieving a competitive edge, and provision of this service of reverse logistics is often outsourced.

The solutions provided by the reverse logistics provider enable Internet-based return requests on forms customized by the merchant. These types of systems offer real-time reports on the products returned, and the reasons and sources of returns. It also provides sellers with valuable information that can be used to develop return-reducing strategies. Customer service representatives (CSRs) issue a return authorization number, reject the request or ask for additional information.

Some solutions have Internet order management components that allow a CSR to act as a gatekeeper. The architecture allows the return to be tied back to the original order so the CSR can understand the return restrictions on that particular product. It also helps ensure that the correct product has been sent back. The gate-keeping function actually provides an opportunity to up-sell and cross-sell.

For some solutions, consumers are advised to go to a designated Web site, generate return labels with the merchant's bar code for returning online purchases and are told where to ship the return. By offering an easy way to return items, customers are more likely to buy and be loyal to an online seller.

Customer satisfaction is also related to a quick money-back facility and once the merchant has agreed to the return, the customer's credit card is credited. The money is returned before the goods are returned, subject to two conditions. Firstly, the goods must be returned within a reasonable time period. Secondly, the goods must be the same as those purchased. If these conditions are not met, the customer's credit card is re-debited. By arrangement with the postal service, customers are provided with bar-coded, postage-paid labels. Finally, e-mails advise customers of the status of their returns.

E-business and Internet technologies nowadays offer an online auction of returned goods or sell them to e-business retailers, which specialize in selling returned goods. There is no single returns solution that is best suited to all industries. Some have achieved success by focusing on 3PLs that service high-tech manufacturers and some solutions seem better suited to Internet retailers.

A fully integrated, Internet-based and centralized supplier relationship management (SRM) solution can manage product returns from a consumer's initial return request, through the entire supply chain to credit/financial reconciliation and final product disposition. It also offers faster transaction speeds, instant communication, flexibility, global availability and completeness of information. It should be based on open content and transaction standards as well as supply-chain reporting rules, which improve the quality, depth and consistent flow of data.

It encompasses the following elements:

- Single-point authorization (gate-keeping technology) and management of goods and refund credit approval throughout the entire returned product cycle
- Simplified consistent return request procedures for consumers through all points of sale, both web–based and brick and mortar
- Centralized processing center and inspection, eliminating the need for multiple shipments of returned products
- Automated rules based disposition of returned goods based on the products physical condition, resale opportunity
- Branded inspection 'factory new' certification seal of approval for saleable returned goods
- E-auctions for saleable returned goods, generating near full price. Liquidation and salvage as necessary via disposition partners

- Exception-based reporting/trend analysis for retailers, e-tailers, distributors and manufacturers
- Automated procedures for detecting return fraud and abuse
- Inventory management
- Shipping and freight management
- Redistribution/fulfillment capability
- EDI, XML, Web and legacy systems integration.

7

Product and plant knowledge management

Learning objectives

- To understand the different operational issues that influence the effectiveness and efficiency of operations.
- To become familiar with the functions performed by different e-manufacturing modules to manage the operational issues.
- To understand how the e-manufacturing modules interact with the enterprise and supply chain.

7.1 Introduction

The manufacturing software market is changing in the same way as the manufacturing market is changing: toward customization, partnership and rapid response. These changes provide opportunities for manufacturers to realize dramatic improvements in performance, competitiveness and profitability. The new generation manufacturing software is moving toward an object-oriented environment, where objects can be linked, shared and re-used to make implementations and roll-out faster, smoother and cheaper.

The historic point-solutions will become redundant and be phased out, unless they adopt and change to the new environment, enabling easy association and integration with other solutions or modules. Vendors complying with a standard like S95 will greatly assist manufacturers, as the data structures will be consistent between vendor products, and will ease integration of best-of-breed solutions.

Figure 7.1 depicts the view of the 1990s into the future, and this is starting to become reality today.

7.1.1 An execution-driven approach

A manufacturing execution system (MES) approach enables manufacturers to schedule work orders, coordinate support functions, manage shop floor execution and communicate work status and problems. It monitors and analyzes work orders based on current shop floor activities, providing immediate knowledge from the proposed schedule. These systems should also include dynamic rescheduling capabilities that allow companies to implement real-time schedule changes and project the impact of schedule variation.

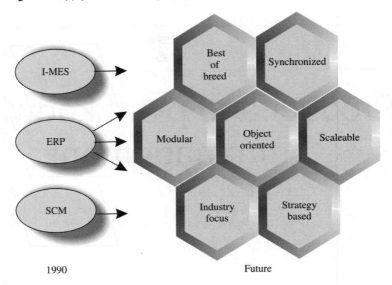

Figure 7.1
New generation solutions

Companies can monitor the shop floor and identify equipment or processes that are unable to perform the scheduled task due to shortage of materials, equipment breakdowns, etc. or employees who are waiting for approval to start a job. It can also assist the manufacturer to identify potential problems or bottlenecks before they occur, and coordinate shop floor and support departments so they execute work as a synchronized operating unit.

Using this approach, the ERP still plans production and materials on an enterprise level whereas the MES provides management with value-added manufacturing activity scheduling and monitoring on a minute-by-minute basis. In this pull-type environment, processes contain only value-added activities, and all support activities can be monitored in parallel to manufacturing activities so they do not become constraints in creating value.

For maximum effect, all non-value-added activities are to be eliminated during the design phase to ensure that inefficient processes are not merely automated, but that real value is released with the implementation.

There are three basic steps to be followed while deploying an execution system along the value-added activities path:

1. Firstly, a manufacturer is to understand the factory operation in detail. Then MES can be applied, providing dynamic finite scheduling and operations monitoring. This provides operators with continuously updated schedules and real-time dispatching of events – when and where it should occur. By managing the execution process, manufacturers gain control of the shop floor and have immediate visibility of any deviation from the scheduled tasks.

2. Secondly, the manufacturer needs to remove all support activities from the critical path of a work order. The collection of all necessary support materials e.g. tools, materials, instructions, prints, etc. to complete a task requires up to 38% of the effective capacity of a plant and can impede the progress of work through the factory. Execution systems enable users to itemize support tasks involved in the fulfillment of customer orders, communicate the requirement of support activities and check their status. This synchronization helps all related functional personnel (other than that of the shop floor) to monitor well in advance their activities in performing a task on the shop floor. This

improved coordination provides the operator with everything required, in the right place and at the right time, to execute a scheduled job.

3. Once support activities are out of the critical path, the factory manager can tightly control which orders are to be released to the shop floor. Only when all supportive resources are available in the shop floor, should new orders be released for manufacturing to make sure that every released order is passed through the factory without delays. This pull environment on the shop floor dramatically improves lead-time, productivity and inventory management.

Apart from the differences in the number of functions a plant needs, there are often variations in the actual focus and type of each function. Software configuration has its limits and the core functional logic is radically different for various industries e.g. scheduling a continuous process line is quite different from scheduling an assembly line. Whatever the type of operation, most manufacturing facilities have the following common components.

7.1.2   Production management

Production management often forms the core to an OMS implementation.

Operational data store

The operational data store (ODS) serves as a primary data source for all OMS applications and has to make provision for production data as well as operational data. It serves as the pipeline of data flowing between the ERP and control systems. Customization within the database is required since the data requirements are determined by the required functionality and changes to the database should be made online.

ODS functionality

- Fast and configurable real-time data acquisition
- Less space for compressed data store
- Real-time access to data through structured query language (SQL)
- Configurable handling of events and summaries
- Connection to third party applications
- Scaleable – capable to handle huge data transactions
- Disaster recovery by strong back-up
- Tabular view of live data and trend analysis.

Production data

Different people required different production data for different purposes. It is therefore important to capture the necessary real-time process data in an easy accessible database, which can be accessed from a number of different tools to extract data. Production data is normally stored in a historian.

Operational data

It is required to store data (history) produced by the various OMS modules on an easy accessible database. Operational data does not have the same time restrictions as production data and an ordinary relational database can be used.

Data exchange

The ODS has to provide this functionality to ensure consistent interfacing between the various clients and other solution components.

Operations reporting

Operations reporting are required to ensure that decisions are based on accurate information through the transformation of raw data into information suitable for decision-making. The primary source of data for the various reports is the ODS that contains manipulated and transformed data.

Standard reports

The use of standardized reports ensures that everybody speaks the same language and uses the same data throughout the enterprise. The throughput in one factory can then be compared to that of another factory on an equal basis, without any additional 'fudge-factor' manipulation to make results look better.

Scheduled reports

A scheduled report is a standard report that is produced at a predefined time or interval. A standard report is available for use at any time or interval. These reports are the windows into all the individual processing plants and data are gathered from the ODS, and reports are made available to specific individuals at specific times as defined during system design.

Event driven reporting

This type of reporting is normally SMSed or e-mailed to the relevant personnel immediately, but an exception report can be printed at specified intervals.

Ad hoc reporting

Ad hoc reports are required in the format of standard reports but for a different period or with different fields. These report are used for problem-solving or budgeting purposes, and needs to be more flexible in terms of format and content.

Availability and utilization

Availability and utilization figures are required to ensure optimal usage of all resources. Data required for these reports are fed manually or via inference using control system tags as inputs.

Balances

Dynamic material, mass, energy, water and gas balances or yield/recovery calculations are required in most manufacturing plants.

Engineering reports

Engineers and managers from various disciplines required data to perform an investigation or trend analysis. Spreadsheets such as Microsoft Excel have become part of the standard tools used by modern engineers. The OMS systems should thus be able to extract and export data to diagnostic tools such as MS Excel.

SQC/SPC

SQC and SPC are statistical tools to detect trends and are essential for plant optimization. They are effectively used and necessary once the plant has reached steady state conditions.

Production planning and scheduling

Production planning has to reconcile market demand with production capabilities and as such, it can also contain replanning triggers. It compares actual production results with the plan and the capacity models through deviation analysis, updates the results and reasons for deviations, and notifies management when action is needed.

Material storage and movement management

Inventory consists of raw materials, consumables, intermediate and final products. These are at various stages in the value chain and are not producing revenue during the period in which they are produced. Hence there is a need for minimized stock at all the levels. Inventory should be accurately planned and controlled as it is potential revenue and a very important asset. Various mechanisms are used during the different processes throughout the supply chain to final delivery at the customer.

Functionality of material storage and movement management:

- Control of weighbridge for all incoming and outgoing vehicles
- Stockpile management
- Calibration of all instruments
- Maintenance and management functions e.g. transaction approval, reconciliation
- Exchange of data with logistics contractors for reconciliation
- Tracking of product through entire supply chain
- Synchronization with laboratory systems.

Energy management

The cost of energy/power is a major operating cost for most companies. It is very important to minimize the power consumption of any production complex.

Functionality of energy management:

- Energy consumption monitoring
- Generation of load profile, load optimization and their effects
- Maximum demand management and power factor management
- Exception reports and alarms, crosschecking the billing.

Real-time cost management

The purpose of any business in existence is to make profit, which is the difference between the selling price of a specific product and the cost to produce that product.

Operational activities are major contributors to production costs. Cost such as consumables, tools, manpower, etc. are fixed monthly costs, and are incurred whether one or one million units are produced. Operational managers control the variable costs, which varies with increasing throughput and needs to be monitored in near real time. Variable costs include raw-material unit costs, energy costs, manpower costs (in some industries), etc. (refer Figure 7.2).

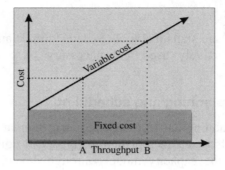

Figures 7.2
Cost structures of manufacturing

The primary purpose of real-time cost management is to manage the cost of a plant in real time. To enable this, it is required to see the effects of decisions on production costs in real time. The real-time cost management module tracks production costs in real time, triggers alarms, raises exception reports and provides suggestions to handle specific incidents. It provides operational personnel with the leverage to actively contribute to the effectiveness of the operational activities.

Miscellaneous

Provision has to be made for a number of miscellaneous functions, which can neither be omitted nor logically grouped with any of the above modules. They are the following:

Data entry

Operators have to manually enter data during certain activities such as availability and utilization, and manual entry should be intuitive and simple to use.

Simulators and plant modeling

Simulation is a very helpful tool to simulate the effect of various decisions before implementation. It is also effective where decisions have to be made in complex situations or where a great deal of uncertainty is present. Simulations can be used to answer various 'what if' questions and to test hypotheses.

Decision support system

Real-time decision support systems (DSSs) are related to simulators in some instances but a DSS is beneficial to operators where ambiguous plant conditions exist. It helps operators to decide what to do when certain situations or emergency conditions occur and ensures optimal decisions. A technology that is best suited for this application is real-time expert systems.

Configuration control and data validation

Configuration control is required to control changes to various data sources used for production optimization, planning and control purposes. It comprises a system that maintains a master configuration database which logs and manages all changes to configuration details of various subsystems.

Data validation is a specific form of configuration control that validates process data online by using predefined rules and relationships within minimum and maximum process boundaries and then automatically/manually corrects data.

Human resource requirement

Production management is about producing products in an optimal way mostly deploying people as an asset for the entire organization. People are required to supervise the process, do maintenance, take preventive actions, etc.

7.1.3 Logistics management

Order management and invoicing

The order management and invoicing process forms part of the business management system (BMS) or ERP layer processes. The main functionality captured in a BMS for outbound logistics, sales order management and invoicing are selling and marketing, order management and customer relations.

Outbound logistics planning

The total outbound logistics planning cycle with integration to production planning, inbound logistics planning, and global supply-chain planning forms an integrated organizational planning system. The outbound logistics planning cycle focuses on dispatch planning and distribution requirements planning.

The logistics planning cycle consists of three main modules – yearly planning, monthly planning and weekly planning, and all these plans are normally rolling plans.

Execution

Outbound logistics execution consists of dispatch management, inventory management and deviation management. The execution cycle converts the planned activities into actual day-to-day activities and monitors the progress and performance of the activities.

Dispatch management

Dispatch management is responsible for generating daily loading schedules, providing resources to meet the schedules, and monitoring the progress and performance of the execution of the schedules.

Inventory management

Inventory management follows a pull system, push system or a combination of the two to accommodate seasonality, market fluctuations and maintenance schedules.

Deviation management

Deviation management provides the essential link between planned actions and executed actions. Deviation information is collected from dispatch management and inventory management for evaluation and then local corrective actions are made for minor deviation. Major deviations can be adjusted for by changing the plan.

Control

The control cycle converts the scheduled activities from execution into actions, controlling and checking the performance of the processes, and providing progress and performance feedback to the execution cycle. Control consists of dispatch control and inventory control.

Dispatch control

All dispatch facilities are controlled by this function i.e. dispatch requests against ordered quantities validates loading requests against transportation constraints, controls all security aspects regarding the loading and dispatch of product, and controls the physical work flow in the dispatch area.

Inventory control

All the inventory facilities are controlled by this function. It includes the actions of physical material handling, the measurement e.g. weights, count or volume, and verification of both scheduled and unscheduled events.

Material genealogy

A combination of dispatch control and inventory control provides the material/product genealogy, tracking materials through the process, combining and separating batches/lots/campaigns to enable forward and backward tracking of materials or products.

Meteorological data

Meteorological data is sometimes required for production planning purposes and all meteorological data is stored on the ODS by uploading it from the various subsystems.

Plant instruction book

Production managers and foremen use an electronic plant instruction book (PIB) to communicate production targets and specific instructions to all shifts. Operators use the PIB to communicate shift specific information to the next shift to ensure that all decisions are traceable. It further eliminates wrong decisions due to improper information sharing.

7.1.4   Quality management

Product quality is a critical success factor for most businesses and the management of the quality inspection laboratory is equally important. To ensure effective and efficient laboratory operations a laboratory information management systems (LIMS) is commonly used. A LIMS normally manage and enforce good laboratory practices and related aspects such as:

- Efficient sample handling
- Seamless integration with analysis equipment for automatic results acquisition, control systems through real-time feedback of analysis results
- Seamless integration with BMS for analysis characteristics, sample approval and invoicing
- Seamless integration with third party systems for contractor feedback
- Planning and controlling of laboratory activities, workload monitoring and expediting, instrument calibration management

- Management of sample preparation, dispatch and analysis process
- Cost allocation, invoicing, automatic publishing of results and customizable reports, SQC charts, online help/manuals
- Consumable management, event monitoring and logging, standard operating procedures, archiving
- Resource planning and tracking, workflow, audit trail, user security levels.

7.1.5   Maintenance management

Maintenance costs are one of the largest operating expenses in capital-intensive industries and they require proper planning and control. It also influences production operations as badly maintained equipment break down more frequently, are unreliable and cause production stoppages and bad quality.

Maintenance planning

It is necessary to plan maintenance according to the production plan, which ensures efficient utilization of human and technical resources.

Equipment performance

The maintenance department has to ensure zero breakdowns but in reality this is not always possible and effective and efficient information capturing and analysis are required to identify areas of opportunity or improvement.

Metrics

To facilitate the above goal, a number of metrics are captured automatically e.g.

- Equipment availability and failures
- Equipment utility and throughput per time interval
- Mean time between failures (MTBF), mean time to repair (MTTR), activity and run-time, etc.

7.1.6   Human resource management

Shift schedule

An integrated shift schedule depicts shift manning and ensures proper rotation of personnel. It also ensures fair and transparent labor practices and the correct level skills on the various shifts.

Standby schedule

These schedules are closely related to shift schedules and are required for standby personnel details such as contact numbers, etc.

Training scheduling

To ensure compliance with some of the regulations and guidelines, regular training and retraining is required to upgrade the skill/knowledge of the personnel and results need to be maintained and performance monitored.

Competency matrix

The competency matrix is a matrix that indicates skills, medical condition, ability, etc. of various individuals. These also include legal issues such as licensing and qualifications validity and status to ensure that only authorized/trained employees operate the affected equipment/machinery or are in charge of safety-critical processes.

Real-time incentives

Production incentives are linked to certain production goals and also integrated with shift schedules and the payroll for automatic processing.

Access control

In order to control access of people and time booked, access control systems are used for security checks and access control. These systems are sometimes integrated with payroll systems, plant instruction books, training schedules, security systems, etc. to ensure that only employees with the correct access authority, training or experience can enter certain areas and notifies the employee when his license/authorization for operating equipment expires.

7.1.7 Safety, health, environment and risk (SHER) management

Incident reporting

Incident reporting logs all production incidents, whether it is an injury, emission violation, accidents, near accidents, breakdowns, etc. Logging can be done manually or electronically, and the main objective for these systems is the capturing of the incident to ensure further investigation and corrective action.

Environment management

Environmental monitoring forms the basis of proper and responsible environmental management. Typical measurements collected are:

- Emissions from stack, quality of water discharged
- Ground water pollution
- Rainfall, Weather station information
- Dust, radio-active pollution, slag deposits
- Rehabilitation monitoring.

Environmental management keeps records of emissions in terms of quantities and durations, and this information can be used to calculate material losses in order to identify improvement opportunities. The information is also used to prove environmental performance and in some cases to trade emission allowances with neighboring companies. These systems can also be used in emergency situations to model emission/pollution movement through the atmosphere or groundwater. This provides valuable information for emergency services to determine potential danger to employees and the community.

Occupational health management

It is necessary to keep track of trends at the medical station to provide an early warning system. Systems also keeps historical records of employee health for legal purposes, including injury statistics and regular medical check-ups, ensuring that each employee is tested on a regular basis to track employee health conditions. It also keeps information on employee exposure to hazardous substances when integrated with environmental monitoring and access control systems, ensuring that exposure periods and frequencies are not exceeded. These systems also keep information on hazardous materials such as material safety data sheets (MSDS) to ensure proper handling of materials and the correct treatment of exposure.

Safety management

A profile of every dangerous task indicates the type of safety equipment is required and individuals are competent to perform those tasks. It can also be used to determine which safety equipment can be issued to which individuals. One of the major objectives is the identification, recording and resolution of potential safety hazards.

Emergency and risk management

For major risk factors and emergencies proper emergency and risk management plans need to be developed and maintained. These can be developed using information from other modules such as simulations, MSDS hazards and environmental data.

Statutory systems

It is required to appoint proper individuals to take responsibility for certain equipment and areas and to ensure that statutory requirements are complied with. It includes tracking of legally required safety inspections and maintenance, equipment lock-out procedures, safety and emergency training and tests as well as personal protective equipment issue and status.

7.1.8 Display and alarm management

While operating a process, machine maintenance personnel and process engineers are able to monitor machine status and alarm displays via an easy-to-read 'dashboard' screens. The gages may display real-time values for user-selected parameters, including:

- Yield efficiency
- Machine speed (e.g. cycles per hour)
- Time to go %
- Production to go %
- Cycle efficiency
- Machine utilization
- Cycle time
- Large number of monitored process parameters.

Furthermore, the user may press a 'job-related' dial (e.g. yield efficiency) to instantly view the related job description and production report. Or perhaps pressing a 'parameter' dial (e.g. any of the monitored process parameters) to see the previous cycles display for that parameter. A text version of the machine status and alarm screen displays additional details about the process parameters that are being monitored. Values for the last machine cycle are color-coded (for example, 'green' if they are within limits, 'red' if they are above limits and white if they are below limits).

The following information is displayed:

- Job-related information including production and time to go
- *Signal:*   The process parameter that is being monitored
- *Current value:*   The current process parameter value throughout a cycle
- *Last value:*   The 'derived value' for the parameter
- *High limit:*   The user-specified engineering high limit
- *Low limit:*   The user-specified engineering low limit.

There are a variety of tools that make process monitoring simple, from process alarm displays, graphs for monitored signals, SPC and SQC, to the last values display. It allows the user to do the following:

- Automatically samples machine parameters for 'automatic SPC'
- Highlights 'RUNS' or 'TRENDS'
- View XBarR charts for SPC and SQC
- View histogram charts for SPC and SQC
- Automatically calculate control limits for processes
- Sample machine parameters for correlation analysis (linear regression analysis) with variable data
- Enter variable SQC data
- Enter attribute SQC data
- Print charts.

7.1.9   Key players of the integrated shop floor

Shop floor operators

Execution driven applications provide operators with the following advantages:

- Real-time dispatch lists of work orders in a priority sequence.
- Continuously adjusted job sequence information as actual manufacturing operations take place.
- Change in priority of job sequence.
- Information of next operation, immediately after the completion of earlier task.
- Detailed work order status including all necessary support requirements and its status with special instructions.
- To communicate actual work order status in real time with production control personnel and the scheduling system.

Production schedulers

Scheduling personnel are able to do the following:

- Change order priority, routings and schedules instantaneously due to changes in engineering, quantity, routings, rework and cancellations.
- Communicate in real time with the shop floor and all related department personnel.
- Download all work order and routings from MRP systems.
- Import all required service resources directly in the shop floor.

Supportive groups personnel

All support groups e.g. stockroom, tool room, maintenance, engineering, etc. are able to:

- Look for work order priority, location, and requirements on a real-time basis which increase productivity.
- All supportive functions obtain detailed plans of what and when to do a task on a minute-by-minute basis.
- Identify problem situations well in advance and respond quickly.
- Indicate in advance when they are unable to meet scheduled requirements, providing information that automatically revises the schedules.

Customer service and supervisors

- The execution system provides customer service, supervisory and other personnel data.
- The work order status for any functional area within the plant.
- Multiple views of production status information including historic, current and projected data.
- To ensure that new customer commitments can be met within the stipulated date.
- To make decisions about the deployment of alternative machines or work centers to clear the shop floor bottlenecks.
- To identify problem orders and material shortages.

Managers and manufacturing engineering

- The system provides detailed data on individual work centers and tools for identification of performance improvement areas.
- Complete backlog analysis and trend, input/output control and throughput calculations. Identifying bottlenecks.
- It also provides a detailed analysis of queue, set-up, run and problem times which in turn provides opportunities to find out the best option for cycle time reductions.

7.1.10   Benefits of execution management

The impact real-time manufacturing management systems can have on the company is quite dramatic (refer Figure 7.3). It can be both operational and financial.

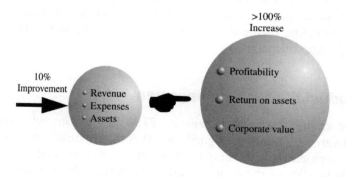

Figure 7.3
Results of real-time management

Operational

- The company is more responsive to its customers.
- Delivering high-quality products in the shortest possible time.
- Achieve true competitive advantage by understanding the process, core competencies and market, and using that information in real-time systems.
- Increased market share and improved profitability.

Financial

- Real-time management systems give manufacturers tight control over the basic processes of their businesses; hence improvements can be achieved quickly by cutting waste, reducing inventories, improving order fulfillment cycle. These translate into huge cost savings for the manufacturers.
- Reduction in working capital requirements, lower levels of debt and improved profitability.
- Improvement in revenue due to reduction in cost, more responsive to the market/customer, and sometimes for premium return on their products.

Effective manufacturing execution

Real-time manufacturing management systems can help companies:

- To re-engineer factory shop floor operations by increased visibility, control and coordination of multiple resources.
- To reduce the inventories and work-in process of each work centers.
- To avoid a backlog of customer demand.
- To deliver high-quality reasonably priced products at shortest possible time.
- Improved productivity and enhanced strategic advantage.

7.2   Product life cycle management

As global enterprises have become more dependent on geographically dispersed teams, and competitive pressures force cost cutting, regularly scheduled face-to-face meetings have become less frequent. The low-powered collaboration tools typically used today, such as e-mail and sharing of design files provide only a basic exchange of information. Without a broader context for the information, remote users are hindered from full understanding of the impact of the design. They are therefore limited in their ideas for design improvements.

7.2.1   Product life cycle management

Product life cycle management (PLM) is a comprehensive and integrated suite of software solutions designed to support optimal decision-making during product development. It is the only market solution that uses advanced planning and optimization tools to address the entire product life cycle – from concept development, to launch, to phase out. Product life cycle management maximizes the speed, productivity and the financial return of the product development process.

Product life cycle management is part of a total strategy to achieve effective and more profitable manufacturing. Manufacturing operations need PLM systems that manage product specifications and recipes, provide production history, create complete product genealogies and track total product quality.

Product life cycle management addresses effectively selected and managed product development investments and resources, in order to ensure rapid product development speed, high development throughput and high organization efficiency. Product platforms, parts, and suppliers for lowest total product and supply-chain cost are to be defined, based on optimal commonality and effective part/supplier selection.

Key development and supply activities are coordinated for optimal product launch-timing and investments in new/existing products. This enables the achievement of consistent high-product launch yields, with reduced product development lead-times. It provides enhanced visibility of resource bottlenecks and schedule impact whilst avoiding high part and supplier complexity with too little commonality, and budget over-runs.

The emergence of creative solutions that solve customer problems and reduce costs requires a rich exchange of information and ideas. Virtual collaboration and PLM technologies increase the variety and frequency of information exchanged throughout the design team to enable creative decision-making. Virtual collaboration technologies are a subset of PLM and include visualization and simulation tools as well as one-to-one scale graphics displays and conferencing. These technologies improve understanding of product designs by allowing teams to jointly interact with a visual model.

Comprising much functionality, including collaborative engineering, sourcing for direct materials, and project and program management, PLM is one of the great emerging territories in B2B technology. Born as an extension of product data management (PDM), the PLM space has huge potential because of the value it offers to manufacturers that master it.

7.2.2   Business benefits

AMR Research estimates that, fully realized, PLM stands to create as much as $106B in incremental operating margins annually for US manufacturers. Traditional PDM just is not enough in a competitive e-business environment. Manufacturers are required to deliver new products better, faster and cheaper. Both customers and suppliers need to be involved in a collaborative effort. Systems without the processes in place to coordinate product data for outsourcing design and manufacturing are destined to fail.

To be successful, systems require additional applications to manage the delivery of product data across the global supply chain. Product life cycle management expands PDM's scope beyond the traditional world of engineering to providing more product-related information to the extended enterprise.

The PLM solution process is unique because it enables companies to link projects, tasks, resources and project pay-offs across the enterprise. This complete, integrated perspective is essential to enable correct trade-offs between schedule/time, features/value and cost/resources.

Product goals

The PLM helps companies to realize numerous product goals:

- Reduction of product development costs
- Reduction of product development time

- Increase development throughput
- Reduction in number and percentage of low-value projects
- Reduction in product material and labor costs
- Reduction in capital equipment acquisition costs
- Closer collaboration with suppliers during product transition planning.

Additional benefits

These include:

- Increased revenues as a result of better product acceptance in the marketplace
- Decreased time to market (30–50% less lead-time)
- Fewer design flaws resulting in lower customer care costs and improved customer satisfaction. Improved product quality and reliability
- Improved value, permitting higher pricing to the customer
- Greater flexibility for mass customization and design-to-order products
- Reduced scrap (up to 25–60% less).

To enable these benefits, businesses need an infrastructure that can foster collaboration and deliver secure access to product data, anytime, anywhere.

The more people throughout the enterprise that have easy access to the product data, the more benefit there is to the business. A large part of the benefit comes from passive access to the data rather than specific work processes.

7.2.3   The PLM components

Product life cycle management applies Internet technologies to allow collaboration among the multiple entities which are involved in the various phases of a product life cycle such as the following:

- Concept
- Design
- Manufacture
- Product Launch
- Field Operation
- Phase out.

PLM allows different dispersed resources like design, sourcing, engineering, testing, manufacturing, sales and aftermarket service to operate from a common database to share information and communicate throughout the product life cycle and function, in effect, as a single organization with shared purpose.

PLM sub-applications

A PLM implementation will include all or some of the following sub-applications:

- Collaborative product design
- Product data management
- Direct materials sourcing
- Customer-needs management
- Product portfolio management.

The PLM technologies also include secure information portals that can aggregate data from a variety of sources to give business partners an integrated view of product data. By utilizing many modes of communication and making them available on a 24 × 7 basis, PLM provides a better exchange of information and greater context within which to understand the impact of the design. The result is higher product quality and more easily manufactured designs.

Other PLM components

These include:

- Portfolio management
- Component supplier management
- Computer-aided process engineering
- Customer relationship management.

7.3 Quality management

Quality control is an effective system for integrating the quality development, quality maintenance and quality improvement efforts of the various groups in an organization so as to enable production and service at the most economical levels thereby extending full customer satisfaction.

Quality control (QC) is based in the concept of variability, whether large or small, which is inevitable in any manufacturing process. Quality control is a systematic and scientific system involving the application of all known industrial and statistical techniques to monitor control the quality of the manufactured product.

The various problems of QC can be grouped into the following three classes:

Engineering The development of a product is basically engineering, including the development of quality evaluation through improved inspection procedures and the knowledge of the causes of defects and their rectification.

Statistical The concept of the behavior of a process, which has brought in the idea of 'prevention' and 'control', including the building of an information system to satisfy the concept of prevention and control, and improving on product quality requires statistical thinking.

Managerial The efficient use of the engineering and statistical concepts is a managerial process. The introduction of quality consciousness in the organization and the effective co-ordination of the quality function with two classes above are management functions.

7.3.1 Benefits

The benefits of QC can be listed as follows:

- Better product design and quality
- Reduction in scrap, rework and consumer complaints
- Efficient utilization of people, machines and materials resulting in higher productivity
- Elimination of bottlenecks in manufacturing process
- Creating quality awareness.

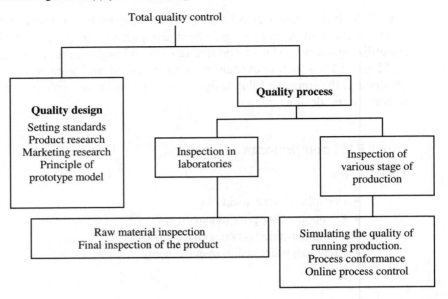

7.3.2   System concept

There are four systems for controlling the quality of a product:

1. Accept–reject
2. Forward control
3. Backward control
4. Process revision.

Accept–reject system

This system of controlling quality involves a quality assessor, who receives the output from process A and sorts out the items that do not fall within his interpretation of the quality specifications. The remainder then passes on to process B (refer Figure 7.4).

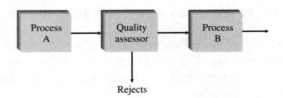

Figure 7.4
Accept–reject system

Controlling the quality by this system involves 100% inspection or may be achieved by sampling.

Forward control system

The forward control system is an effective system in situations where the process itself is liable to uncontrollable variations. Figure 7.5 shows the forward control system. The flow of the products and information must be synchronized with the movements of the product to which it applies. Process B is adjusted to rectify variations which have been introduced in process A. Food processing for example, is one industry in which forward control can be very effective.

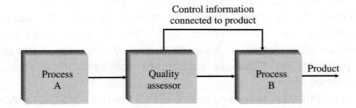

Figure 7.5
Forward control system

Backward control system

This is the most common type of control. In this system the assessment of quality of a particular product is communicated back to the person operating the process. On the basis of this information the person is able to decide upon the appropriate action, which will maintain or improve the quality of subsequent work. This system is shown in Figure 7.6.

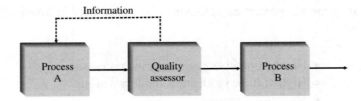

Figure 7.6
Backward control system

Process revision system

Production equipment or machinery and the people operating it may be unable to achieve the required level of quality in the product. If it is not possible to gain control of the quality either by forward or backward control, then the process may possibly be changed. This system is shown in Figure 7.7.

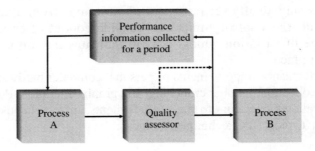

Figure 7.7
Process revision system

7.3.3   Statistical process and quality control (SPC and SQC)

The SQC is the application of statistical techniques to measure and improve the quality of processes. Statistical quality control includes SPC, diagnostic tools, sampling plans and other statistical techniques.

The SPC and SQC have an effective part in continuously improving a manufacturing process. When measurements are accurately collected and analyzed, improvements are identified and implemented, and controls established to ensure improvements are permanent; a process is well on its way to meeting quality requirements. The SPC is the application of statistical techniques for measuring and analyzing the variation in processes.

The statistical process control (SPC) (based on SQC) methodology is one of the most important analytical developments available to manufacturing. The SPC is used as a real-line tool providing close-up views of a process at a specific moment in time. The SQC provides support analysis and decision-making tools to help determine if a process is stable and predictable from shift to shift, day in and day out, and from supplier to supplier. Using these tools together, users can view the current and long-term picture about processing performance simultaneously.

The idea behind continuous improvement is to focus on designing, building and controlling a process that makes the product right the first time. The key to improving a process is removing as much variation as possible. Manufacturers applying SPC and SQC techniques rely on a variety of methods, charts, and graphs to measure, record and analyze processes to reduce variations.

In general, processes achieving the most benefit from SPC and SQC are products with:

- Highly repetitive manufacturing processes
- High-volume production and low margins
- Narrow tolerances.

The SQC methods are used to analyze recorded data and establish which variations are a natural part of the process, and which are unusual variations caused by external factors (such as variations in raw materials).

The SPC and SQC tools

Control charts are a fundamental tool of SPC and SQC and provide visual representations of how a process varies over time or from unit to unit. Control limits statistically separate natural variations from unusual variations. Points falling outside the control limits are considered out-of-control and indicate an unusual source of variation, unless a trend indicates that no unusual jump in variables have taken place.

Performance improvements of personal computer hardware and software permit real-time data collection, number crunching, and graph generation. Providing SPC information in real time allows operators to make adjustments, or to schedule maintenance on an as-needed basis. Typical SPC techniques includes:

- Control charts
- Check Sheets

- Pareto Charts
- Cause and Effect Diagrams
- Defect Concentration Diagrams
- Scatter diagram
- Histograms.

Once the process-monitoring tools have detected an out-of-control situation, the person responsible for the process makes a change to bring the process back into control.

7.4   Laboratory information management systems

Most laboratory information management systems (LIMS) are configurable systems that provide laboratories with various functions. These systems normally integrate to other systems such as MES and ERP systems, passing test and quality results to these systems.

The LIMS allow users to connect to the application in a variety of ways. It operates over phone lines, remote cellular phone connections, remote frequency LAN and WAN connections and the Internet. This technology provides features for remote connectivity, platform and operating system independence and simple single point LIMS application deployment.

The LIMS are database applications that are used to store and manage information associated with the laboratory, such as customers, sample matrix, tests, results, methods, parameters, employees, control limits, passwords, etc. Open database connectivity (ODBC) is a database standard that provides the ability to link the LIMS with different databases (such as accounting, MES, ERP, etc.). Laboratories are in the information business. Those that can deliver quality information to their clients ahead of their competitors will emerge as market leaders.

Before selecting a LIMS it is important to have a clear understanding of exactly what the laboratory's data management requirements are, in addition to the benefits that the laboratory can expect to gain from a LIMS and automation. Perhaps the most common reasons for acquiring a LIMS are to decrease turnaround time, enhance reporting and to improve overall data quality. Turnaround can be greatly enhanced with a rapid sample login process.

A successfully implemented LIMS will increase laboratory productivity, improve data accuracy and increase the laboratory's overall effectiveness. A LIMS can organize all the information that is pertinent to the laboratory and allows for rapid data retrieval and reporting. It also allows data to be accessible to others, promoting collaboration among different departments. In addition, many laboratories utilize either a LAN or WAN that allows users to share network printers and information.

Most LIMS are set up in a client/server configuration. In this configuration, the database tables reside on the server and the graphical user interface resides on the client machines. The advantage of this configuration is that data processing occurs on the server. As a productivity enhancement, the LIMS can provide the means to connect laboratory instrumentation, reading results from analytical quality tests directly into the LIMS. The figure below depicts the typical functions of a LIMS system.

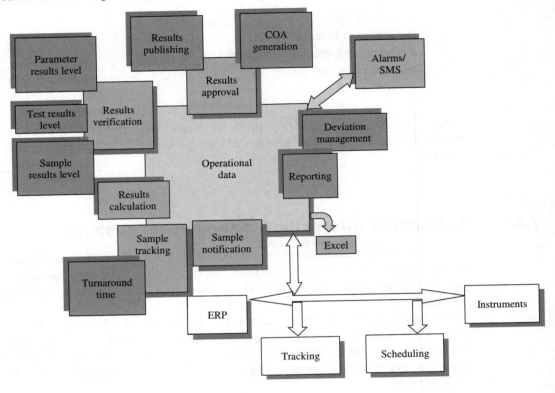

7.4.1   Sample login

Users can create and modify login information using single or multi-sample logins. A multi-sample login is a powerful means of logging in numerous samples in one simple data-entry sheet. It permits copy and paste functionality as well as fill-down by increment. Login can also be streamlined by using pre-assigned login groups that define existing sample types and assigned tests. Increased efficiency can be gained by employing bar coding in high throughput laboratories. After a sample is logged in, only a password-privileged person can modify sample information and tests, complete with audit trail.

7.4.2   Results entry

Results can be entered either manually or by transferring data from instrument files through specialized software interfaces. Samples can be selected for data entry by QA batch, pending analysis, sample ID, query, login or sample delivery group. Screens can be set up by the user without programing to display:

- All tests for a sample or groups of samples
- Only a single test for a group of samples
- Only a single sample
- Or a sample and group samples of tests both assigned and unassigned.

7.4.3   Security and privileges

It includes several security elements beyond the network and operating system security. User creation tools allow the LIMS administrator to identify users of the system and assign functional user privileges.

7.4.4   Calculations

The LIMS offers a variety of solutions to automate the calculations. Some systems use the Excel Calculator module to provide flexibility by sending results to an Excel spreadsheet, performing calculations or logic operations and returning the results back to LIMS. Convenient user interfaces provide a seamless link to Excel. Other systems have built-in calculator engines where calculation logic can be built during system configuration.

7.4.5   Validation

The LIMS provides complete validation procedures. A sample remains active until all tests are complete. Upon validation, the results are sent to a reporting queue. Complete validation keeps data secure.

A wide variety of validation procedures can be specified:

- Validation between parameter results – where more than one result of the same parameter are produced by instruments, and needs to be compared and validated
- Forced validation of all tests – where different tests are done for the same parameter (e.g. instrument and wet chemical) that needs to be compared and validated
- Validation of total sample – where the sample is broken or split in two or more samples that is tested in parallel, and results are compared and validated
- Validation of sample custody
- Validation through use of secure electronic signature – manually by an authorized person.

7.4.6   Exception reports

In order to take a pro-active role in assuring the quality of a product or service, LIMS allows the user to produce automatic exception reports for specification quality deviations. The reports may be set up to automatically e-mail the responsible party, speeding up the correction of problems and reducing production costs.

7.4.7   Certificate of analysis

The Certificate of analysis (COA) function allows user-defined specifications, ingredients, nomenclature and units for product quality management. 'As shipped' data is maintained that allows users to create reports and accurate SPC historical charts. Result evaluations can be made to find matches of all customer specifications. The COAs can be printed on request, as well as multiple COAs where required.

Where long/big campaigns are produced, samples can be taken on a frequent basis throughout the campaign and tested to provide backward control during the process. At the end of the campaign, the results are collated and averaged to calculate a total campaign quality from the combined results for COA purposes.

7.4.8   Trend charting

Trend charting is an invaluable tool for graphically reporting LIMS information. Single and multiple charts may be displayed simultaneously. Simple trends can be displayed, mapping parameter or test results per sample, or alternatively X:Y parameter comparison trend charts can be displayed, permitting cause and effect troubleshooting. Laboratory information management systems can also allow users to create SQC charts and calculations

7.4.9 Sample process tracking and scheduler

The process scheduler provides a comprehensive view of samples in the system. It is ideal for production laboratories with multiple samples, where it acts as a communicator for sample requests/notifications from people outside the lab. The scheduler saves time by offering sample-oriented workflow and barcode entry of login information. Instead of choosing a task and then finding the samples associated with it, users may choose any sample and view where it is in the process.

It can also be used in high-volume laboratories with large routine sample schedules, where the scheduler provides a graphical schedule for each month, day and hour to schedule samples and tests for login. It supports periodic sampling schedules and auto-login of samples.

7.4.10 Instrument maintenance and calibration

Most laboratories require records of instrument maintenance and calibration. In addition to record keeping, the system tracks the instrument test qualifications, and controls which tests can be assigned to currently qualified instruments. It also ensures that instruments are calibrated regularly, and that results are within calibration limits before tests are performed. The calibration results can also be used to determine instrument drift and identify when maintenance is required.

7.4.11 Personnel training

All personnel can be tracked for education, training schools or certification. Each person may be tracked by LIMS test code certifications, date and time and when re-certification is required. Tests may not be assigned to or performed by non-certified personnel.

7.4.12 Hand-held data entry devices

Hand-held devices used in other high throughput industries such as package delivery services or meter readers add a degree of automation to laboratories. This can include hand-held barcode readers to palm top computers with GPS sample point validation.

The interfacing of lab instruments increase efficiency and reduce manual entry errors. Studies show that transcription error rates of 3–5% are common in non-automated laboratories. The interfaces provide fast results transfer from instruments, virtually eliminating errors. Figure 7.8 displays sample flow through a typical laboratory.

7.4.13 Benefits of LIMS

Productivity

- *Increase throughput:* Increase the quantity of samples analyzed and tracked, without hiring additional lab personnel.
- *Track productivity:* Generate productivity reports which allow lab managers to track turn-around times, number of samples or analyzes completed per unit of time, expired samples and exceptions.
- *Schedule routine tasks:* Increase efficiency by using LIMS' scheduling features to control routine sampling, analysis, instrument calibration and maintenance tasks to adjust for peak sample volume.

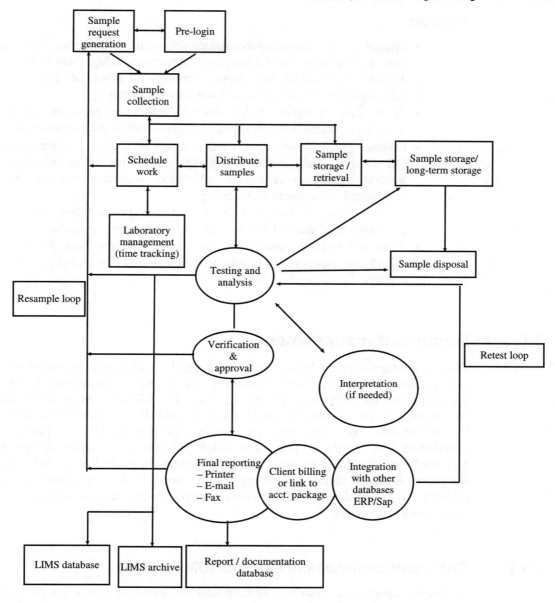

Figure 7.8
Schematic representation of a typical LIMS workflow

- *Spend time effectively:* Eliminate many tedious paper-handling tasks, so technicians may spend more productive time at the bench.
- *Document lab output:* Realize increased levels of documentation without an increased level of effort.
- *Provide access to data by outside personnel:* Establish read-only access for personnel outside the lab. This feature allows authorized personnel to directly pursue their own questions dynamically, effectively eliminating non-productive data search, retrieval and copying time.
- *Interact directly with LIMS system developer:* Ensure prompt and accurate responses to technical-support questions by interacting directly with the system developer.

Accuracy

- *Reduce errors:*   Significantly reduce data transcription error, since manual data entry is limited to specific sample characteristics at login and to those results not available from direct data import. Virtually all other data entry is done using selections from predefined pull-down lists and menus.
- *Verify data at input:*   LIMS enforces validation protocols on many fields, allowing technicians to catch data input errors before they are entered.
- *Enhance quality control:*   Document data's accuracy and precision by using the QC facility. Lab managers set QC limits for both individual analytical methodologies and instrument calibration. Associated control charts can be maintained and updated as necessary.
- *Avoid missed deadlines:*   Use the automated scheduling prompts to notify personnel in advance of sampling events and sample expiration.
- *Avoid the embarrassment of false results:*   Screen data thoroughly through val-idation protocols to eliminate illogical data conditions prior to generating reports.

7.5   Document management

Documents represent the knowledge of industrial corporations and are much more than just formatted data. Largely, document processing is paper-based. One obvious drawback to this is the long processing time. In addition, certain information that is embedded in paper documents can be partly or entirely lost after their active processing has terminated. The documents are usually stored in insufficient paper archives or transferred only partly to persistent online databases. In order to operate a plant efficiently, one of the most important requirements is control of the information about the plant. A clearly defined set of procedures needs to be implemented to ensure that information are readily available, accurate, easy to locate in an emergency and relevant to the work performed.

7.5.1   Document management systems (DMS/EDMS)

Engineers, operations personnel, or maintenance personnel may often find it difficult to locate the right information at the right time. In addition, current and accurate documents, which reflect the current plant configuration may not be made readily available.

The systematic management of drawings and documents can have a dramatic effect on project and operations efficiency at the plant. Failure to properly manage these critical documents can result in delays in projects and problems with consistent plant operation. Many companies are employing electronic document management systems (EDMS) to organize this critical set of information.

An EDMS ensures that the documents are accurate and up to date whether it is:

- An update to a drawing to correct an error or record a change to the 'as built' plant.
- An update to an operating procedure or maintenance instruction, reflecting a change in plant configuration or vendor specification.
- Changes affecting process chemicals, technology, equipment or facilities.

The management of change process is intended to ensure that changes are:

- Properly reviewed by personnel who are cognizant of the issues, to ensure additional risks are not introduced.
- Incorporated into relevant documents such as piping and instrument diagrams, operational and training manuals.
- Communicated to affected employees via coordinated training programs prior to implementation of the change.

An EDMS is a computer-based system that controls:

- The creation/capture and storage of documents
- Distribution and user-access of those documents
- The process for document updating.

The EDMS also includes control over document checkout, check-in and revisions. A typical system contains a document repository for the physical storage of the documents on where to find hard copies of documents not stored electronically. The system also operates on a database that contains relevant information about the documentation being managed. This data is important for finding required documents and for tracking the relationships between documents.

The system also contains the documentation's evolution history, which includes information about each document and its status:

- Who created the document?
- What is the current revision level?
- Is the document currently checked out for modification?
- Which subsystems and assemblies are affected by the document?
- Who must approve any changes to this document?

The history of all changes to the document is available, including who changed the document at which time and for what reason. Also noted are the people who reviewed and approved each change. Users connected on a network can retrieve documents and perform markups and check documents into the system.

7.5.2   The DMS and workflow solutions

A DMS manages not only document types, but also aids in understanding the inter-relationships between documents, and contains the complete procedure for which the documents are required. The DMS would determine for example, if someone using the maintenance procedure module should need to access the latest vendor manuals, related material safety data sheets and other such documents for the entire procedure.

Documents needed to operate a process operation can be very diverse, and these typically include:

- Plot plans facility drawings
- Process flow diagrams
- Standard operating procedures
- Engineering drawings
- Electrical diagrams
- Training records
- Correspondence
- Purchase orders

- Incident reports
- Emergency action plans
- Specifications
- Maintenance history files
- Work orders
- ISO procedures.

To be effective, a DMS should not only manage all of the document types, but it should also make it easy for a user to understand the inter-relationship between the individual documents.

7.5.3   Enterprise-wide document management solutions

A DMS software provides the most economical way to access and share information. Document imaging alone cannot adequately solve the total document management problem. An integration strategy is needed for managing images and drawing from various packages and technical documents in a single, logical DMS. The DMS software encompasses the following:

Documents capture

The source drawing or document is converted into a computer-compatible format. Even if a computer first generated the document, it is made compatible with all other manually generated and computer-generated documents in the system.

Document storage

All documents are stored in the system in such a way as to allow access by a common user interface without regard to original medium, format, size or original location.

Document version update

Automatic control is maintained over which functional departments and specific individuals have authority to revise the documents. In addition, a workflow mechanism can be put in place that allows specified personnel to review, comment and suggest revisions before they are approved and released for general use.

Document retrieval

Document access is limited to those with a need or authority to view, make hard copies, retrieve and distribute documents. There are multiple levels of rights like view, modify, print, etc.

Document distribution

Only those with the proper authority are allowed to retrieve documents in any desired output medium, including paper hard copy, exportable electronic file formats, computer monitor display, color transparency, etc. Document retrieval is instant, giving the user optimum value.

7.5.4   Benefits of DMS software

- On-time project completion.
- Plant availability, which is tied to efficient, scheduled and unscheduled maintenance practices, as well as minimization of time to perform plant reconfigurations.

- Plant productions target is directly tied to the plant availability.
- Compliance with critical quality credentials like ISO 9000.
- Safe operations in compliance with regulatory mandates.
- Easy to manage both papers based and electronic documents using one common computer desktop.
- DMS software allows opening over 200 file formats without need of the application software.
- Readily available paperless office.
- DMS software provides the platform for instant information access.
- It presents a complete audit trail for complete tracing and accountability.
- Faster new product development and cost reduction (e.g. eliminating the cost of storing paper or microfilm archives, decreased cost of online distribution vs paper copies, blueprints, etc.).
- Better customer satisfaction due to faster resolution of complaints.
- Reduction of plant downtime; when necessary engineering documents are readily available online for remedial maintenance or an emergency response.
- Decreased turn-around times for interdepartmental comments on an engineering document.
- Improved document security as imaged documents are never taken out of storage, as paper documents would be.

8

Production capability management

Learning objectives

- To understand organizational areas that influence production capability.
- To become familiar with capability functions such as equipment, materials and labor management.
- To understand how the different capability management modules can benefit the organization.

8.1 Introduction

Various capabilities exist in an organization. The 5 Ms (materials, manpower, machines, money and market) must be deployed in the most effective and efficient manner to ensure maximum return on investment. A lack in one or more of the critical capabilities in a manufacturing organization can have disastrous implications for any company.

The role of labor is continuously growing, especially in the context of union and legal activities. Total cost of employment represents a very significant portion of the total cost of any company, as it includes employee time, health, safety and statutory requirements. For effective control of labor cost, all these areas need to be managed more effectively and opportunities for improvement be identified and implemented.

Today, many companies view maintenance as the last controllable function in which they have an opportunity to reduce costs. However, arbitrarily reducing the maintenance budget can lead to lower levels of operating capacity and reliability. The goal of maintenance management though remains the same as always – optimize capital investments by reducing the costs associated with owning and operating assets.

To maintain a competitive edge, however, more than cost cutting alone is needed. Complex equipment and facilities must be maintained by employing asset management strategies such as total productive maintenance (TPM) and reliability-centered maintenance (RCM). In addition, a company is also required to comply with mounting regulations, environmental laws and international standards, while at the same time coping with constant change.

In the age of cut-throat competition and multiple products, no industry can maintain a favorable position if the cost of production is not maintained at a competitive level. The present concept of material storage has increased in its stature because of changing tastes, technical complications, attitude of buyers and increasing specialization in industries. Proper materials storage and management can contribute greatly to the bottom-line of a company by ensuring minimum stock-holding and maximum asset turns.

8.2   Labor management

Various solutions exist to assist with the management of labor issues, such as time and attendance, occupational environment, health and safety and accident management systems.

8.2.1   Time and attendance systems

As companies cut costs and seek ways of positively impacting on the bottom line, time and attendance technology can provide immediate and tangible benefits.

Companies that have installed time and attendance systems claim an immediate 1–2% increase in attendance levels, presenting the business with serious cost savings and efficiency.

Originally created as punch-to-pay cheque systems, Time and attendance systems (T&A) and workforce tracking packages have evolved to handle hourly labor, budgeting, scheduling, human resources and forecasting functionality. As products evolve, however, organizations are beginning to use them in new ways:

- Employees use T&A to record time and labor.
- Supervisors use T&A as a 'real time' tool to monitor attendance, labor and material movement.
- Finance uses T&A to keep control of labor and product costs.
- Payroll uses the solution to keep track of hours, pay rates and to account for overtime and other pay rate changes.

Time and Attendance systems can provide an immediate and long-term return on investment and the benefits can be seen across an organization. The T&A systems are normally flexible enough to ensure that the solutions can be configured to suit most business environments. The T&A systems are designed to reduce the administrative burden from software-only solutions (where employees clock in/out using PCs) or hardware-based solutions (with a choice of magnetic swipe, barcode and proximity readers). The benefits of using working time monitors are many, namely:

8.2.2   Benefits of T&A

The most advanced time, attendance and labor tracking software delivers measurable benefits across all industries, including: professional organizations, FMCG companies, office environments, service industries, health sector, manufacturing, retail, hospitality, construction, education, government and any organizations operating sales and marketing teams.

Considerable benefits can be found in the areas of:

- Time and attendance
- Real-time management information
- Labor distribution
- Scheduling
- Leave management
- Activity monitoring and costing
- Employee self-service
- HR and payroll integration.

Time and money savings

These systems automatically calculate gross pay and overtime, and allow employees to focus on the important aspects of their business. The T&A systems also highlight problems as they occur, and data is fed through in real-time.

Job costing information

The T&A systems ensure that labor hours are captured and associated to a specific job or activity. The hours in turn can be used to calculate a specific cost, enabling activity-based or job-costing.

Eliminate errors

With manual time-keeping systems, the opportunity to introduce inaccurate data is enormous. Users can easily mistype information or simply not look up correct codes or not have the timesheet add up properly or simply enter work as 'miscellaneous'. Errors increase work downstream in the time-keeping process and create work and heartache for both the payroll staff and the employees.

Increased speed

With data entered directly into an automated system, turning data into useful reports or transactional files for moving into other organizational systems becomes much faster and easier than before. One of the biggest attractions of automating the time-keeping environment for managers is access to *actual labor* data in a more timely fashion.

Reduce workload

With an automated system, controls could conceivably be created to trap typographical errors and other inaccuracies in the timesheet while it is being completed. Also, there is no need to re-enter manual time sheets into a computerized system. This ensure that all billing is done accurately and that hours are not lost

In any manual time-keeping system, it is difficult if not impossible to establish checks and balances to ensure that billing information is done completely and accurately. An automated system can provide reports that compare expected billings with actual billings and identify any unbilled hours for scrutiny by the billing department.

Provide the *actual* element of variance reporting

Project-oriented systems often provide some type of variance report. This kind of report shows the actual progress to-date against the original plan. Automating the time-keeping environment provides the project system with the progressed element of the budget vs actual report.

Department time savings

To manually add up time cards takes roughly 6 minutes per card. The T&A software can easily reduce that to just a few seconds. One to three percent of the total payroll imprecision is caused by human error. Long lunches, breaks, tardiness and early departures represent on average a time loss of 4 h and 5 minutes per employee each week.

8.2.3   Management of environment, health and safety

Companies around the world are carefully considering not only products and processes, but also the growing number of regulations focusing on occupational health, safety and environmental issues.

EH&S Solutions manage the entire range of environment, health and safety (EH&S) processes across the company, throughout the product life cycle and are efficiently integrated. It helps the user to track vital information, optimize EH&S processes and comply with regulations accurately and efficiently which in turn control costs, enhance company image, and ensure the safety of employees, customers and the community.

EH&S Solutions provide the tools and capabilities required to manage today's sophisticated EH&S processes, in all areas of a company and all relevant areas of environmental impact. Not only the effective storage and handling of EH&S data, but also the tight integration into existing ERP-systems empower the user and improve usability of information for decision-making purposes.

EH&S functions

EH&S Solutions help companies to:

- Store and manage all EH&S-related data in an easy and flexible way.
- Manage hazardous substance data for regulatory compliant documentation and reporting.
- Produce and maintain a variety of documents, such as material safety data sheets, transport emergency cards, standard operating procedures and labels.
- Keep track of dangerous goods, create accurate dangerous-goods documentation integrated into the supply chain.
- Manage the disposal of hazardous and non-hazardous waste efficiently and comply with all relevant national and international rules and regulations.
- Enhance the safety of employees by identifying and minimizing risks to employees health.
- Track employee healthcare treatments and examinations, and monitor the results and diagnoses.
- Report all environmental-related data in a flexible way to comply with regulatory requirements.
- Track, store and report all environmental data via logger, monitoring and reporting.
- Manage and host all solutions in appropriate technical environment (IT-Services).
- Integrate these solutions on demand directly in ERP-system, for full leverage through integration.

8.2.4 Dangerous goods management

Customers, suppliers and carriers are involved in the dangerous goods transportation process. Checks as a result of legal requirements, depending on the means and routes of transport are automatically performed at any time in the supply chain. Dangerous-goods-related data is automatically made available and printed out with the delivery note. Road and inland waterway transportation cards may be printed and handed to the carrier or dangerous goods information can be exchanged electronically with the carrier in advance. The key benefit is the tight collaboration and the savings in time and money due to faster and legally reliable procedures.

8.2.5 Accident management

Potential accident identification and elimination requires interaction between industrial hygiene and safety (IHS) and internal occupational health (First-aider) professionals, asset life cycle management (ALM), human resources and the authorities (insurance company or workers compensation association). For maximum effectiveness, the recording of legally required data, the triggering of safety measures and the creating of accident reports need to be integrated in support of a companies' accident prevention program.

When an accident occurs in an enterprise, any person injured are generally given first-aid, and the legally required data about injuries can be entered in the injury/illness log. The data relevant for processing the accident is transferred to the incident/accident log and this information is made available directly to the industrial hygienist for further accident processing.

An industrial hygienist researches the events in connection with the accident on the basis of the data from the injury/illness log. An accident is often triggered by a technical fault, and as a result of the accident, damage is often incurred to technical protective equipment. The industrial hygienist is required to ensure that this damage is repaired. To do this, the industrial hygienist can create a plant maintenance notification that is automatically sent to plant maintenance. The industrial hygienist can check the status of the progress of the notification at any time. When the plant maintenance work has been completed, the industrial hygienist receives a message of completion.

The industrial hygienist must send an accident report containing the information gathered about the accident to the supervisory authorities or accident insurance company within a certain time frame.

The time management department records the period of absence of the person injured in the accident accordingly, so that it can be used for human resources management processes such as settlement. This data (only the absence times caused by the accident) is available to the industrial hygienist to complete the time data in the incident/accident log. Once all the data for the accident report has been gathered and saved in the system, the final report can be sent to the supervisory authorities or accident insurance company electronically. Internal accident reports are also sent to HS&E committees and accident investigation teams to identify corrective and preventive actions.

8.2.6   Occupational health management

The interaction between the occupational physician in the internal health department, the IHS professional and the HR department process are described below. Within the scope of risk assessment, measurements are made of hazardous agents in work areas within the enterprise. These hazardous agent measurements are often repeated at regular intervals. If the measurements show a significant change with respect to the existing measurements, the exposure rating for the work areas affected is changed accordingly.

Human resources often move employees to new positions as a result of modified shift planning. These positions often involve new tasks with related triggers for health surveillance protocols. To keep the assignment of persons and health surveillance protocols in the enterprise up-to-date at all times, industrial hygienists generate a report at particular intervals to automatically generate a proposal list. This report takes all changes into account that result from the change to a risk rating or the transfer of employees. The industrial hygienist checks the content and then opens medical services for the employees. On the basis of these medical services, the physician's assistant makes appointments at the health center.

Once an appointment has been arranged, the invitation letter and any questionnaires linked with the health surveillance protocols can be sent directly to the person to be examined. In the medical service, the physician can immediately enter any necessary information such as diagnoses, restrictions, medical test results and medical measures at the time of the examination. This information is used to create medical certificates that can be issued to the employee examined or the HR department.

8.2.7 Health, safety and labor/management committees

Health, safety and labor/management committees recognize the importance of harmonious relationships achieved through joint problem solving and ensuring a safe healthy work environment. The purpose of the committee is to promote mutual understanding and goodwill between management and employees. The committee discusses ways and means of improving working methods, safety, operating efficiency and eliminating waste in labor and materials. The employer ensures that health and safety representatives receive the prescribed training in health and safety.

HS&E committees

The committee will:

- Apply the relevant health and safety legislation and regulations when making decisions or recommendations.
- Establish subcommittees, as it deems necessary and will set their terms of reference.
- Ensure proper training of committee members.
- Take minutes, distribute copies to committee members and post on relevant bulletin boards.

HS&E representatives

The powers and duties of a health and safety representative are several. The health and safety representative will:

- Consider and expeditiously dispose of health and safety complaints.
- Ensure that adequate records of work accidents, health hazards and the disposition of health and safety complaints are kept, and regularly monitor the data.
- Meet with the employer as necessary to address health and safety issues.
- If there is no policy committee, participate in the development, implementation and monitoring of programs to prevent hazards in the work place, which also provide for the education of employees in health and safety.
- Participate in all inquiries, investigations, studies, and inspections pertaining to the health and safety of employees.
- Cooperate with health and safety officers.
- Participate in the planning of the implementation of changes that may affect occupational health and safety, including work processes and procedures.
- Inspect each month all or part of the work place, so that every part of the work place is inspected at least once each year.
- Participate in the development of health and safety policies and programs.
- Assist the employer in investigating and assessing the exposure of employees to hazardous substances.
- Participate in the implementation and monitoring of a program for the provision of personal protective equipment, clothing, devices or materials and, where there is no policy committee, participate in the development of the program.

8.3 Equipment management

Machines and equipment wear out all the time. The greater the number of parts in a machine, the bigger the possibility of wear, tear and hence breakdowns. Breakdowns are very costly, not only to repair. Depending on industry, maintaining plant, fleet facility and equipment can be up to 25–50% of total operating cost. The costs that occur due to breakdown are:

- Downtime of machines resulting into loss of production and sales
- Idle time of direct and indirect manpower
- Increase in the scrap and rejections
- Dissatisfaction of the consumer due to delays in delivering on a commitment
- Actual cost of the repairs.

A typical split of direct cost is depicted in the following graph.

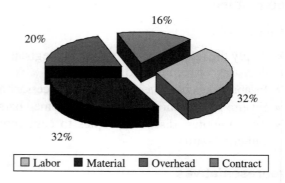

In general, breakdown work is two to five times the cost of planned work. Unplanned corrective work is approximately one-and-a-half times more costly than planned/scheduled work. Emergency work is three times more costly than planned/PM work. These differences in costs are due to various problems related to unplanned activities.

If the reliability of the system is improved, the frequency of breakdown time can reduce and hence the costs too will come down. The total downtime of a machine is the summation of downtime multiplied by its frequency. To increase the reliability of the system, either the size of the repair facilities (services) has to be increased (to reduce the machine downtime in order to attend it quickly as and when breakdown occurs) or the system of preventive maintenance needs to be improved and followed.

The typical split between planned (preventive) and unplanned (breakdown) maintenance is depicted in the following figure.

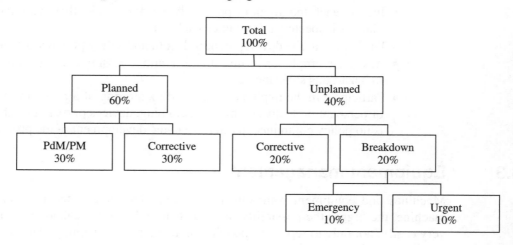

In world-class companies, the ratio is as depicted in the following figure:

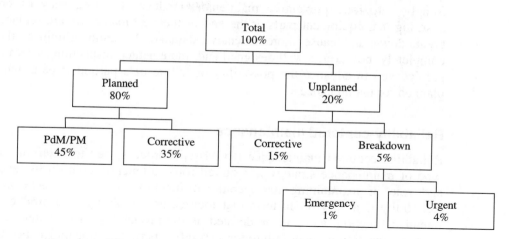

8.3.1   Maintenance concepts

Various maintenance concepts are used to manage the maintenance function, and maintenance professionals have developed various calculations to assist in the prediction and management of equipment breakdowns.

Frequency distribution

Breakdown time distributions indicate the distribution of maintenance-free performance of equipment. Depending upon the nature of the equipment, the frequency distribution curves take different shapes (refer Figure 8.1).

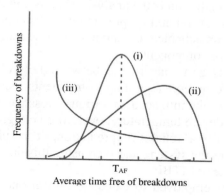

Figure 8.1
Frequency distribution of time free of breakdowns

A simple machine breaks down nearly at regular intervals of time with least variation in time distribution. This type of distribution resembles normal distribution in a sense, such that a major portion of the breakdowns happen near/on an average value of downtime (i).

In a complex machine, each part of the machine will have its own frequency distribution of breakdown time. Therefore even after the repair of one part, the machine can break down at any time due to the wear out of any other part of the machine. The variability in the time of breakdown is much more when compared to the earlier case. Such breakdowns will follow skewed distribution and for the same average breakdown (ii), wider variability of breakdowns is noticed.

Preventive maintenance

Simply put, good preventive maintenance reduces the frequency of breakdowns by ensuring that equipment parts at the end of their estimated life are replaced before they break down and cause more extensive damage. It cannot eliminate the breakdowns completely; however, a well designed and implemented inspection system can indicate or predict well in advance the possibility of failure of a particular part, and hence can be planned for maintenance.

Reliability-centered maintenance

Reliability-centered maintenance (RCM) is a process used to determine why and what type of maintenance strategy is required for equipment, based on its level of criticality and role in maintaining its operational function at the best achievable reliability. Reliability is a very broad term that focuses on the ability of a product to perform its intended function, and can be defined as the probability that an item will perform its intended function without failure for a specified period of time under specific conditions.

Performing a reliability analysis on a product or system can actually include a number of different analyses to determine how reliable the product or system is. Once these reliability and maintainability (RAM) analysis have been made, it is possible to anticipate the effects of design changes and corrections in order to improve reliability. The different RAM analyses are all related, and examine the reliability of the product or system from different perspectives, in order to determine possible problems and assist in analysing corrections and improvements.

Reliability prediction

A reliability prediction is the analysis of parts and components to predict the rate at which an item fails. A reliability prediction is usually based on an established model for electronic and mechanical components. These models provide procedures for calculating failure rates for components.

Failure rates and 'mean time between failure' ratings (MTBFs), are calculated by gathering information regarding components using predefined standardized equations. The equations take into account various stress parameters of the components, and include data such as device temperature, operating voltage, rated voltage and power stress ratios. In order to do a reliability prediction, it requires gathering information about the components in the system, and then using this data in mathematical equations to compute the failure rate or MTBF.

The models employed to calculate failure rate do not contain listings of failure rate values for specific devices, but include equations to calculate the failure rates for any conceivable device. The complexity and required parameters of these equations vary depending on device type.

Failure rate

Failure rate can be defined as the anticipated number of times that an item fails in a specified period of time. It is a calculated value that provides a measure of reliability for a product. This value is normally expressed as failures per million hours, but can also be expressed as a FIT (failures in time) rate or failures per billion hours. For example, if a component has a failure rate of two failures per million hours, then it is anticipated that the component fails two times in a million-hour time period.

A calculated failure rate is generally based on an established reliability prediction model (for instance, MIL-HDBK-217 or Telcordia). Calculations are based on component data such as temperature, environment and stress. Calculated failure rates for assemblies are a sum of the failure rates for components within the assembly.

A component manufacturer may sometimes provide a specified failure rate usually based on field or laboratory test data. Similarly, a manufacturer can also provide a specified failure rate for an assembly.

Failure rate is also the inverse of the mean time between failures (MTBF) value for constant failure rate systems. For example, if a component has an MTBF value of 500,000 h, and the failure rate is desired in failures per million hours, the failure rate would be:

$$\text{Failure Rate} = (1\ 000\ 000\ \text{h}) / (500\ 000\ \text{h}) = 2 \text{ failures per million hours}$$

For an existing product MTBF can be found by studying field failure data, but for a new product or if significant changes are made to the design, it may be required to estimate MTBF before any field data is available. In some cases, failure rates for previous products can be used if changes to a design are unlikely to affect reliability. In that case reliability prediction technique is required to estimate reliability.

Maintainability analysis

Maintainability analysis provides calculated information regarding various aspects of maintenance. The goal of performing maintainability analysis is to determine the amount of time required to perform repairs and maintenance tasks. In other words, if a system does fail, how long will it take to fix it? Before maintainability analysis, the components/parts in a car engine were positioned in such a manner that some of the parts requiring more frequent maintenance such as the oil and air filters were located beneath the main body of the engine, effectively requiring the engine to be pulled when changing the filters.

A common maintainability calculation is mean time to repair (MTTR). Mean time to repair (MTTR) is the most common measure of maintainability. It is the average time required to perform corrective maintenance on all of the removable items in a product or system. This kind of maintainability prediction analyzes how long repairs and maintenance tasks will take in the event of a system failure.

Mean time to repair also factors in other reliability and maintainability predictions and analyses. It can be used in reliability prediction to calculate the availability of a product or system. Availability is the probability that an item is in an operable state at any time, and is based on a combination of MTBF and MTTR.

Failure mode effect analyses

An FMEA is a bottoms-up method of analysing a system design or manufacturing process in order to evaluate the potential for failures. It involves identifying all potential failure modes, determining the end effect of each potential failure mode, and determining the criticality of these failure effects. The FMEAs are sometimes referred to as FMECAs (failure mode, effects and criticality analyses). They are based on standards (both military and commercial) in the reliability engineering industry. These analyses can take many forms, but they are all used to study a particular system and determine how that system can be modified to improve overall reliability and avoid failures. Once this is completed, an FMEA or FMECA can be used to determine the most critical failure modes and then determine how these critical failures can be minimized or eliminated.

To begin an FMEA, the lowest levels of the system are outlined. This can be the individual components (piece part FMEA) or the lowest level assemblies in the system (functional FMEA). For each lowest level, a list of potential failure modes is generated. Effects of each potential failure mode are then determined. For example, one component in a television may be a capacitor which has two potential failure modes i.e. capacitor may fail 'open' or it may fail 'shorted'.

If the capacitor fails in an open state, the effect might be that the screen appears with wavy lines and if the capacitor fails shorted, the effect might be that the screen goes completely blank. If the capacitor fails shorted and the screen goes blank, that failure mode could be considered more severe or critical than if the capacitor fails open and wavy lines appear. In this case, it requires finding ways to prevent these failures from happening or lessen their criticality.

An FMEA can use failure rate calculations, which were performed during the reliability prediction portion of an analysis to determine probability of occurrence. In an FMEA, failure rate is used to compute mode criticality, or the probability that a particular failure mode is actually going to occur.

Life cycle cost analysis

Manufacturers may take reliability data and combine it with other cost information to illustrate the cost-effectiveness of their products. Life cycle cost (LCC) analysis can prove that although the initial cost of a product might be higher, the overall lifetime cost is lower than that of a competitor as the product requires fewer repairs or less maintenance.

Life cycle cost analysis is a method of analysing the cost of a system or a product over its entire lifespan. The LCC enables you to define the elements included in the lifespan of a system or product, and assign equations representing the calculation of the cost of that particular element.

The objective of performing LCC analysis should be to choose the most cost-effective approach for using available resources over the entire lifespan of the product or system. The LCC provides a systematic process for evaluating and quantifying the cost impacts of alternate courses of action. It can be used to support trade-off analysis between several product design configurations, or as a measure of sensitivity of a specific product design to changes in selected performance parameters (such as reliability, maintainability and testability). Product quality can affect the distribution of costs between up-front program/manufacturing costs and field operation, and repair costs. Higher-quality products can minimize associated scrap and rework costs.

The LCC analysis of a typical system could include such costs as system planning and concept design, preliminary system design cost, design and development costs, product costs, maintenance costs and disposal costs. This type of analysis often uses values calculated from other reliability analyses like failure rate, cost of spares, repair times and component costs.

A company could use LCC analysis to determine, for instance, warranty costs. This type of analysis could be based on anticipated failures, repair times and costs of repairs. Many companies are finding that LCC analysis is a valuable tool during the design phase of a project in order to determine the most cost-effective solutions before substantial costs are incurred.

8.3.2 Maintenance systems

The following conditions indicate that a computerized maintenance management system (CMMS) can be of assistance.

- Planned maintenance work is less than 90% of the total maintenance work load.
- Craft productivity is less than 80% of capacity.

- Craft overtime is more than 10%.
- Finished product quality is consistently less than 95%.
- Equipment availability is less than 95%.
- Maintenance inventory cost, including holding cost, is more than 30% of the annual maintenance budget.

Most (CMMS) provide at least the functionality as depicted in the following figure.

Planning and scheduling

Maintenance of planning and scheduling should dramatically improve the productivity of maintenance. This is where the most gains in execution can potentially be made and acted upon. In some larger organizations planning and scheduling are split, allowing more adequate resources for each role. The role of the planner needs to cover the full range of the work-order system, from input into coding, prioritization and a degree of autonomy in execution. As such, these roles need to be staffed by skilled and versatile people.

The difference between planning and scheduling needs to be clear within each company. These are differing areas worthy of differing measurement and improvement initiatives.

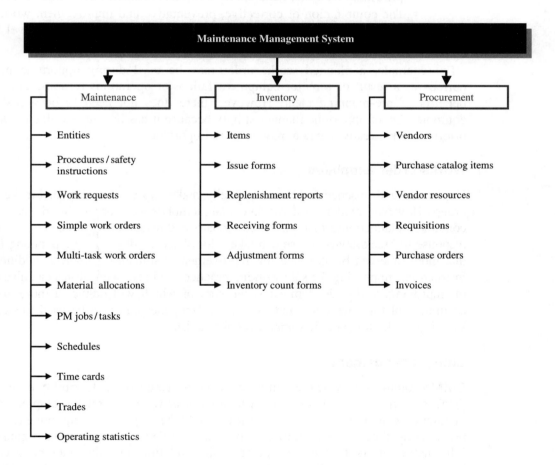

Planning

Planning can occur at any stage during the life of a work-order. An electronic indicator in the work-order systems needs to be able to identify the work-order by the status of planning. In this manner work orders requiring parts, procedures, documents, skills or equipment can easily be focused upon. A work-order can only be considered 'planned' until all of these items have been considered. In addition, exception reporting needs to highlight:

- Number of resources
- Number of quotation estimates
- Incorrect coding.

Scheduling

Scheduling is the function of coordinating all of the logistical issues surrounding the execution phase of the work. This can also uncover some areas of planning deficiency that need to be captured. Scheduling is best performed in a capacity-scheduling manner. The following needs to be scheduled:

- Overhead labor hours such as safety and toolbox meetings, break times and training times are gathered, along with holidays and scheduled work-orders (Permitting future analysis of these items).
- Addition of corrective and approved improvement actions as dictated by the prioritization system and operations plan (Planned work-orders only).
- The combination of corrective, preventative and improvement work constitute the most effective use of labor and resources, and a workable level is 70–80% in the initial stages.

For example, a planned works order may be used during opportune maintenance periods as a result of a major equipment failure or operations reasons. This makes the benefits of pre-planning clear. However, there does not need to be a rush to repair equipment in an opportune manner simply because it has become available. Other higher priority work already planned may be more important.

Works-order templates

Unplanned and unscheduled work generally makes up the majority of breakdown work-orders. However, modern systems do contain template work orders. Works order templates contain all planned information, including parts and resources requirements, ensuring faster response to breakdowns. These can take a lot of the work out of the planning function so that more time can be allocated for improvement. Estimation variances, additional tips or instructions, improving the safe working practices and reviewing stores can also offer areas of improving work-order templates, the sum of which will deliver a more efficient and accurate tool for scheduling and execution. Templates can also be used to store trouble-shooting guides for specific symptoms/fault modes.

Entity management

CMMS solutions provide the 'entity' concept to maximize the flexibility of an equipment database. An entity is anything for which the user wishes to retain complete costing and maintenance activity records. An entity could be a piece of equipment, component, process line, factory, room, railway, division or any other element in the organization.

It enables users to identify 'parent–child' relationships that can be used to build hierarchies for cost rollups and operating statistics. A maintainable entity includes anything that needs to be repaired or serviced by maintenance staff: e.g., a piece of equipment, assembly line, truck or a repairable spare. An organizational entity refers to any arrangement of an enterprise that is used to collect cost, statistical, budget or backlog information such as a division, cost center or department.

Work management

Work management ensures that maintenance personnel have control of incoming work, while personnel requesting work can track work-order status. The planning function ensures that labor, material, tools, drawings and subcontractor requirements as well as safety information can be identified on work-orders to support pro-active maintenance activities. In addition, a library of predefined standard jobs can be established to aid in the creation and execution of work-orders. Work management provides the tools to empower plant floor personnel to enter and view equipment-related activities.

It provides a mechanism to make notes and track activities related to a particular piece of equipment. In addition, integration to the inventory and procurement functions ensures material availability, improved productivity of maintenance staff, better communication for inventory and procurement, and complete costing of all activities.

Preventive maintenance

The preventive maintenance (PM) function offers the ability to create a library of standard jobs with the automatic work order generation (based on any combination of user-defined triggering criteria: operating statistics, elapsed time, calendar date), inspection checklists and PM routes. Some solutions provide applications that are fully integrated to real-time operating statistics from the plant floor. The preventive maintenance function helps users manage maintenance in a proactive and planned manner, rather than just treating maintenance as reactive or repair work.

Safety

Providing personnel with proper and timely safety information is a critical element of an effective maintenance management program. Solutions automate the collection and dissemination of procedures such as stoppage, shutdown and startup information.

Reliability analysis

Solutions can provide detailed history of equipment information based on day-to-day maintenance activities. Failure history including symptoms, the cause of failure, and action taken can be easily reviewed and analyzed. In addition, the reporting of MTBF and MTTR is available to determine proper fine-tuning of equipment maintenance requirements.

Maintenance, repair and operations inventory

Maintenance, repair and operations (MRO) inventory addresses the main challenge of maintenance, repair and operations inventory management, controlling a large number of unique and low unit value items that are subject to unpredictable demand. The system automates the re-order process by using calculated safety stock levels, replenishment lead-times and sophisticated 'available-to-promise' logic based on expected receipts (open purchase orders) and issues. The ability to establish multiple storerooms and locations as well as flexible categorization capabilities allow user-defined grouping of items for searching, analysing and reporting purposes.

Procurement

The main objective of the procurement function in a maintenance environment is to minimize the cost of buying high volumes of MRO inventory items. Solutions focus on fully automating the entire procurement process, including requisitions, purchase orders, expediting, receiving, request for quotations and invoice matching.

Extensive analysis capabilities are available to streamline the procurement process and allow more time for the procurement professional to add value in contract negotiation and vendor relationship management.

The real-time equipment information coupled with the coordination between production schedules and maintenance activities are used for selection. Seamless integration allows the information to flow from meter readings and alarms to the automatic generation of work orders and statistical information.

Condition-based monitoring

Real-time condition-based monitoring (CBM) uses information gathered from instruments such as temperature, vibration, current and voltage to determine the status of equipment. Excessive vibration or temperature can be used to predict imminent equipment failure, and maintenance can therefore be planned and scheduled. The CBM routines can be implemented in real time, or at predetermined frequencies using PM schedules.

Rotable spares management

Rotable repair management enables the repair of rotable spares and equipment, such as engines, gearboxes, pumps or pump motors, and rollers. These are items that are repaired and reused and rotate through stock. Generally, rotables are taken out of operation and replaced with another from stock. It is then placed into stock in an unserviceable form, repaired/reconditioned under a work-order and returned to stock in a serviceable form.

Rotable repair management includes the manufacturing modules required to manage the build and rebuild of spares and equipment. It also keeps track of repair costs (LCC) and determines the point where the repair cost starts to outweigh the cost of a new piece of equipment. Rotable systems also keep track of the location of rotable equipment, adjusting MTBF figures when refurbished equipment is installed.

8.3.3   CMMS benefits

There are several asset management solutions, which provide tightly integrated Maintenance, inventory management and procurement components as well seamless connectivity into the real-time plant floor information. Asset management solutions also have the flexibility to address continuous improvement strategies in business processes and have the following benefits:

- Improves competitiveness and return on investment
- Provides a single, scalable asset management solution to fit the business
- Can be tailored to changing business environment.

Research indicates that a well-executed maintenance improvement plan can produce solid results:

- 5–30% reduction in maintenance operation cost
- Asset reliability improvements up to 98%
- Inventory reductions up to 75%
- Maximized production
- Better resource utilization
- Pro-active, instead of reactive, maintenance operations
- An economic balance between production and maintenance.

8.3.4   Overall equipment effectiveness

In an ideal condition, equipment would operate 100% of the time at 100% capacity, with an output of 100% good quality.

The difference between the ideal and the actual situation is due to various losses in the process and factory. There are six types of wastes that reduce the effectiveness of the equipment and these six big losses are grouped in three major categories: downtime, speed losses and defect losses.

Downtime

Downtime refers to the time when the machine should be running, but does not. It includes two main types of loss: equipment failures, and setup and adjustments.

Equipment failures

Sudden and unexpected equipment failures, or breakdowns, are an obvious cause of loss, and an equipment failure means that the machine is not producing any output.

Setup and adjustments

Changeover of products may require some period of shutdown so that internal tools can be exchanged. The time between the end of production of the last good part and the end of production of the next good part is downtime. This downtime loss often includes substantial time spent on making adjustments until the machine gives acceptable quality on the new part.

Speed losses

A speed loss means that the equipment is running, but it is not running at its maximum designed speed. Speed losses include two main types of loss: idling and minor stoppages, and reduced speed operation.

Idling and minor stoppages

When a machine is not running smoothly and at a stable speed, it will lose speed and obstruct a smooth flow. The idling and stoppages in this case are caused not by technical failures, but by small problems. Operators can easily correct such problems when they occur, but the frequent halts can dramatically reduce the effectiveness of the equipment.

Reduced speed operation

Reduced speed operation refers to the difference between the actual operating speed and the equipment's designed capacity. The goal is to eliminate the gap between the actual capacity and the designed capacity.

Defect losses

A defect loss means that the equipment is producing products which do not fully meet the specified quality characteristics. Defect losses include two major types of loss: scrap and rework, and startup losses.

Scrap and rework

Loss occurs when products do not meet quality specifications, even if they can be reworked to correct the problem. The goal should be zero defects – to make the product right the first time.

Startup losses

Startup losses are yield losses that occur when production is not immediately stable after starting the equipment, so the first products may not meet specifications. This is a latent loss, often accepted as inevitable.

OEE calculation

The OEE calculation provides information about how effectively the machine is running and which of the six big losses needs improvement. Overall equipment effectiveness is not only an indicator to assess a production system, but is also important for improvement. Overall equipment effectiveness is calculated by combining three factors that reflect these losses: the availability rate, the performance rate and the quality rate.

The *availability rate* is the time the equipment is really running vs the time it could have been running. Low availability rates reflect downtime losses.

$$\text{Availability rate} = (\text{operating time} - \text{down time}) / \text{total possible operating time}$$

The *performance rate* is the quantity produced during the running time, against the potential quantity, given the designed speed of the equipment. A low performance rate reflects speed losses.

$$\text{Performance rate} = \text{total output/potential output at rated speed}$$

The *quality rate* is the amount of good products vs the total amount of products produced and a low quality rate reflects defect losses.

$$\text{Quality rate} = \text{good output/total output}$$

$$\text{OEE} = \text{availability rate} \times \text{performance rate} \times \text{quality rate}$$

8.4   Material storage and availability management

The problem of storage is of great concern to most companies, as a substantial amount of a company's working capital is normally invested in stores. The objectives of material storage are the following:

- Ensures adequate supply of stores to the production and service departments under proper authority
- Prevents under-stocking and over-stocking
- Identification and location of items without delay
- Prompt issue of materials
- Protection of materials from loss and wastage due to defective storage
- Protection of stores against pilferage, theft and fire
- Minimizes the storage cost.

8.4.1   Material storage concepts

Stores location and layout

The location of stores department in a company should be carefully planned. Stores should be near to receiving or dispatching departments so that transportation charges are minimal. Stores should be easily accessible to all other departments in a factory and facilitate an easy flow of materials through the manufacturing process. A planned location of stores departments can thus save time and money in the long run.

Inside the store, the materials/products should also be stored in a logical order, with high turning materials closest to the dispatching points and low turning materials further away. In high-volume-throughput stores, bin-location, picking and packing operations can be automated to increase stores efficiency even further.

Stock levels

The *maximum stock level* represents the upper limit beyond which the quantity of any item is not normally allowed to rise. The main object of establishing this limit is to ensure that unnecessary working capital is not blocked in stores. Theoretically, maximum stock level is the sum-total of minimum stock level and economic order quantity.

Maximum level may be expressed as follows:

Maximum level = reorder level + reordering quantity – minimum consumption.

= reorder level + reordering quantity – (minimum consumption per period

× Minimum Recording period)

The maximum stock level for a particular item is fixed after considering the following points:

- Rate of consumption of material
- Storage space available
- Lead-time from the date of placing the order
- Nature of material
- Working capital required
- Inventory carrying cost
- Market trends
- Fashion habits
- Government restriction
- Economic order quantity, etc.

The *minimum stock level* is the lower limit below which the stock of any item should not normally be allowed to fall. This is also technically known as safety or buffer stock. The main object of determining this limit is to protect against stockouts of a particular item. The prime considerations in fixing the minimum stock level or safety stocks are:

- Average rate of consumption
- Lead-time i.e. time required for replenishment.

Minimum level = reordering level – (normal usage per period × average delivery time)

The *re-order level* is fixed between the minimum and maximum stock levels. When stock of a material reaches this point, the purchase process should be initiated. The re-order level is slightly more than the minimum stock level to guard against abnormal usage, abnormal delay in supply, etc.

Re-order level = maximum consumption during the period
× maximum period required for delivery

The *danger level* is generally fixed below the minimum stock level. Normally stock quantities should not be below the minimum level. If it reaches the danger level at any point in time, urgent action for replenishment of stock must be taken to prevent stockout.

The *economic order quantity (EOQ)* can be defined as the purchase order size, which takes into account the optimum combination of stock-holding costs and ordering costs. Economic order quantity helps to achieve the lowest unit cost for stored material. The concept of economic order quantity is primarily based on the consideration of acquisition cost and possession cost, which has been discussed below.

$$EOQ = \sqrt{\frac{2 \times \text{annual consumption} \times \text{buying cost per order}}{\text{cost per unit} \times \text{storage and carrying cost rate}}}$$

The EOQ calculation is based on the following assumptions:

- Known stock-holding cost
- Known constant ordering cost
- Rate of demand are known
- Constant price per unit is known
- Replenishment is made instantaneously.

The *acquisition cost (S)* is used in determining the EOQ. It is necessary to know the cost of placing a purchase order, the cost of material units and the quantity discounts available for the items being purchased. It is the incremental cost rather than the average cost per order that is important when determining these costs.

$$S = \text{Cost per order} \times \text{number of orders}$$
$$S = \text{Cost per order} \times \text{anticipated usage/order quantity}$$

The inventory carrying cost is referred to as *possession cost (I)*. Carrying inventory carries a substantial cost. This cost is represented by items like rent of a storehouse, cost of insurance, and opportunity cost of tying up large working capital in inventory. The major components of carrying cost are as follows:

- The cost of money invested in the stocks.
- Cost incurred on the physical storage facilities such as storage space, racks, handling equipment, etc.
- Cost added by deterioration of items with low self-life or losses due to evaporation etc.
- Salaries to stores personnel.
- Losses in stores due to pilferage, wastage, breakage, etc.

$$I = \text{Average inventory quantity} \times \text{unit cost} \times \text{Inventory holding cost\%}$$

$$I = \text{Order quantity/2} \times \text{unit cost} \times \text{Inventory holding cost\%}$$

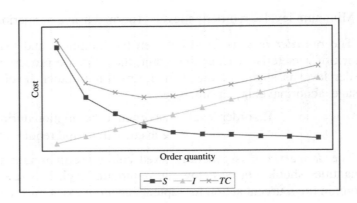

The EOQ is used to determine the most optimum order quantity in terms of total cost. Smaller orders mean less possession cost (I), but more frequent orders with the associated acquisition costs (S). Bigger orders mean higher possession cost (I) but less frequent orders and associated acquisition costs (S). The EOQ is where the ordering costs equal the inventory holding cost ($S=I$), as this is the point where the total cost (TC) would be at its lowest. The following figure depicts this situation.

In a situation where the following is applied to the above formulas, $S = 80.00$, $I = 0.3$, unit cost = 5.00 and the anticipated usage = 6000, the following table can be developed for possible order quantities (Q).

Q	S	I	TC
10	48000	7	48007
100	4800	75	4875
300	1600	225	1825
500	960	375	1335
600	800	450	1250
700	685	525	1210
750	640	563	1203
800	600	600	1200
850	565	638	1203
900	533	675	1208
1000	480	750	1230
1200	400	900	1300
1500	320	1125	1445

It can be seen from the table that the lowest total cost is at 800 units where $S = I$.

In theory, the EOQ and re-order level will operate together as indicated in Figure 8.2 below.

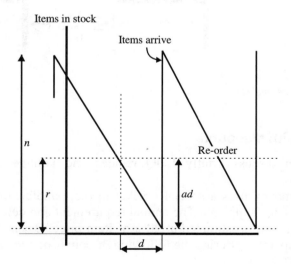

Figure 8.2
EOQ and re-order level

The following is assumed:

- Average rate of use/demand = a
- Time to get a delivery/lead-time = d.

A = the product of the average rate of use, a, and the average time to deliver, d. $A = a \cdot d$

A is therefore the number of units which one would expect to use during the delivery period. We assume that the delivery is always in a batch of size n, which is fixed for all time.

Inventory holding conflict

There are two other cost parameters to consider. The first is the cost of holding one unit in stock for a unit time. Inventory-holding cost will have to include all the costs such as rent of shelf space, security, cost of obsolescence, insurance, cost of capital and so on. As Inventory-holding cost increases, it becomes more likely that the optimum strategy is to reduce the average stock level and risk running out of stock.

The other cost parameter is the cost of being empty and failing to have stock when it is needed. As stockout cost increases, it is important to have adequate stock to reduce the chances of running out of stock.

Therefore the two cost parameters, inventory holding and stockout costs are working in opposite directions and we can see the need for an optimization (refer Figure 8.3). It is thus critical that a company do not evaluate EOQ in isolation, but that safety-stock and stockout cost is also considered before deciding on the appropriate order quantity.

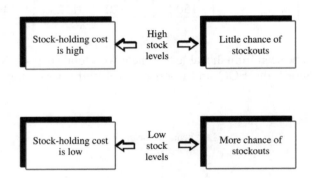

Figure 8.3
Stockholding conflict

EOQ with discounts

We can discuss briefly EOQ rebates and volume discounts with the help of an example:

A company uses a special bracket in the manufacture of its product, which it orders from outside suppliers. The annual requirement and delivery cost per order is $20. Stock-carrying cost is estimated at 20% of item price. Basic item price is $10 per bracket and the company is offering the following discounts on the basic price:

400–799 less 2%
800–1509 less 4%
1600 and over less 5%

It is required to establish the most economical quantity to order. This problem can be answered using the following procedure:

$$\text{Basic EOQ} = \sqrt{\frac{2 \times 200 \times 20}{10 \times 0.2}} = 200 \text{ brackets}$$

The savings from the lower price and ordering costs and the extra stock holding costs at each discount point with the costs associated with the basic EOQ are given in the following table:

Order Quantity	200	400	800	1600	Note
Discount	–	2%	4%	5%	
Average no. of orders p.a.	10	5	2 1/2	1 1/4	Demand of 2000/order quantity
No. of orders saved p.a.	–	5	7 ½	8 3/4	
Ordering cost saving p.a.	–	$100	$150	$ 175	
Price saving per item p.a.	–	$ 0.2	$ 0.4	$ 0.5	
Savings for 2000 No.	–	$400	$800	$1000	
Total gains	–	$500	$950	$1175	
Stockholding cost p.a.	$200	$392	$768	$1520	(Order quantity/2) × cost per item × carrying cost %
Additional costs incurred by increased order quantity	–	$192	$568	$1320	
Net gain (loss)	–	$308	$382	($145)	

From the above table it is clear that the most economical order quantity is 800 brackets, thereby gaining the 4% discount

Inventory records

Two types of records are often maintained for material received, issued and transferred.

Bin card   A bin card is used by storekeeper to keep quantitative records of all items of materials and goods in the store. There is normally a bin card for each material. All bins, racks or shelves, etc., should be numbered consecutively in order to indicate their location to the storekeeper.

Store ledger   It provides a continuous record of materials and stores received, issued and returned.

Inventory systems

There are various ways to keep track of materials in the store to ensure accurate inventory figures.

Perpetual inventory system   It is a method of recording store balances after every receipt and issue to facilitate regular checking and to prevent closing down for stocktaking. A bin card can be viewed as a perpetual inventory record. A program of continuous stocktaking can be run in parallel with a perpetual inventory system. Perpetual inventory means a system of records, whereas continuous stocktaking means the physical checking of those records with actual stocks on a continuous basis.

Periodic inventory system   In the case of small value items or slow moving items, a periodic inventory system can be adopted to determine the physical movements of stock and their closing balances as on particular dates. This is normally associated with stocktaking exercises at defined intervals.

Always better control concept of classifying goods in an inventory is used for exercising effective inventory control where the items are classified according to value of usage. The higher value items have lower safety stocks, because the cost of protection is very high with respect to higher value items. The lower value items can carry higher safety stocks, as the cost of protection is lower.

Stock value pricing

The first-in-first-out (FIFO) method assumes that items first received are first to be issued and that the requisition is priced at the cost at which items were placed in stock. The oldest costs are used for accounting purposes first regardless of actual material flow. This is ideally used for slow moving and comparatively high unit cost.

Advantages	Disadvantages
It is not based on approximations and estimates Conforms to sound principles of economics and business Convenient method to be used for pricing of material issues under any circumstances	Involves a lot of calculation work For pricing one requisition, more than one price may have to be adopted Prices of different lots of materials are different – this may lead to distorted cost In case of frequent fluctuation, the costs of issues do not represent the current market price

The last-in-first-out (LIFO) method is based on the assumption that the last items purchased are the first to be used. The balance in hand is priced at the cost of the earliest purchases. The LIFO method does not mean that material purchased last is always first to be used. It simply means that the prices of the last purchase are used for accounting purposes regardless of actual material flow.

Advantages	Disadvantages
Cost of material is closer to market price Conforms the principle that cost should be related to current market price Neither profit nor loss is made using this method (unlike FIFO method)	Considerable calculation work The balance stock is valued at the oldest price – this may not correspond with the prevailing market price Under falling prices, issues are priced at lower prices and stocks are valued at higher rates. Suitable for few items.

The moving average method is the simple average of prices over a period of time. It makes use of number of periodic prices to calculate a simple average price. It is useful for materials in a highly fluctuating market.

The weighted average price method is the total cost of an item of material in stock divided by the total quantity in stock.

Inventory turnover

Inventory turns are receiving considerable attention in the market today, as it gives an indication of the asset utilization of a company. Inventory is an asset, and unused assets drain the organization of valuable resources. Money has been spent to purchase the inventory, but if it has not been used it did not add value or generate wealth. It is thus important for most companies to ensure that stock (asset) turnover for their particular company is in line with industry standards or benchmarks. Inventory turns can be demonstrated with the following example:

	Material A	**Material B**
Opening stock 1/1/02	$10 000	$9 000
Purchases during the year	$52 000	$27 000
	$62 000	$36 000
Less: closing stock 31/12/02	$6 000	$11 000
Materials consumed	$56 000	$25 000
Average inventory {(Op.stock + Cl.Stock)/2}	$8 000	$10 000
Inventory turnover ratio (Consumption/avg. inventory)	7.0 times	2.5 times
Inventory turnover (days) (365 days/IT ratio)	52 days	146 days

In the example above, material A is moving faster than material B.

8.4.2   Warehouse and inventory performance measures

The tables below indicate some of the best-practice performance measures used to measure the effectiveness and efficiency of warehouse and inventory processes.

Order fulfillment performance

Measure	**Definition**	**Calculation**	**Current**
On-time delivery	Orders delivered on time as per customer-requested arrival date	Total orders on time/total orders shipped	%
Order fill rate	Orders filled completely on first shipment to customer	Orders filled complete/total orders shipped	%
Order accuracy	Orders picked, packed and shipped perfectly	Orders shipped w/o errors/total orders shipped	%
Line accuracy	Lines picked, packed and shipped perfectly	Lines shipped w/o errors/Total lines shipped	%
Order cycle time	Time from order placement to customer shipment	Actual ship date (minus) customer order date	Hours
Perfect order completion	Orders delivered without changes, damage or invoice errors	Perfect delivery orders/total orders	%

Inventory management performance

Measure	Definition	Calculation	Current
Inventory accuracy	Actual inventory quantity vs system-reported quantity	Actual Quantity/SKU Reported Qty. by SKU	%
Damaged inventory	Damage measured as a % of inventory value (cost)	Total Damage $ Total Inventory Value $	%
Days on hand	Average sales days of inventory on hand based on historical sales	Average inventory value ($)/ average daily sales during past month ($)	Days
Storage utilization	Occupied square footage as a % of storage capacity square footage	Average Inventory sq. ft/storage capacity sq. ft	%
Dock-to-stock time	Average time from carrier arrival until product is available for order picking	Average dock-to-stock hours per receipt	Hours
Inventory visibility	Time from physical receipt to customer service notice of availability	Time of host system Receipt data entry –(minus) time of physical receipt	%

Warehouse productivity

Measure	Definition	Calculation	Current
Orders per hour	Average number of orders picked and packed per person-hour	Orders picked /packed/ total labor hours	Ord/Hour
Lines per hour	Average number of order lines picked and packed per person-hour	Total Lines picked/packed/ total labor hours	Lines/Hour
Items per hour	Average number of order items picked and packed per person-hour	Total items picked/packed/ total labor hours	Items/Hour
Cost per order	Total warehousing costs fixed: space, utilities and depreciation. Variable: labor/supplies	Total warehousing costs, total orders	$/order
Cost as a % of sales	Total warehousing cost as a percent of total company sales	Total warehousing cost, total revenue	%

8.4.3 Warehouse management systems

Warehouse management systems (WMS) provide a bridge between enterprise-level purchasing, manufacturing planning, manufacturing execution, customer service systems and the warehouse or distribution center. With real-time visibility of available inventory, the WMS marshals people, space and equipment to efficiently receive, store, pick and ship components and raw materials to production, and finished goods to wholesalers, distributors and end customers.

A high-performance warehouse management system dramatically cuts costs and increase warehouse productivity. A WMS optimizes all the resources in the warehouse by automating the materials handling process and providing productivity tools to ensure that businesses stay competitive, allow cost cutting, increase productivity, fulfill orders accurately and efficiently, and improve customer service.

WMS opportunity identification

Challenges most frequently cited as the basis for WMS investment include:

- Receiving, picking and shipping errors
- Long search times due to misplaced or lost stock
- Manual transaction recording
- High direct and indirect labor costs as measured by cost per money-unit shipped
- Inventory accuracy below 99%
- Low inventory turns (if calculated at all)
- Shrinkage
- Lot tracking and shelf-life management issues
- Lengthy-order cycle times, low fill rates and related customer service problems
- Poor space utilization, increasing use of outside storage and related shuttle costs
- Performance measurement issues
- Customer demands.

Efficient and accurate order fulfillment

Most WMS solutions integrate barcode and wireless data collection with powerful rules-based client/server software and with multiple work zones or multiple sites. The WMS systems are versatile and scalable to handle a variety of logistic needs and complete paperless movement of materials in the warehouse. Productivity and profitability are achieved by optimizing all the resources within the warehouse including labor, equipment, storage, space and inventory.

WMS functions

Receiving Some systems are completely paperless and integrated with planning and scheduling systems, eliminating all the errors associated with paper being misplaced or misfiled. Every move of materials in the warehouse is interactively verified using barcodes and scanners or tags and readers to guarantee accuracy. Order entry provides available-to-promise (ATP) information online with immediate item allocation, including price confirmation for standard and custom orders.

- Paperless check-in
- Verification and labeling
- Cross docking
- Notification of orders pending for this product
- One-step put-away to bins
- Stock immediately available for picking
- Cross-referencing of supplier part numbers
- Incorrect shipment identification
- One or more purchase orders can be received at once
- One or more receivers can work on the same order.

Picking/packaging To optimize the efficiency of the order fulfillment process, it directs warehouse staff to the correct pick-bin location. Searches for goods are eliminated and travel within the warehouse minimized. With multiple warehouse zones, picking strategies can be modified using simultaneous and sequential zone picking, pick-and-pass picking, and touch screen re-pack workstations. Conveyors, carousels and a shipping system can be integrated to further accelerate order fulfillment.

- Direct picking to shipping carton(s)
- 'Order verification' and price ticketing in multiple formats
- Concurrent replenishment and picking
- Group replenishment
- Tracking of serial numbers and lot codes
- Picking based on expiry dates
- Slot management
- Customer-specific labeling
- Bin assignment with reaches codes and size codes, etc.

Inventory control Integration of order entry requirements into the materials functions for accurate ATP scheduling enables the system to distinguish firm orders from forecast orders, the tracking of customer order throughput status and reporting shortages into the customer service function for customer notification.

- FIFO or LIFO stock rotation
- Date/product /bin location
- Audited product relocation
- Inventory tracking of product through the warehouse
- 'Fill rate management' capability provides a flexible 'predictive' allocation process
- Tracking of both picking and overstock
- Issues replenishment requests based on actual orders (vs min./max.)
- Paperless replenishment requests issued as bins run down
- Stock adjustments
- Slotting recommendations
- Dedicated, non-dedicated and random bin locations
- Quarantine management.

Shipping Most systems provide information about due date commitments, can schedule carriers, provide pick lists by location, by order, by customer and can produce 'picked complete' information.

- Integrated multi-carrier shipping systems
- Packing slips, carton content labels
- Shipping labels produced at start of picking process
- Tracking of vital shipping information
- Estimates shipment cube and weight
- Compliance labeling
- Automatic customs and dangerous goods documents available
- Rate shopping
- Prints compliant freight labels
- Automatic manifest generation
- Customer service views
- E-mail order notification of order status

- Assigned incident reports
- Tracking incidents with views by order, client, customer, product, buyer and vendor
- Customer lists with shipping history.

Order entry automation and integration into production and delivery functions of the organization can reduce order cycle times, improve the accuracy and integrity of the planning and scheduling processes, and improve order promise commitments.

8.5   Lean manufacturing

An integrated set of activities designed to achieve high-volume production using minimal inventories (raw materials, work in process and finished goods).

Just-in-time (JIT) philosophy is dedicated to the elimination of waste. In the context of JIT, waste is anything that does not add value. In the ideal JIT system, the throughput time exactly equals its processing time. When the JIT philosophy is implemented, throughput time is minimized. Inventory holding costs are almost eliminated and large gains are realized by improvement of quality and productivity. The throughput time is the aggregate of processing time, inspection time, conveyance time and waiting time.

Throughput time is made up of value-added time and non-value-added time.

The value-added time is the time during which work is actually performed on the product. In JIT, inventory is viewed as a form of waste, cause of delays and a form of production inefficiency, and thus should be eliminated. Both TOC and JIT operate on the basis of a pull system (JIT using the just-in-time concept and TOC using the rope concept). The JIT differs from TOC in that it views all inventory as wasteful and unnecessary, whereas TOC realize the need for inventory buffers around the constraint (drum) to avoid schedule and production disruptions.

Waste in operations

- Waste from overproduction
- Waste of waiting time
- Transportation waste
- Inventory waste
- Processing waste
- Waste of motion
- Waste from product defects.

JIT production philosophy

- Acquire material just in time for fabrication
- Produce fabricated parts just in time for sub-assembly
- Produce sub-assemblies just in time for final assembly
- Produce finished goods just in time to be sold.

JIT purchasing

The JIT purchasing is the purchase of materials and supplies in such a manner that delivery immediately precedes the demand of use. This will ensure that stocks are as low as possible. The JIT purchasing is implemented by developing closer relationships with suppliers so that the company and supplier work together cooperatively. With JIT, arrangements are made for more frequent deliveries of smaller quantities of materials in order to maintain a minimum of stock. Making suppliers responsible for the quality of their supplies can have considerable savings in material-handling expenses.

The improved service is obtained from fewer suppliers who provide high quality and reliable delivery. Suppliers are encouraged by instituting long-term purchasing orders that ensures their revenue. The JIT purchasing substantially reduces the investment in raw materials and WIP stocks. It also saves factory space, eliminates the need for large quantity discounts and reduces paper work. As a blanket order is placed, ordering costs are also reduced. In JIT purchasing, the policy is not always guided by EOQ decision models (refer Figure 8.4).

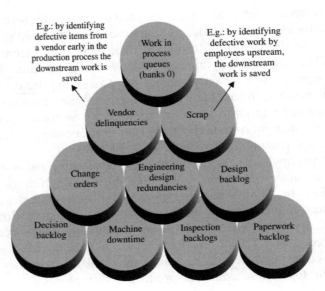

Figure 8.4
Elimination of buffer inventory

Benefits of JIT System	JIT – Implementation	Design of JIT System
Inventory Production lead-time Space Worker involvement Coordination Product quality Process quality	Design a flow process Use total quality control Stabilize schedule Use kanban pull control Work with vendors Reduce inventory Improve product design	Container/lot size Number of containers/load Average demand during lead time Safety stock Average waiting time Average processing time

9

Production scheduling, management and control

Learning objectives

- To understand how production capability is exploited through scheduling, management and control.
- To become familiar with production functions such as scheduling, dispatching, allocation and data-collection.
- To understand how the different scheduling and production modules can benefit the organization.

9.1   Introduction

For an enterprise to succeed in the current global fast-moving business environment, it needs to have efficient production facilities that are agile in order to fulfill changing customer requirements. Unless the majority of revenue within a complete supply chain is generated through long-term contracts with reputable customers providing steady demand, any company will have to adapt to changing requirements at a fast pace.

In order to become agile requires quite a few prerequisites:

- The company has to be in complete control of the production process, ensuring consistent quality and throughput. Inconsistent quality and throughput introduce variability into the process and makes planning and scheduling more complex and unreliable.
- The company has to be able to schedule production in such a fashion that it produces schedules optimized for cost, meeting customer delivery deadlines and maximizing production capacity.
- The company also has to operate with a minimum of stock-holding, both in raw materials and as WIP, whilst being able to meet customer demand. In order to do this, it needs to know what materials are available in the plant or store, whether it has been allocated to a specific production order, or if it can be used for a new order.
- The company has to know when the materials move between production processes, where it has been dispatched to, and if it has been used in the next process already or not. If companies do not know this, WIP tends to increase,

and production is a 'black box' where material move into, and after some time, products move out, with no knowledge of where materials or production orders are in the plant.

- The company has to know how many resources (besides materials) are required and available to carry out the work, including manpower and equipment. If equipment and manpower are inadequate, materials move into the production process, get into a WIP stage, and then wait for further processing, increasing the WIP inventory very quickly.
- The company also has to control the physical production process in terms of material efficiency, quality and capacity utilization in order to identify process or quality deficiencies using SPC/SQC or other methods. The information from the process control function is then used for corrective actions and for identifying and preventing potential deficiencies.

Without these production management tools, less management information is available, making it difficult to introduce changes quickly. The effect of changes will be difficult and more time-consuming to identify, as the information is not readily available and would require investigation. Agility requires speed, and without the right tools to provide accurate, decision-making information, speed is lost.

9.2 Enterprise scheduling

All over the world, competition is increasing and customers are looking for a faster turn-round and on-time delivery, and this rather than price, is becoming a key differentiator. This pressure to be more agile is often seen in make-to-order environments but even where a company is in a make-to-stock environment, there is pressure to reduce stocks to a minimum whilst keeping a wider range of products in stock. It leads to smaller lot sizes and any issues that reduce the efficiency of the plant become important.

Many companies concentrate on the engineering issues such as commonality of parts and assemblies, changeover issues using techniques such as single minute exchange of dies (SMED), design issues such as design for ease of assembly, and reducing 'floor to floor' processing time by using bigger, better and faster machines. This forces the manufacturers to concentrate on issues that maximize throughput and minimize lead-times, and this may not be the same as maximizing resource usage. Hence sequencing of lots to maximize throughput and minimize lead-times is the key to increasing service levels whilst reducing the cost of inventory.

9.2.1 Enterprise scheduling

Most ERP/MRP systems have a capacity requirements planning (CRP) module, to indicate capacity overloads. This does not take into account sequencing of the work and gives no facilities for taking account of late or new orders. This is why many MRP suppliers have added some form of graphical and interactive tool to replace CRP, either by offering tools as an integrated part of their own package or developing their own. Finite capacity scheduling (FCS) produces better sequencing and achievable schedules, which allows companies to become more agile and responsive and still maintain customer service levels.

The MRP systems take customer orders and break them down into individual parts using a bill of materials (BOM), then aggregate the requirements for the parts into works and purchase orders, but the relationship between a works or purchase order for a part and the customer orders is lost during this process. To provide APS functionality, the scheduling system should understand these relationships in order to know how to sequence the works/purchase orders that have to be made. The APS systems often duplicate MRP functionality, re-blowing the BOM to understand the links between the works/purchase orders.

It also duplicates ERP functionality including forecasting systems, distribution software, etc. The latest material control modules of APS include the concept of dynamic material control (DMC) whereby the linking between works/purchase orders is carried out during the scheduling run. During re-scheduling the planner can decide whether to keep the existing links or re-allocate the materials because a problem has occurred somewhere in the production facility. As a by-product this gives traceability of which materials and components go into which customer orders.

To extend the APS and DMC principles to the whole of the supply chain, the schedules of the manufacturer and those of their suppliers and subcontractors need to be synchronized. Using a single high-level SCM model for the entire supply chain is not accurate enough to take into account the current and future workloads of the entire team, since much of the work of suppliers and subcontractors is not related to the manufacturer. The solution to this problem is to make the APS system of each member of the supply chain available to all the other members.

Supply-chain scheduling (SCS) provides this functionality and it also provides capable-to-promise (CTP) or make-to-order functionality, which is suitable for subcontractors, while the DMC module provides true available-to-promise (ATP) functions, taking existing stocks into account.

9.2.2   Supply-chain scheduling

The advent of finite capacity scheduling (FCS) has enabled companies to provide an accurate prediction of delivery taking into account all the constraints of their plant such as materials, machinery, labor and tooling.

Electronic document interchange (EDI) and supply-chain management (SCM) foster closer relationships between the companies within a supply chain. Both EDI and SCM are mechanisms for companies to place their demand (requirements) on their suppliers, without taking much account of the supplier's capacity.

Supply-chain scheduling reviews the supplier's capacity and then determines the delivery requirements. The SCS has the potential to improve supply-chain efficiency for companies in the make-to-order sector, but to work in its most effective way, these companies will have to work more closely together. Benefit for multi-level companies e.g. mineral resources group can be a good example.

Multiple mines within the group produce various products (like coal, bauxite, iron ore, etc.), which is sent to metallurgical processing plants (refineries, coking plants, etc.). This material in turn is sent on to metal smelters/casting houses producing final products (aluminum, steel, stainless steel), from where it is sent on to warehouses for storage, distribution or exporting to customers or distribution centers for final delivery.

All of the above can potentially take place within the same company, with transport between multiple plants located in multiple countries with own or contracted transport between them. Supply-chain planning and optimization tools assist in the scheduling of material movement and logistics and can also assist in determining the most profitable location to get materials from and the most profitable place to manufacture a product for a specific customer order given production costs, transport costs, stock-holding costs, distribution costs, etc.

9.3 Finite capacity scheduling

Finite capacity scheduling (FCS) emerged, as a response to the limitations of MRP (MRP assumed infinite capacity and material availability). The MRPII started establishing a relationship between material requirements and production capacity, with added functions such as master scheduling and business planning, but not at the detail and specialized optimization functionality of FCS.

FCS itself is not a new idea, and for many years, versions of FCS emerged to enhance the accuracy of the production plan to better manage inventory, resources and customer satisfaction. Initially, the FCS systems were simple, involving the measure of only the primary resource to do work.

Multiple constraints were later added to provide more accurate modeling of the production resources. About the same time, backward and bidirectional scheduling began to appear in some packages. Most of the larger ERP systems have started incorporating some FCS functionality into their product suites. The new FCS systems are flexible enough to accommodate all the production philosophies to support job-based, resource-based and event-based modeling. By applying any one, or a combination of these three views of the production floor, an FCS model can produce the accuracy that adds value to the business goals of the enterprise.

9.3.1 Finite capacity planning

Finite capacity planning and scheduling is a process whereby a production plan consisting of a sequence of operations to fulfill orders is generated based on the real capacity and sequence of resources. Resources can be machines, tanks, piping, labor, storage areas, forklift trucks, delivery trucks, tooling or anything that could constrain production processes.

Traditional planning systems consider that sufficient resources are available to produce batches when required. These planning systems take the orders, break them down into component parts or ingredients and then calculate when to launch the batch based on the individual lead-times, and when raw materials are required taking into account current stocks. The real capacity of resources and its utilization to produce batches in a timely manner is not considered and the same lead-time is used to calculate the launch time for any batch.

The materials required are also ordered to arrive in time for work to start and in case of delay in production upstream of a particular operation, the material is accumulated and work in progress increases. With no concept of bottlenecks available to the planning system, resources become overloaded and queues of work get longer. Batches join queues for resources at each process step, orders take longer to make progress through the process route, expected lead-times are too optimistic and deliveries are late.

In finite capacity scheduling, operations are only scheduled when resources are available and materials are ordered only when they are needed for the operation to be carried out.

9.3.2 FCS benefits

The value of FCS is in the accuracy and detailed process and equipment sequencing not obtainable in traditional infinite capacity planning. With this accuracy, raw material levels can be synchronized with demand from the production floor. It helps in reduction of inventory levels and key resources are utilized better and flow of work is more controlled. The WIP remains at a relatively constant level, lead-times are more predictable and delivery dates are more reliable. The FCS allows management to concentrate on the balancing of variable demand with the capacity available, instead of spending time on fire fighting and expediting.

There is an optimal point in time and cost for a given need to produce goods (refer Figure 9.1). In manufacturing there are two means to advance the delivery time:

1. Expedite the order, or
2. Create WIP inventory to have excess products to deliver against.

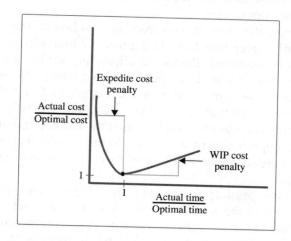

Figure 9.1
Cost vs time

One obvious drawback with WIP is that the value-add is not recovered quickly and therefore the manufacturer has lost the time-value of the money. What is not obvious is that expediting costs much more than WIP, because of the ripple-through impact of moving other orders around and otherwise disrupting the optimal point for all the other orders, accumulating a much greater financial impact than creating excessive WIP.

A good FCS can compute scheduling solutions that set these points for all orders, within the context of the production philosophy and the business goals to be achieved. However, the scheduling solution cannot exist in a vacuum, and the output of a scheduling system needs to provide useful data in order to fulfill its other role as a decision support tool.

9.3.3 Scheduling concepts

There are a number of very simple heuristic rules used to select the order for loading jobs on the planning board. One rule that can be used is to sort the jobs by the priority value that has been assigned to each job. Other sorting criteria may be earliest due date, earliest release date and smallest remaining slack time. Each rule represents a different strategy and focus in planning the jobs. For example, the due date-related rules (earliest due date, least remaining slack) focus on reducing the number of late jobs, whereas the priority-based rule strives to complete the most important jobs as soon as possible.

In some plants, a given operation can be performed on two or more different resources. For example, a batch operation might be performed in either of two reactor vessels. In such cases, the schedule is first determined by the ordering of jobs, then by the rule that is used to determine to which resource a given operation is assigned during the loading process. Again, simple heuristic rules (e.g. assign the operation to the resource that will complete the operation first) can be used to determine the schedule.

Although schedulers typically load each job working forward in time by starting with the first operation and continuing through to the last operation (forward sequencing), it is also possible to use this same sequencing scheme but reverse the procedure to work backward in time (backward sequencing). In essence, a forward sequencer fixes the start times for a job and determines the end times (which may violate the due dates), whereas a backward sequencer fixes the end times and determines the start times. However, the backward schedule may require start times that are infeasible, that is, the jobs must start before the current time.

Although the idea of constructing a schedule that has no late jobs is appealing, backward sequencing has some practical limitations, even in cases where a feasible solution is generated. Backward scheduling shifts all jobs to the right on the planning board so that they start as late as possible while still meeting the due date. This means that there is no time buffer in the system and any disruptions that occur (machine breakdowns, late material arrivals, etc.), will typically create late jobs. In addition, by postponing the use of available capacity and waiting until the last minute to start each job, the opportunity to consider additional jobs that may arrive later and need to be added to the schedule, will be lost. For these reasons, many schedulers prefer a forward sequencer.

It is also possible to schedule in a bidirectional mode. In this case, an operation somewhere in the middle of the job sequence is selected, the remaining operations is scheduled using forward sequencing and the preceding operations using backward sequencing. This is useful in cases where there is an available time slot on a critical, highly utilized resource. One can then assign an operation to this critical resource and then load its upstream and downstream operations around this operation. This is accomplished by fixing the operation on the critical resource and bidirectionally sequencing the remaining operations for this job.

9.3.4 FCS functionality

The FCS systems offer the user the opportunity to try different methods of scheduling or different dispatching rules, routings or constraints that are in use in the company. This is because a truly interactive FCS system not just gives the 'result' of a scheduling run but provides a mechanism by which, with the aid of analytical tools, alternative schedules can be generated and compared.

Depending on the complexity of the process and sophistication of the FCS tool, schedules can also be optimized based on different variables such as cost, production capacity, delivery dates, material availability, order priority, etc. Using these variables and developed algorithms, FCS can create schedules that optimize for maximum throughput, or least manufacturing cost, or least late orders, or maximum capacity utilization, or a combination of the above. Normally the optimization engine provides users with a choice of a primary optimization variable and one-or-more secondary variables to use as a base for optimization.

There are a number of ways by which scheduling systems offer tools to display and analyze the results of scheduling runs. These include:

Gantt charts and schedule performance metrics

The Gantt chart is the earliest and best-known type of planning and control chart. It is the most common way of displaying the proposed loading of jobs onto individual resources over time (i.e. the schedule) and comparing it with the actual performance of the facility in meeting that scheduled start and finish times for each job. Normally the vertical axis displays the names of the resources that have to be scheduled. The horizontal axis normally represents time. The bars normally display the order number, which is colored to highlight the product type. Other systems use colors only to differentiate between different orders, regardless of product types. The figure below displays a typical scheduling Gantt chart.

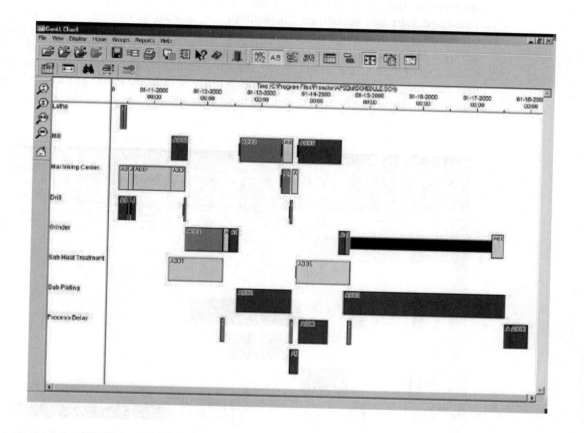

Tools are usually available in FCS systems to highlight the process route that a particular job has taken. There are a number of ways by which schedules can be measured, for example, resource utilization may be an important performance parameter in one plant, due date compliance in another and total changeover time in another.

Typical rules when scheduling operation sequences:

- Shortest setup time
- Shortest processing time
- Earliest due date
- Lowest value of critical ratio
- In a preferred sequence, etc.

Many key-performance indicators are often in conflict with one another. For example, it may be possible to obtain reduced set-up times by sequencing like jobs together on resources (e.g. material type, color or flavor), but this may make some other orders late.

This conflict is one of the major reasons for the development of primary and secondary scheduling rules.

Job analysis

Another tool that can be useful in analysing schedules is to track the progress of individual jobs, and in particular to compare the scheduled start and finish times for each operational step with the actual times achieved. As each operation is completed the scheduled start and scheduled finish times are compared with the actual values. Also a CR value is calculated. The CR is a ratio found by dividing the remaining process time by the time left to the due date at the actual finish time of the operation. The closer that the value approaches 1, the more urgent the job becomes and by looking at the values, the user can judge whether a change in job priority should be made at the next reschedule. The figure below displays a typical job tracking chart.

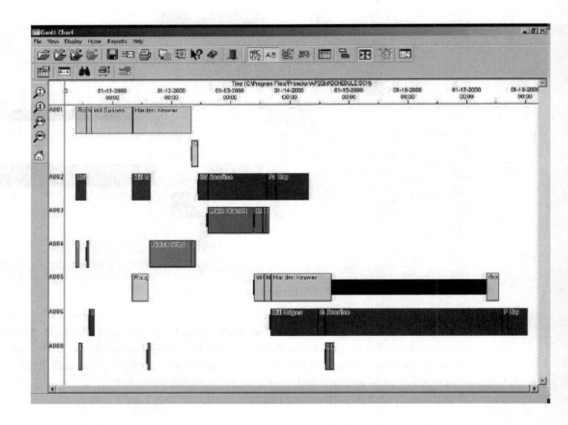

Comparative analysis

When comparing two schedules with each other, it is difficult to decide on the better schedule without graphically seeing the effect of the schedule change. This is particularly useful to determine the effect of order priority changes, expediting and the introduction of unexpected urgent orders. In the example below, two order tracking schedules are compared based on order due dates. It can be seen that one schedule is worse than the other, as one order has moved out to past the due-date for that order.

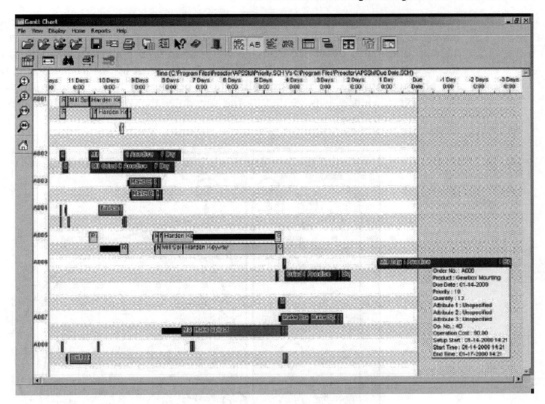

Material allocation

More advanced planning and scheduling systems will use not only resources such as machines, labor, tooling, etc. to constrain the loading of demand, but also the availability of materials. Some systems are able to take the BOM structure of a product and allocate materials for each level of the BOM. These additional links between jobs can be used to constrain the schedule based on the availability of these materials. The process of material allocation is often referred to as material pegging and is most used when the scheduling system is used along with an MRP/ERP system in a make-to-stock production environment. This can obviously only be done with good effect if the material availability is accurate and the FCS is integrated with the MRP/ERP system.

Reactive scheduling

Reactive scheduling enables concurrent scheduling and schedule execution. To remain competitive, a manufacturing system needs to react adequately to disturbances in its environment (e.g. rush orders) and uncertainties in the manufacturing process (e.g. defects, delays and variable yields). Reactive scheduling has combined good reactivity with high performance. Shop floor control is performed by both an online control system that reacts to disturbances immediately and a reactive scheduler that does not react as fast, but uses this larger time span to adapt the existing schedule to optimize global performance. Both subsystems cooperate to complement each other's abilities.

Scheduling and online shop floor control

Scheduling and online shop floor control (refer Figure 9.2) require several modules: orders, workstations, an online shop floor control system and a reactive scheduler. These modules may all be autonomous entities, operating together to provide business benefit. On the lowest level, resource allocation is achieved by the linking of an order and a workstation. This is coordinated by advice from the online shop floor control (workstation available) and the scheduler. The online shop floor control has to react to disturbances in real time and is not able to optimize the resource allocations because of the computational complexity of this problem. Hence, it is guided by the scheduler that reacts to disturbances periodically and on an event-basis.

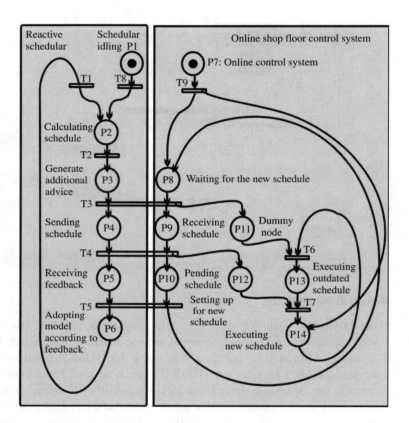

Figure 9.2
Scheduling and online shop floor control

9.3.5   Master scheduling

Master scheduling reflects the overall, long-term production plans, taking into account the company's sales forecast, material availability and plant capacity. It provides on-line and current available to promise (ATP) information for individual inventory items, and is a powerful tool to aid order processing, sales personnel and production planning. It virtually eliminates manually reviewing the master schedule required by other systems.

Master scheduling allows three separate schedules to be maintained to do what-if analysis. Each one of these schedules allows analysis of load vs capacity using rough cut capacity planning and also allows flexible, period-based demand vs supply analysis.

In the present situation it is important to reschedule frequently to adapt to new market conditions and customer demands. It requires an agile scheduling process that enables manufacturers to incorporate changes and produce a realistic schedule that maximizes production throughput.

9.4   Dispatching production units

Dispatching production units involves the management of the flow (routing), of materials to resources, in order to produce product. It directs workflow of production units in the form of jobs, orders, batches, lots and work orders according to production plans and detailed schedules. Dispatch information is presented in the sequence in which the work needs to be done and changes in real time as events occur on the factory floor. It has the ability to alter prescribed schedules and/or production plans on the factory floor. Additions and alterations may include material preparation and handling, and process operations such as rework, recovery and salvage.

Dispatch also has the ability to control the amount of WIP at any point with buffer management. Dispatching always ensures that the best resource is automatically assigned to the right job by matching skill sets, territories, customer requirements and resource costs, ultimately decreasing costs up to 25%. It maintains all customer records in one place, including a complete history of all work performed, customer communications, and notes. It can quickly and easily add new customers and work to be performed in order to track and profitably service that customer. All customer and work-related information could be imported from external databases if it already exists.

Dispatching reduce manual dispatch and scheduling time by 75% or more and increase accuracy by automating the process of resource allocation and workload balancing, even with emergencies and other unforeseen schedule changes.

Function of dispatch:

- Account for time spent needed to accomplish work on the job site.
- Automatically match the right equipment or person to the job.
- Provide automatic scheduling and optimization of work to technician.

9.4.1   Dispatching system functionality

Dispatching systems assign each work order to the right resource for getting the job done using automatic routing and assignment capabilities. It can quickly and effortlessly modify a schedule to accommodate new work or changes as they happen. All customer information, work time and work- or service-performed data are automatically captured for easy invoicing and other management reporting needs.

Tracking

It is important to keep a track of every resource which goes into a production area each time, in order to stay a step ahead of the competition, run business efficiently and keep customers happy. It understands which resources are busy during different times of the day. It helps to:

- Specify inventory locations, minimums, current levels and reorder points
- Keep track of equipment/asset location and management by customer locations
- Use both stock number and serial number identifications
- Maintain 'standard' installation times for each part/unit
- Produce valuable asset management and location reports.

9.4.2   Real-time dispatching vs cycle time

A comprehensive dispatching system is essential to improving efficiency. It considers process and equipment limitations, critical layer concerns, line balance, lot priority, ontime delivery and more. Daily process condition changes in the tools further complicate the difficulty of dispatching.

These problems were addressed by combining an MES with a dispatching system and recipe management system. The integrated dispatching system developed in which all the production data of lots, products, routes, operations and tools are stored or updated in the MES and the recipe distribution management system (RDMS) allows users to query available steppers by a particular product and check the limitations of products.

As the MES updates data to its database, a filter program called 'repository' sends a copy to another database for use by the real-time dispatcher (RTD), a rule-based dispatching program (refer Figure 9.3). Also, the RDMS sends tool limitation data to the RTD every time the RDMS is updated. When an operator triggers dispatching through the MES on any operation or tool, the RTD calculates lot priority and informs the operator which operation is to be next.

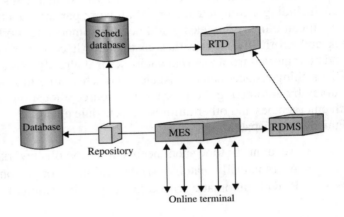

Figure 9.3
Integrated dispatching system

The RTD system uses a copy of the process history data and tool limitation data from the recipe distribution management system (RDMS) to calculate lot priority when dispatching is requested.

9.4.3   Dispatching rules

The use of dispatching rules has been proven to have significant impact on cycle time and has been an important topic in production control. The shortest remaining processing time (SRPT) rule sums the processing times of the remaining steps along the route of each queued lot and picks the lot with the shortest remaining processing time first. The SRPT is one of the best rules for reducing cycle time, but the resulting cycle times have unacceptably high variation. For this reason, slack policy is applied instead as a general rule of dispatching. Many issues had to be taken into serious consideration during the rule design phase for any company. An example of a multicriteria rule as implemented in an RTD system is provided below (Figure 9.4).

Step 1   Lots available. When a stepper becomes idle, an operator triggers dispatching. All available lots are searched by the dispatcher and filtered by RDMS limitation table.

Step 2   If the status of the lot is on hold, it will have the lowest priority. Hold lots will appear at the end of the dispatching lot list.

Step 3   Hot lots and super hot lots will have highest priority. Super hot lots take priority over hot lots, and if more than one hot lot is in the dispatching list, they will be ranked by slack value.

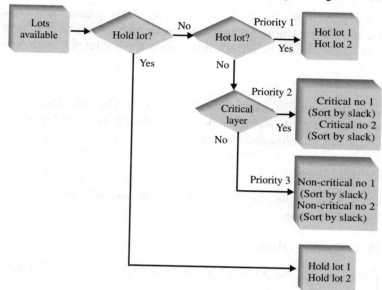

Figure 9.4
Dispatching rule flow

Step 4 Normal lots will be divided into two groups. Lots needing a critical layer process have higher priority.

Step 5 Dispatcher sorts lots for critical layer operations according to NQ ratio. The least slack rule is applied among lots with the same NQ ratio.

Step 6 Step 5 is performed on lots for non-critical layer operations.

Step 7 The list is output to the terminal for the operator.

Slack rule

A slack rule performs well in reducing cycle time and with variation. Though slack rule dispatching is designed to minimize the deviation of lot cycle time, it also performs well in reducing overall cycle time. For this reason, a slack value is generally chosen as the default-ranking rule.

The formula is:

$$\text{Slack} = \text{due date} - \text{today} - \text{remaining cycle time}$$

Where remaining cycle time data for all operations of a particular route are calculated using MES history data. A negative slack value obviously means the lot is behind cycle time. A smaller slack value indicates more urgency for that lot.

Next queue rule

A next queue rule takes WIP of the next main machine into consideration. The 'NQ ratio' for each entity type is given by:

$$NQ = \frac{\text{CWIP}}{\text{SWIP}}$$

Where CWIP denotes the current WIP of the entity type, and SWIP is standard WIP level of the entity. For some pre-bottleneck machines, SWIP is replaced by safety WIP level. The SWIP values are updated monthly according to changes in the product mix.

A low NQ ratio means a low WIP level for the next main machine and high priority for the lot. These values are updated automatically every few hours. Using the next queue rule keeps the line balanced and protects the bottleneck from 'starvation'.

Lot grade

Lots are generally divided into three priority classes: super hot, hot and normal. There are few super hot lots and more hot lots, and they have the highest priority. The lot grade rule dominates all other dispatching rules.

Critical layer

Critical layers also need high priority because of tight process specification and lack of backup machines.

Prioritization flow

Calculation of lot priority incorporates tool limitations and variations with the four rules listed above.

The RTD implementation for lot dispatching in one company continuously reduced the mean and variance of cycle time. The trend chart with weekly data is shown in Figure 9.5, illustrating a cycle time reduction of 31.3% from May 1997 to July 1998 and 52.3% reduction in cycle time deviation for the same period. In addition to cycle time reduction, the RTD simplified procedures for choosing which lot to run next. As more of the dispatching system was implemented (bars), cycle times became shorter and more consistent. The error bars represent one standard deviation in each direction.

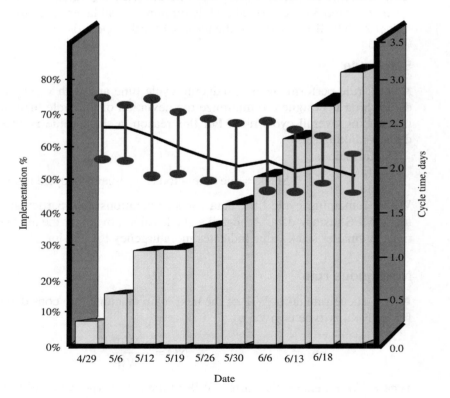

Figure 9.5
Cycle time improvement

9.5 Resource allocation

Resource allocation is an aggregation of the functions required to track and manage all resources related to production. These resources include: labor, machines, tools, fixtures, materials and other entities such as documents that must be available in order for work to start at the operation. It provides a detailed history of resources and ensures that equipment is properly set up for processing and provides status in real time. The management of these resources includes reservations in support of the operations/detail scheduling function.

9.5.1 Resource allocation components

A unique material ID identifies all system materials and is linked to each of the other material attributes. These other attributes include material name, description, material type, retest and expiration rules, required sampling and unit of measure. An additional link ties each material to a supplier or vendor that can provide the material.

For each defined material that can be manufactured, packaged, or otherwise produced on the plant, a BOM is created. The BOM defines the materials to be produced, the standard quantity produced, and each of the materials used in the production process.

Once the BOM is created, a recipe is also created. The recipe defines the minimum, standard, and maximum material quantities that can be produced, since every manufacturing and packaging process has limitations. The recipe also links the production material with a BOM, a bill of equipment (BOE) and a production area.

A site handles more than just materials. Equipment, instruments and physical locations can be defined as resources. For manufacturing and packaging operations a BOE is created to indicate equipment or instruments required for the production process. Physical locations can also be defined and are usually limited to production areas such as a bottling or a packaging line. These areas can be listed in a recipe as a specific area to produce a product and can also be allocated toward the fulfillment of a particular order.

Every material item within the inventory is either provided by a supplier, vendor, or it is produced. In each case the supplier and vendor information is stored within the database for use during receiving and manufacturing operations. An additional database table provides a connection between suppliers and the materials they can provide. This connection aids in providing material traceability in the receiving process.

In a similar way to supplier and vendor records, an additional set of record entries is maintained for shipping destinations. Specific materials should be assigned to a destination to avoid shipping sensitive or inappropriate products to an unauthorized recipient.

In order to maintain precise records for material and resource locations a location hierarchy is defined. Each place where a material or resource can be stored is defined by a site, an area and a location. The location hierarchy allows for multiple sites, areas and locations to better divide and subdivide a physical area into logical divisions. Each site-area-location designation is combined into a coded location value that is placed on a bar-coded label at each location. Bar-coded location tags are used during container movement events to note initial and final container locations.

While the sources may be located at the same site or facility, there is no site, area and location designation for the requester. If a usable quantity is available within the inventory, a material request can be fulfilled either by sending or shipping the requested material, otherwise the material needs to be produced and then delivered.

For materials requiring manufacturing or packaging, a production order is created. Each material request is assigned a unique ID value for tracking and fulfillment purposes. This unique ID is referenced for shipping any required production operations.

When either manufacturing or packaging is required to complete a material request, a production order also referred to as a batch-order is created. The production order associates a new product to be produced with a material request in order to distinguish the completed product from general inventory. A lot number for the finished product is created as the batch ID for the order. Once the finished product is entered into inventory, it receives a lot number with the same ID as the batch ID. The continuity between these two numbers allows for correspondence with the requester to include a finished product lot number. Once a production order has been created, material allocation, resource allocation and production can begin.

Material allocation is used to distinguish between available material inventory from total material inventory and aid in material resupply. Material allocation allows fixed quantities of material from a specific lot and container to be allocated for a specified purpose, such as the completion of a production order. By allocating materials before their actual use, quantity conflicts are resolved electronically rather than by physical inspection.

Similar to material allocation, the resource allocation module allows the operations planner to reserve people, equipment, instruments or physical locations for use in producing materials. Resource allocation associates selected resources with production events to ensure that equipment, instruments and locations will be available and dedicated to the production process. Stand-by equipments/instruments are arranged well in advance to avoid any set back during operations. Human allocation includes arrangements of right person for any particular assignment in terms of knowledge and experience on the same task. It also ensures the exact number of employees and the required duration for the same. Alternative arrangements are also kept ready for any contingency, for example, absenteeism, etc.

9.5.2   Bill of materials

The key to a successful manufacturing software implementation begins with the bill of materials. A BOM defines the products structure in terms of materials and provides an optional connection to plant resources such as machinery, tooling and labor defined by a bill of routing. The BOM application provides a solid base for production activity to be defined, tracked and reviewed.

The BOM is a set of ingredients, which identify specific requirements to construct a finished product. Following table is an example of the BOM recipe. To produce the top-level 0 or finished product, the material requirements are exploded through to the lowest level 3 which comprises raw material and purchased parts.

Level	Item	Quantity	Type
0	Calculator	1	Assembly
1	Processor	1	Subassembly
2	Casing	1	Subassembly
3	Keypad	1	Purchased
3	Fasteners	6	Purchased
3	PCB assembly	1	Purchased
3	Inserts	4	Raw materials

A BOM example (calculator assembly)

In this example of a BOM, a series of levels show what material is required and where in the process it is required. Level 3 items, comprising raw material and purchased parts, are assembled into level 2 items. Level 2 items are assembled into level 1 item. Then level 1 item are assembled into the level 0 item to form the finished product. The complexity of the BOM will largely depend on the product and type of manufacturing environment. The degree of complexity can be illustrated by comparing the balanced demand BOM and the structured demand BOM. The balanced demand BOM (refer Figure 9.6) shows a situation where very few subassemblies are required to form the final product. The components are assembled in a short space of time, except that is for subassembly, which may have been assembled some time in advance of assembly.

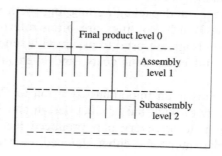

Figure 9.6
Balanced demand

The balanced demand BOM is a JIT BOM – as the throughput speed of the manufacturing processes increases, the BOM becomes flatter, that is process is more assembly than manufacture.

The structured BOM (refer Figure 9.7) comprise a number of subassemblies prior to assembly into the final product. The process of producing the subassemblies through to completion of the final product may take several days or weeks. The variety of subassemblies increases the need for planning and logistics movement and storage.

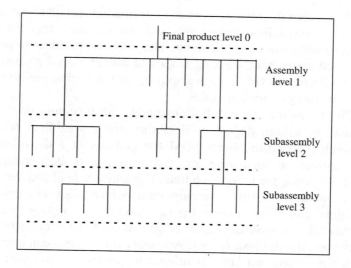

Figure 9.7
Structured demand

BOM in manufacturing

Bill of materials is used extensively in the manufacturing process, to assist with material requirements and to detail the exact formula or recipe for the finished goods. Some manufacturing industries simple use the BOM to provide a benchmark to which production is compared, some use the BOM for exact manufacturing instructions where component quantities and mixtures are critical such as the health industries. The most important part of a computerized manufacturing process and MRP is the BOM. Accuracy of the formula/recipe is crucial for material requirements planning and accurately projecting costing. Inaccurate BOM cost numerous manufacturers a lot of money in expediting raw materials, down time, confusion and stress.

Some computer systems extend the BOM by adding specific manufacturing details, scrap percentages and packaging/labeling methods. Most provide the ability to add routings to the BOM. Routings are often referred to as work centers or equipment areas. These routings are used to assist with scheduling the manufacturing processes, adding labor and equipment costs, and even adding start-up and overheads to the BOM.

Bill of materials is an important extension to CAD programs for tracking and listing parts all the way through the product design and manufacturing process. In the product manufacturing industry every part produced has some type of distinguishing mark – typically the part number – either stamped, etched, embossed or silk screened onto the part.

The manufacturer also needs to document the design and track the characteristics of each part like material, cost of manufacturing and out-side vendor code during the product development process. Most manufacturers use sales orders, inventory minimums or forecasting (MRP) to trigger the manufacturing. Some manufacturers just trigger manufacturing directly. Whichever method is used to trigger manufacturing, they all use the BOM to break down each routing into scheduled jobs. Raw material, equipment and labor requirements are then made available to the production managers, purchasing departments and the entire information system.

9.5.3   BOM systems

Bill of materials software contains information related to the structuring of a product. The BOM modules link components and raw materials together into assembled or finished subassemblies or complete items. The systems provide subassembly of up to 99 levels per finished component, tracking labor and machine costs needed to produce a finished item. The systems provide complex inquiry and reporting facilities and are directly linked to the inventory control modules.

The BOM module is fully integrated with inventory control and sales order modules, providing a high degree of flexibility and a wealth of features for managing a light assembly operation. Each BOM file consists of a list of component parts, which can themselves be subassemblies, raw materials or description-only items (i.e. royalties, shipping cost, etc.). For each item, a quantity is held and the cost of materials for the bill is accumulated. Any inventory item, which does not exist in the inventory control module, can be created, as the two modules are tightly integrated. There are large text fields for extensive descriptions, so that user can detail where to place components and any special build instructions. Assembly instructions can also be added to the item file for machines used and labor requirements, so that a total cost of the finished unit can be determined. Additional information, such as engineering drawing numbers and build standards, can also be kept with each assembly.

Items are usually entered in the sequence in which the finished product is built, but this order can easily be resequenced as specifications change or components substituted. Components within a BOM file can quickly be copied to a new bill number when producing a modified version of an existing item. Costs can be adjusted any time current material prices change, and this change can be done for one or all finished items. Different costing methods can be used for each item as required to suit varying operations and build sequences.

Each level of the subassembly can be viewed with the current inventory status, and detailed information displayed as to whether enough on-hand stock is available to produce a given number of finished items. It will display the maximum number of finished items that can be produced from the existing, on-hand inventory. The BOM module provides a sophisticated part substitution procedure, which enables replacement of one component with another for every assembled item.

The history of movements of units through the system can be viewed on-screen at any time, so that a full history of substitutions and issues is readily at hand. The BOM module maintains a history of issues for each assembled item, enabling hard copy reports to be produced on demand to ensure traceability.

Reports

Information displayed in hard copy (paper) form.

- *Single level/multi-level BOM:* A report of the next structures for one level/through all levels.
- *Single/multi-level where used:* A report of the next level assemblies in which the part is used/all next level assemblies the component is used as well as all higher-level assemblies it is used in.
- *End item by contract:* Report of end item pegging.
- *Summarized parts list:* A report that lists all parts through all levels in the BOM but only lists each component part once with its associated total quantity.
- *Pending engineering changes:* A report that lists all engineering changes that will be implemented in future periods, based upon effectivity date. This display may be broken into date ranges, e.g. 1–5 days, 6–15 days.
- *Engineering change history:* A display of engineering changes going to history and out of active status.
- *Loose item exceptions:* A display that lists all parts that have no relationship to other parts. That is, parts that have no BOM and are not a component in a BOM (nowhere used). Used to help manage data integrity.
- *Dangling relationship exceptions:* A display that lists all parts that is not coded as being end items and is not a component in a BOM. Used to help manage data integrity.
- *User library report writer:* The ability to store a complement of reports, which are configured specially for the user.

9.6 Process management

A process is a sequence of activities (tasks) that are intended to achieve a particular result – typically, to create value for customers (refer Figure 9.8).

To ensure that processes are controlled adequately, some management tools are required. These tools are used to control processes in real time, collect data and generate information for decision-making. The systems are also used by operators to control the process and view the process and parameters graphically. The systems also give warning of out-of-control processes, production stoppages and equipment breakdowns.

Figure 9.8
Value creation in process management

9.6.1   Process monitoring, fault detection, location and diagnosis

Fault diagnosis has become an area of primary importance in modern process automation. It provides the prerequisites for fault tolerance, reliability or security, which constitute fundamental design features in complex engineering systems. The process under consideration is monitored and the data is passed to fault detection algorithms or procedures. The basic task of a fault detection scheme is to register an alarm when an abnormal condition develops in the monitored process. Once a fault is detected, procedures may also be subsequently used to identify or diagnose the cause of the abnormality.

Fault detection and diagnosis techniques are based upon the use of process models. Data from the plant is fed into these algorithms and the outputs are compared with the corresponding plant outputs. If there are discrepancies, it is an indication that at least one fault has occurred. The location of the fault is determined using a representative model (not necessarily the one used in fault detection). In some instances, the type of fault may indicate the location of the fault. Genetic algorithms and rule induction systems can be used to classify the fault.

9.6.2   Process supervision via artificial intelligence techniques

Human beings are able to make judgments in the face of subtle nuances and ambiguities. Number crunching data processing algorithms cannot match these knowledge-processing capabilities. Although the human decision system may not be precise, the results are often of sufficient accuracy for quick and effective problem solving.

It has been the goal of computer scientists for many decades to build systems that mimic the decision-making powers of human beings, i.e. artificial intelligence (AI) systems. The AI techniques are also model based. Some would regard neural network-based techniques to fall into the AI category.

Such real-time knowledge-based systems (RTKBS) have been used to tune controllers, supervise the performance of adaptive controllers, perform fault detection and diagnosis, perform alarm management and even provide direct online process control. In the process industries, they have been applied to reactors, separation processes and power-generation systems.

9.6.3   SPC and SQC

Statistical process control and statistical quality control methodology is one of the most important analytical developments available to manufacturing in this century. Statistical process control provides close-up online views of what is happening to a process at a specific moment.

Statistical quality control provides off-line tools to support analysis- and decision-making to help determine if a process is stable and predictable. When SPC and SQC tools work together, users see the current and long-term picture about processing performance (refer Figure 9.9). Quality check points measure the state of the process and quality control points measure the process result.

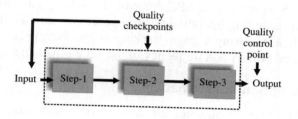

Figure 9.9
Quality measurement

The idea behind continuous improvement is to focus on designing, building and controlling a process that makes the product operate correctly the first time. A process can be improved by removing as much variation as possible to meet customer requirements and expectations by delivering products and services with minimal variation.

Manufacturers applying SPC and SQC techniques rely on a variety of methods, charts, and graphs to measure, record, and analyze processes to reduce variations. Processes are measured through intermittent or batch testing as well as with in-line analyzers. In-line analyzers measure product or WIP product quality in real time, the same as temperature and pressure sensors measure process quality. The SPC/SQC are used with in-line analyzer results to determine total batch/campaign quality, and to display quality data to plant operators and management in real time.

9.6.4 MES and process control

An MES provides all the necessary and correct information to operators or assemblers at the correct time. Quality, manufacturing and engineering data, stored in separate databases, is accessible across the network for combined reporting. An MES also allows operators to request resources from other department databases linked within the system.

A business can take advantage of MES on several fronts:

Tie-in to manufacturing

An MES in conjunction with an ERP system informs management about jobs in production, in queue or waiting to be scheduled. Jobs can be scheduled against inspectors and operators. This allows a production manager to view in real time any shop floor bottlenecks and make appropriate changes.

Tool and gage management

An MES can easily show gages distributed throughout the facility that are due for calibration. Because MES is an electronic system, related documentation such as gage prints, specifications and pictures are immediately available.

Cost control

An MES can track the quality costs associated with inspection, operation and data collection. It also provides a mechanism for inspectors, supervisors and quality engineers to track their time. Using Pareto charts, department managers can easily compare predicted costs vs actual costs.

SPC online

An MES system with SPC imbedded tracks production time and quantities, and can electronically flag operators when a process is due for inspection. In addition, when tied in with a company's gage tracking or management system, gage calibration can be scheduled based upon frequency of use rather than an arbitrarily specified time frame.

Tracking non-conformance

An MES can display an electronic non-conformance document as soon as a non-conforming piece is identified, which forces the operator to identify the problem. The system can then automatically e-mail a notice that a disposition decision needs to be made regarding the non-conformance.

Real-time monitoring

An MES can send an alarm throughout the network about a problem on the shop floor. The notice will even include specific information such as process, part, operation, job number, serial number and operator. With this information, the quality control can identify the problem right after it has occurred, rather than hearing it later from manufacturing.

9.6.5   Equipment breakdown and process

Equipment reliability directly impacts production capacity for the following reasons:

Equipment outages and unscheduled downtime result in the loss of an opportunity to sell a product that could have been produced during that time period. This results from lost run time and also from the reduction in quality as a process is brought back up to full product specifications. Such quality losses can add up significantly over a period of just 1 year.

If each time equipment stop, even for just a few minutes, it may take lot of time to return it to full production within quality specifications. Equipment stoppages, either because of direct failure or the failure of facility support systems, impact on bottom-line profitability by creating product waste or scrap. The product cannot be sold, and, in addition, overtime is used to make up for the lag in the production schedule.

In most companies, waste is measured, but is not necessarily viewed as a direct loss of profits. The losses due to excessive waste product, overtime, power consumption and reduced equipment life are sometimes not measured. Seventy percent of equipment failures are caused due to lack of skilled work force, operator errors, reactive culture and not performing planned maintenance properly.

Replacement of equipment, item or machine is considered whenever the machine or item deteriorates in its function or fails to work. When the efficiency of the equipment and machines declines over a period of a time, the following things happen to the process:

- Forced idle time is caused with loss of production
- Scrap in the process increase
- Rate of rejections increase

- The frequency of repairs increase
- Causes accidents, or rate of accidents increase.

The operating costs of equipment increase as the life of the equipment reduces or as the age increases. It is necessary to predict the time when the cost of maintenance and operation of the old equipment is higher than the cost of replacing old equipment with new one so that the old equipment can be replaced with new one.

Some equipment do not deteriorate markedly with age yet suddenly fail after some time. The lifespan of such items is not constant but will follow a certain frequency distribution. The probability of mortality or failure increases as the age of the machine or item increases. Replacement of usable items, which is still functioning, will be justified if the cost of replacement is higher after failure than the cost of replacement before failure.

The cost of replacing failing equipment include:

- Cost of the item itself
- Cost of labor
- Cost of loss of production that can occur due to delay
- Cost of damage to the material
- Cost of damage to the equipment
- Cost of a possible accident.

The right instrumentation on critical equipment to measure equipment condition can provide data that enables management to identify potential equipment deviation or imminent equipment breakdown. This will assist to prevent quality deviations as result of failing equipment and equipment performing suboptimally. The data can also be used to calculate equipment running hours, number of starts and stops and other life cycle information useful for maintenance packages.

9.6.6   Alarm and event management

To prevent production upsets, unplanned outages, unnecessary equipment problems and lost production, manufacturers need to implement a cost-effective alarm and event management solution.

Alarm and event management solutions collect and store alarms and events from any control system in a long-term archive. Standardized alarm and event analysis and monitoring functions instantly reduce problematic alarms and alert operators instantly based on changes to key control system parameters that may damage the process or equipment.

Alarm and event management allows the operator to instantly identify and address problematic alarms and continuously enhance operator knowledge and increase the visibility of equipment health. Operators often set process values that can affect the life cycle of equipment, and when this happens, an alarm event is generated – to inform the operator and also for history purposes. These events can then be used to identify knowledge gaps and training requirements

Alarm and event management can vocalize alarms over speakers, intercom systems, radios, multimedia speakers, pagers, telephones, PCS phones and cellular phones. Operators can access and acknowledge alarm information and data from any telephone or cellular telephone – saving valuable time and money. It also turns a PC into a communications command center that monitors an entire network 24/7. The software normally ties into plant historians, and all analysis can be performed with a minimum of user interaction and customization.

Alarm conditions are normally set on SCADA and DCS systems, but these are normally parameter-level and individual event alarms. These normally result in a flood of alarms (specifically during breakdown conditions) to be acknowledged by operators, and some are deactivated or their deviation levels adjusted after a time.

Good alarm and event management software are able to identify a sequence of events before generating an alarm, and automatically deactivate identified alarms under breakdown conditions.

For instance, a pump stoppage alarm, a low-flow alarm and a temperature-high alarm individually will most likely not catch the attention of an operator, especially when mixed with other flow, temperature and equipment condition alarms. Pumps start and stop, temperatures fluctuate and flow-meters may stick or deviate, so individually these alarms may not mean a real emergency situation.

But given the situation that the alarms all relate to a smelting furnace, and a high-shell temperature indicates potential burn-through, these alarms become significant. In the event that the coolant-water pump stops, the flow in the coolant line is low, and the temperature of the furnace shell is high, an alarm should be generated, as this is an emergency situation requiring action.

It should thus be possible to link alarm events and conditions, and to establish rules to eliminate nuisance alarms.

Benefits

Alarm and event management product allows the operators to instantly identify and address problematic alarms and continuously enhance operator knowledge and increase the visibility of equipment health.

- Minimize impact of alarm showers
- Improve plant safety
- Eliminate nuisance alarms
- Integrate alarm and process data
- Archive alarms and events
- Integrated with data analysis tools.

9.6.7   Downtime tracking and production monitoring system

This enables manufacturers to continuously monitor productivity and efficiency from any desktop browser and easily interfaces with applications such as ERP or MES, or human machine interfaces. It automatically collects and analyzes the detailed real-time and historical information that is required to continuously optimize all sources of manufacturing losses. Leveraging the existing plant automation system, it tracks downtime and monitors production with accurate, comprehensive, real-time data, such as:

- Status of the running plant and process
- Production details – type of product, quantity, further production plan, etc.
- Equipment status – equipment required for the present and future production, last maintenance done and due date, equipment replacement status, spare parts position, etc.
- Deviation measurement, fault analysis and corrective measures.

Features and benefits

- Capture multiple, and even simultaneous, causes of downtime at a system and area level.
- Tap into plant's hidden capacity by reducing downtime, bottlenecks, changeover time, maintenance problems and product flow problems while pushing equipment to run at a higher rate.

- Reduce the cost of collecting downtime data by using real-time data instead of clipboards.
- Optimize maintenance by monitoring reliability instead of operating hours.
- Continuously monitor production performance
- Understand the costs associated with out-of-spec consumables or unreliable equipment.

9.6.8 Data historian

Most applications require recipes, data logging, and other means of reading and writing from and to databases. Data historian systems can log incredible amounts of data to disk for later review. This assists in problem solving as well as providing information to improve the process. The critical capabilities of historians are the management, storage and movement of the data.

Whilst most SCADA systems have the ability to store historical data, historians have been able to optimize data gathering and compression techniques.

9.6.9 SCADA and data historian

The availability of SCADA data to the corporate IT system is frequently done via a data historian.

The position of Data historian systems within system hierarchies differ from company to company, depending on the comparable power between the control and instrumentation department and the IT department. The two possible levels could be either on the control or on the MES level.

A data historian is a suite of programs designed to efficiently store large volumes of time series-based data, and which provide easy end user access to this data. The SCADA data is normally automatically fed into the data historian, and stored for many years. The feed can be in real time or in batches. The data historian will typically allow data to be sent onto further applications, sometimes on the basis of 'triggers'. It will allow data to be retrieved on an ad hoc basis to the desktop of the user where automated reports can be generated based on the data (refer Figure 9.10).

There are savings in the cost of data collection for the reports, savings in the cost of producing the reports, but these savings are negligible compared to the value of efficiently running the plant on the basis of this information.

Simple dumping of SCADA data into a data historian is not advisable. Such data is unstructured and difficult to interpret without detailed knowledge of the individual sites. It will have a variety of units, levels of accuracy, meanings and descriptions. An understanding of a complex tag naming system is normally necessary to find any data from the data historian.

A data model is required to ensure that everyone has a common understanding of the data. Device object models can be used, but still need some of the concepts of data modeling – good definition of the data to be collected, derived data to be calculated, consistency in units of measure, consistent descriptions and tag naming, definition of required accuracy, automated quality tagging and so on.

If this is not done the data historian will be a mass of useless data without a correct retrieving facility. The aim should be to collect only the useful data, of sufficient quality, so that it can be easily retrieved and used for simulation or analysis. To do this, someone is needed who takes responsibility for the data, and edits/cleans and ensures the integrity of the data when necessary.

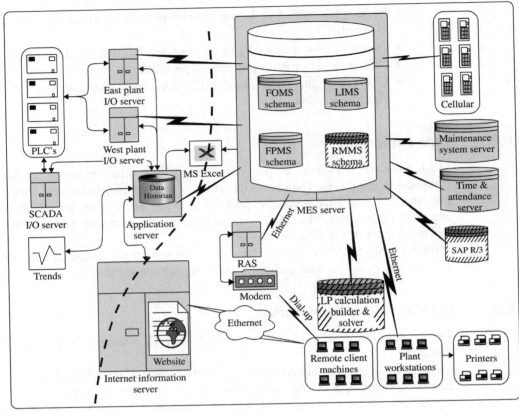

Figure 9.10
Data historian on plant level

9.7   Production systems collaboration

Production systems focus on the value-adding processes, helping to reduce manufacturing cycle time, improve product quality, reduce WIP, reduce or eliminate paperwork between shifts, reduce lead-time and empowering plant operations staff.

Production systems deliver information enabling the optimization of production activities from order launch to finished goods. Using current and accurate data, production systems guides, initiates, responds to and reports on plant activities as they occur. The resulting rapid response to changing conditions, coupled with a focus on reducing non-value-added activities, drives effective plant operations and processes.

By design, production systems manage customer order and product status information as well as information on the use of material and process efficiency. Data points from SCADA systems feed the floor scheduling and resource allocation modules in real time for detailed tracking of materials, settings, and events to and from the process management system. It is designed to pass job setup information to and capture data-point information from those loops. Production systems perform job control and data collection by directly feeding data to and collecting data from SCADA systems, distributed control systems or PLCs.

Process optimization techniques are not restricted to the design of predictive constrained controllers. Process optimization is a task in its own right. Unlike local controllers, which seek to maintain unit-operating conditions at desired levels, the plant optimizer utilizes a model of the plant to adjust operating conditions of the process so as to minimize raw material usage and maximize profits. The outputs of the optimizer therefore set the targets for the local controllers, taking into consideration the operational limits of the plant. This effectively bridges the gap between the plant's true business objectives and its actual operations.

Production systems typically operate together as depicted in the figure below.

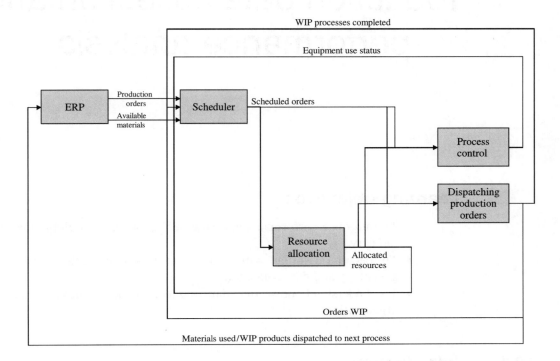

Incomplete production orders and available material stock information is passed down to the scheduler. The scheduler checks orders already in progress, free materials (available and not already allocated) and equipment available to execute the schedule. The scheduler then generates a schedule based on the scheduling rules and business priorities. The schedule is downloaded to resource allocation that allocates materials to specific production orders.

Once equipment is free to execute a new order, the material already allocated is dispatched to the plant or equipment and goes into WIP status. The material quantities used data is sent to the ERP system to adjust available material quantities.

The equipment goes into an 'In use' status and the process parameter data is collected and associated with a specific production order. Once a WIP process is complete, the dispatch production orders system sends the complete WIP product information to the ERP increasing WIP stock. It also sends WIP process complete data to signal process completion to the scheduler for schedule update.

10

Production data collection and performance analysis

Learning objectives

- To introduce performance measurement concepts and global measures used in industry.
- To become familiar with production functions such as product tracking, genealogy and data collection.
- To understand how the data collection and analysis can benefit the organization.

10.1 Introduction

Market factors are forcing manufacturers to rethink their business and operational strategies on a continuous basis in order to maintain or grow their position in industry. Without adequate information about the strategic and operational performance of the company, it becomes increasingly more difficult for enterprise executives to decide on the best strategic direction of their operations.

For an enterprise to succeed in the current global business environment, it not only needs to execute processes effectively and efficiently, but also has to collect performance data, manipulate it into decision-making information and analyze the information to identify areas of potential improvement.

The data collected depends on the specific processes employed by the company to manufacture/produce its products, but the specific measurements depend very much on the strategic positioning of the company (e.g. lowest cost, highest quality, process flexibility, new technology, research and development, product mix, throughput maximization, etc.). The process data collected thus have to be in a format that is easy to access and manipulate in order to provide meaningful information that can be used as a strategic measure.

Apart from data on the efficiency of specific processes and equipment such as throughput time, yield and capacity utilization, companies also measure various product attributes such as quality and quantity. Apart from these, companies must also track products and WIP materials through the manufacturing process in order to identify specific processes where improvements can be made. This includes the identification of bottlenecks, the WIP inventory and activity costs. If done accurately, the collected information will enable the development of a product genealogy for each completed product sold.

The data collected needs to be translated into a format where it can give an indication of the performance of a company toward its strategic goals. Various tools and measurement concepts are available that enables management to view the same data from different perspectives, as different departments focus on different aspects of the business. Some tools and concepts to assist in the collection, structuring, analysis and representation of data and related decision-making information are discussed in the rest of this chapter.

10.2   Changing face of manufacturing strategies

Historical key performance indicators (KPIs) focused on cost, production throughput, and machine uptime, but failed to enhance manufacturing performance related to customer satisfaction, service levels, product returns and product market penetration.

This situation was historically not a problem, but as market factors and technology changed, this is becoming more of a problem.

10.2.1   Shift from manufacturer to customers and consumers

Historically, the market power base rested with the producer and owner of manufacturing capacity. As few local suppliers of manufactured goods were available, the manufacturers could more or less determine the price of their good and the structuring of their deals with retailers and distributors. The KPIs were centered on mass production practices to leverage-installed capacity to the maximum level, as the factory with the lowest production cost had an advantage over its competitors.

Over a time the power base moved from the value chain to the supply chain. With the breakdown of trade barriers and the globalization of the marketplace, customers had a wider choice of suppliers, and the retailers and distributors owned the relationship with the customer. They compelled manufacturers to make lower-cost deals, and in response, manufacturers concentrated on mass customization, supply-chain relationship management, and supply-chain process efficiency to ensure market share and profitability.

Currently, customer service is the highest priority KPI for most companies, and with the evolution of e-business strategies, all partners in the supply chain are linked and need to cooperate to ensure the continued existence of the chain. The need for better coordination and synchronization of supply and demand data between trading partners is becoming more apparent in many industries and supply chains.

Higher transaction speeds with increased access and sharing of information is driving the power base to the customer who is going to expect even more efficient support and responsiveness, and a wider range of options from manufacturers.

This change in power base has altered the business focus toward supply chain and customer service effectiveness, away from internal manufacturing efficiency. The shifting of the market power base, the changing focus of KPIs with their related business priorities and the various IT applications that have supported information requirements during the shift is depicted in Figure 10.1.

Easy access to information gives customers more choice and a lower cost of switching between suppliers. Accurate information is required to support strategies that address customer needs and expectations, and provide the ability to drive market segmentation continuously. Manufacturers are grappling with the cultural and technological transition from push-driven efficiency to pull-driven customer service.

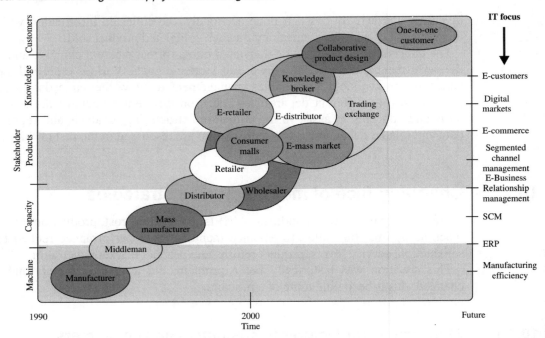

Figure 10.1
The market power base moves to customer and consumer

10.2.2   Move from push-to-pull market requirements

Figure 10.2 represents the two main phases of the change in manufacturing and business practices developed as result of the Internet and the improved access to information. Many manufacturers are in the process of planning this transition as part of e-business strategies. This will ease the challenge of converging data requirements of supply chains to calculate and view KPIs and other common data, rather than physically moving the data between businesses.

New Web-based technology in emerging e-business applications allows cross-functional processes and access to information that is oblivious to traditional organizational, geographic, functional and departmental boundaries, and allows companies to separate the flow of physical goods from the flow of information in the extended supply chain.

The improved availability of information and IT-based processes helps leading manufacturers execute global consolidation strategies, and manufacturers are implementing KPIs based on optimization of global inventory, orders and materials.

Now manufacturers can capitalize on the tremendous opportunity to improve the global supply-chain performance by gaining visibility of raw materials, product performance, inventory and orders using a standard and integrated global foundation of ERP and supply-chain planning systems. The information links between manufacturing, customer service and customer-facing processes support cross-functional KPIs and decision support. These links focus on several factors:

- Coordination of plant and supply chain processes
- Synchronization of processes between the supply chain and manufacturing
- Optimization of the manufacturing plant as part of the supply chain.

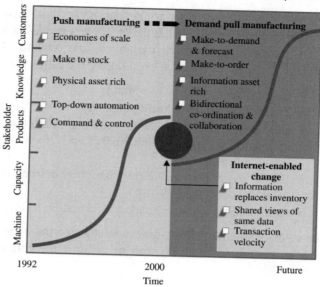

Figure 10.2
The changing focus of manufacturing strategies

10.2.3 Make-to-stock manufacturing strategy

In make-to-stock manufacturing strategies, manufacturers have three primary problems:

1. Perishable goods manufactured in excess of market demand have to be discounted or disposed of (which impact on profitability). Local optimum manufacturing KPIs are not adequate to identify and address these problems.
2. New product development and introductions (NPDs and NPIs) to meet emerging customer needs were slow, inefficient and complicated. Traditional KPIs failed to measure performance for these processes.
3. Consolidating retailers and distributors represent a growing power base that enforce a high demand and restriction on manufacturers i.e. shelf-space restrictions, retailers' own brands, reduced number of products and brands, and trade and promotion discounts limited overall growth and profitability of manufacturers. Traditional KPIs are unable to drive improved responsiveness in product and customer relationship management functions.

Manufacturing strategies in most process industries are moving away from the inefficient make-to-stock processes. The trend is toward more responsive make-to-order and make-to-demand, which allows consolidation of assets and increased return on investment. Businesses seek globalization, increased market share and profitability, reduced costs, and improved customer and market responsiveness.

10.2.4 Move from make-to-stock to make-to-demand/order strategies

As manufacturers are becoming more responsive to market change, it is natural for them to make an effort to increase the ratio between products made to order or to demand, and products made to stock.

Manufacturers with more accurate and predictable forecasts are developing ways to coordinate, synchronize and optimize manufacturing output across all production operations. This allows them to define KPIs that drive make-to-demand practices. Manufacturers want to make what the customers need, but also want to be flexible to deal with any changes in market dynamics. To ensure their competitiveness in the marketplace, manufacturers have been taking the following measures:

- Consolidate products and brands
- Consolidate manufacturing capacity
- Deal more effectively with growing retailer and distributor power bases by exploring alternative channels and focusing on customer relationship management
- Being more responsive to changing market dynamics by introducing new products continuously
- Collect more accurate market information to increase forecast accuracy.

Make-to-demand and make-to-order has changed the focus of KPIs across the business.

The new manufacturing KPIs aim to achieve the following objectives as opposed to traditional KPIs:

- Ensure maximum customer profitability through segmented channel management vs lowest cost of production
- Meet customer needs and market demand vs highest throughput quantity
- Satisfying customer expectations vs highest product quality
- Segmented channel management effectiveness vs highest capacity utilization.

Where traditional bottom-up KPIs aim to improve local optima rather than overall business performance and customer service, top-down KPIs focus on coordinating business processes across the extended supply chain and aligns manufacturing goals with business priorities e.g. market share, profitability and customer service.

10.3   Performance analysis strategies

The Baldridge Criteria (1997) booklet reiterates this concept of fact-based management: 'Modern businesses depend upon measurement and analysis of performance. Measurements must derive from the company's strategy and provide critical data and information about key processes, outputs and results. Data and information needed for performance measurement and improvement are of many types, including: customer, product and service performance, operations, market, competitive comparisons, supplier, employee-related, and cost and financial. Analysis entails using data to determine trends, projections, and cause and effect that might not be evident without analysis. Data and analysis support a variety of company purposes, such as planning, reviewing company performance, improving operations and comparing company performance with competitors' or with 'best practices' benchmarks'.

It is clear that organizations that manage performance through strategy-focused performance measurement do better than those that do not, as performance improvement frequently occurs if business units and staff are well informed about their current and likely future levels of performance. Business performance management ensures a management style that plans and acts to achieve strategic and operational objectives by measuring and monitoring outcomes and drivers.

But, not all measurement is good. Many organizations are in chaos because of a flood of data that is irrelevant, too detailed, poorly integrated, difficult to access and of little value in making decisions. Some measures bear little relationship to what the organization is trying to achieve – they are not relevant to objectives. Other measures are misleading because their meanings are poorly understood, unclear or ambiguous. Irrelevant or misleading measures lead to poor or even disastrous decisions.

Knowing the performance gap is not enough to make a decision and take action. To take effective action, the reason for the gap needs to be known. Measuring and understanding the reasons for gaps will lead manufacturers to make the right decisions. Tools are thus required that will enable not only measurement, but also analysis of information and problem-identification.

A major consideration in performance improvement involves the creation and use of performance measures or indicators. Performance measures or indicators are measurable characteristics of products, services, processes, and operations the company uses to track and improve performance. The measures or indicators should be selected to best represent the factors that lead to improved customer, operational and financial performance. A comprehensive set of measures or indicators tied to the customer and/or company performance requirements represent a clear basis for aligning all activities with the company's goals. Through the analysis of data from the tracking processes, the measures or indicators themselves may be evaluated and changed to better support such goals.

10.3.1   Key performance indicators (KPIs)

Companies trying to reduce costs, increase return on assets and profitability, and improve market responsiveness are consolidating capacity and moving toward make-to-order manufacturing but they often fail to integrate market demand, capacity, and order status information to support enterprise-wide KPIs. The KPIs combining manufacturing, supply chain, and customer service processes are not readily available in a packaged solution e.g. ERP, supply-chain management, or fragmented plant IT systems. Some IT groups have built complex, inflexible data warehouses that are unaligned with integrated business and manufacturing strategies.

Manufacturers try to make more accurate product, pricing, promotion, supply-chain decisions by using segmented channel management KPIs such as profitability by product, brand, pack, manufacturing site and customer segment, but there is a need for new responsive KPIs to link manufacturing and supply-chain processes that encourage better customer service.

Companies are linking marketing, order execution, and fulfillment process information to achieve better responsiveness and reduce inventory levels. The KPIs in the plant support new business goals such as customer responsiveness and service and enterprise-wide quality management. Fragmented IT architectures, complex point-to-point integration, and isolated manufacturing information make it difficult to access the performance management information that enables capability-to-promise (CTP), available-to-promise (ATP) and profitable-to-promise (PTP) KPI processes.

10.3.2   Customer first

Key performance indicators should facilitate enterprise-wide coordination, synchronization and process performance. The Internet bridges traditional functional, departmental, and geographic boundaries and make it possible to exchange views of the same data using browser-based technology. Web-based business processes reduce the need to physically integrate systems to move data and view KPI information.

The Internet has brought about a fundamental shift in information exchange, giving customers more influence over supply-chain and manufacturing decisions. Collaboration is a most important aspect in emerging e-business application and trading exchange developments. To support these new trends, manufacturers use KPIs to link customer service, manufacturing and business performance, and drive new behaviors and capabilities.

Manufacturing KPIs no longer focus on local production optimization only, instead new-paradigm KPIs such as order fill rate, production and order lead-time, planning cycle time, number of customer returns, time-to-market for new products, ATP, CTP and PTP are used increasingly. It focuses on overall coordination, including synchronization and optimization of the extended enterprise supply chain for the benefit of the customer.

10.3.3   Performance management and change

Key performance indicators are tools for change, performance management and sustainable business improvement. The KPIs measure business performance and drive change as well as monitor and sustain ongoing business performance. For instance, a manufacturer with a goal of reducing waste in manufacturing by 10% may use waste and yield KPIs to track first-batch yield and variability. These KPIs can assist companies to identify the root causes of problems that can be prioritized and corrected.

The KPIs can be discrete measurements such as measured defects or quality results, but are often related to business process performance such as time-to-market for a new product or time-to-first-batch for control. The KPIs can be manual process-based or supported by IT. Many companies use multi-level KPIs, which consist of real-time shop floor systems, transactional business systems and systems that access historical data.

A KPI measurement in production may trigger a physical action, such as starting a pump, or it may help identify a root cause and change a business process; so accurate analysis and mapping is vital. To get the full value of the KPI as a tool to drive and sustain change, companies need to define the KPIs from a top-down perspective. Change and performance management leveraged from KPI initiatives depends on the links between KPIs, personal goals, business priorities and performance management processes.

Traditional manufacturing KPIs supported the drive for efficiency from assets, sometimes at the expense of customer service. The focus of traditional KPIs was on production cost and efficiency and local optimization around functional and departmental goals. No cross-functional process KPIs were implemented that could drive practices that ensure that plants make predictable, conforming and high-quality products to meet market needs.

Figure 10.3 presents the process of deriving KPIs from one integrated business strategy, which ensures that KPIs are derived from the strategy and focus on driving practices that enable manufacturers to meet the changing demands of the market.

10.3.4   Business drivers and KPIs

Only a few manufacturers have actually devised a business strategy that defines specific KPIs and prioritizes initiatives that have the highest impact on business performance.

Manufacturing KPIs align with the business and supply-chain priorities and impose practices and behaviors in manufacturing that promote business change, responsiveness and agility. Focus on the strategic manufacturing priorities and links to the supply chain allows leading global manufacturers to consolidate product brands, products and manufacturing capacity.

Responsive, customer-oriented manufacturing is a key part of an e-business foundation. Manufacturing KPIs should encourage practices and behaviors that build agility and responsiveness in customer support as required by the changing marketplace.

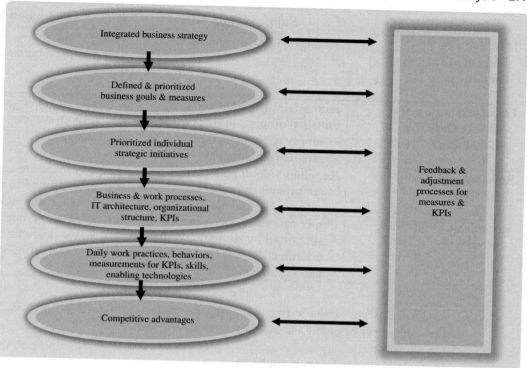

Figure 10.3
Business strategies drive top-down KPIs

When KPIs in a company do not relate to a specific strategy or objective, they are meaningless and cause unnecessary work and confusion. The KPIs can be used to drive company direction, and as such should relate directly to company strategy, goals and business drivers. The table below shows the links between business drivers and KPIs in manufacturing.

Business Drivers	Supply-Chain KPIs	Production KPIs
Shareholder value	Segmented channel management, market penetration	Responsive make-to-order and make-to-demand manufacturing
Highest quality (Brand image)	Meet customer order and product specification throughout fulfillment process	Reduce production variability
Lowest cost	Optimized supply-chain processes, consolidated and balanced capacity, demand	Eliminate production waste, inventory reduction
Highest margins	Optimized and reduced supply-chain costs, profitable to promise	Optimized production costs, Optimum yield and transformation
Increased agility (product availability)	Responsive delivery to plan, available-to-promise and capable-to-promise capabilities	Produce-to-schedule, capable-to-promise, predictable quality

Business Drivers	Supply-Chain KPIs	Production KPIs
Effective new product development and introduction	Minimum time to peak sales	Minimum time to first packaged batch in control
Expanded customer relationships, like e-business	Increased fulfillment velocity Fulfilled orders and meet extended supply chain demand and satisfied customers	Produce to specification responsiveness, agility to change
Reduced cost of compliance	Regulatory compliance: full and continuous traceability, auditability and records	Conformance to regulated procedures like cGMP, HACCP and SOP
Sustainable competitive advantage	Organizational effectiveness supported by technology-enabled coordination, synchronization and optimization	Coordination, optimization and synchronization of manufacturing with supply chain

The following strategic manufacturing priorities identify KPIs and measurements that drive customer-oriented manufacturing improvements:

Reduced production variability

Reducing variability in production ensures predictable and consistent manufacturing output to the supply chain. Plant variability KPIs support responsive supply-chain performance.

Conformance to global product specifications

Products and supply chains can be globally coordinated and synchronized with customer needs on a right-first-time basis. Consistent quality between different facilities makes it possible to satisfy orders from multiple manufacturing facilities within the same company. Consistent quality and manufacturing methods also make it possible to compare actual activity costs between facilities and enable the enterprise to source products from the most cost-effective manufacturing facility where stock is available. Various manufacturing options and flexible operations must also be available to support global customers and requirements.

Eliminated waste

Eliminating material and product waste reduces production cost, maximizes yield and optimizes capacity utilization. Even when materials can be reused and defective products reworked, they still introduce inefficiency into the process, as other resources such as capacity and manpower are consumed in the manufacture of the defective product.

Delivery to schedule

Consistently not meeting ATP delivery schedules can cause customers to lose trust in an enterprise, leading to canceled orders and reduced revenue. The word-of-mouth effects are more difficult to quantify, but are potentially even more harmful. Measuring and managing on-time delivery ensures that manufacturing output is delivered on time to support the supply chain and meet customer delivery requirements.

Compliance management

A systemic approach to compliance management ensures consistent regulatory compliance across the organization. Minimized cost of compliance is an emerging KPI in regulated industries.

Production agility

Minimized changeover and lead-time in production supports production agility. Agility ensures the best leverage of assets within the scope of market needs. This results in a high return on net capital assets rather than maximum utilization of capacity assets. Maximum utilization may not be profitable as it often leads to goods of a certain type in excess of market need and other required goods that are out of stock.

10.3.5 Customer-focused vs traditional KPIs

The lack of balanced integration between compliance, production and supply chain KPIs has allowed enterprise inefficiency to escape attention. The lack of process-based KPIs to track overall manufacturing performance within the confines of strictly enforced procedures causes inefficiency and invisibility to overall order fulfillment performance. In many regulated organizations the overall cost of doing business, including items such as compliance costs, is neglected as a KPI. Such costs are often simply disregarded as possible sources of improvement. The following table shows how enterprise efficiency can be measured by combining customer-focused and traditional KPIs.

Supply-Chain Process Measure	Metric	Manufacturing KPI
Customer service goal: Planning cycle time. Adherence to plans.	Order fill rate Line item fill rate Dollar fill rate Cycle time Variability On-time delivery Back-order duration Perfect order fill rate Customer satisfaction survey results	Production costs per unit Product quality Number of defects Set up and changeover times Quantity produced
Asset management goals: Inventory turns	Inventory days of supply Inventory accuracy Inventory turns Cash-to-cash cycle time	Manufacturing efficiency Factory efficiency Machine uptime Waste and scrap Capacity utilization
Forecast accuracy goals	Orders vs sales forecasts Shipments vs forecasts	Not applicable
Costs goal: Componentized and traceable	Various costs Unit production costs	Cost per unit manufactured
Value-added goals: Economic value-added (EVA)	EVA	Trended manufacturing costs

Supply-Chain Process Measure	Metric	Manufacturing KPI
Manufacturing resource planning goals: Economy of scale, finished goods, decreased inventory, predictable conforming product	Percentage of complete shipments on time Actual sales vs plan Actual production vs plan Warehouse receipts vs orders Percentage of items completed Manufacturing cycle time Conformance to product specification Percentage of schedule changes Schedule adherence Actual production to plan Conforming product released for shipment Balanced inventory and production costs	Completed orders Scheduled batches completed on time, to specification

10.3.6   REPAC KPIs

The REPAC model can be used as a guide to identify specific measures in the manufacturing environment. The table below gives a few examples of how REPAC can be used to derive KPIs specific to manufacturing.

REPAC Process	Focus	KPI Measure
READY	Product life cycle management, recipe management, and specification management	Product recipes and manufacturing standards used to drive practices and behaviors at all manufacturing levels Time-to-market for new products decreased
EXECUTE	Recipe and word order execution	Business and customer requirements accurately translated into manufacturing procedures. Minimum disruption from changed requirements
PROCESS	Production machinery control	Machine uptime Performance to specification Waste and variability in control
ANALYZE	Performance analyses	Non-conformance detected in real time Problems solved before product reaches next process stage Real-time performance visualization used to continuously improve process
COORDINATE	Coordination of plant resources and logistics	Aligned schedule execution with supply chain and customer service plan

10.4 Performance analysis systems

Most manufacturers have best of breed plant IT architectures and depend on a collection of point solutions to run plant operations. Contrary to conventional wisdom, the key for these manufacturers is to avoid building tightly integrated architectures from heterogeneous islands of technology and information. Manufacturers with fragmented information architectures must rather build loosely coupled, modular software components linked by event-based messaging processes.

Linking processes such as planning and quality management improves overall manufacturing responsiveness, but bringing them together requires a close evaluation of the IT architecture. E-businesses must offer collaborative, unconstrained and responsive information flows for organizational effectiveness. User access to enterprise-wide KPI information should be seamless. Loosely coupled architectures with evolving portal applications can accomplish these goals.

Portal applications use combinations of Java, HTML and XML to access and display KPI information. Cross-functional KPI information and monitoring processes are difficult to build in a heterogeneous IT architecture and require detailed analysis by business and domain functional specialists. Building flexible KPI monitoring applications from tightly integrated, closed, best-of-breed application architectures are a difficult task.

Manufacturers underestimate the extent of analysis needed to build a solution to sustain KPI monitoring in the long term. Many leading manufacturers have built software tool-based modeling processes to map business processes and identify KPIs that offer the biggest return for the money spent. The primary goal is to build an information management architecture that fuses business and IT changes, and to identify strategic KPIs that need to be monitored to improve performance.

10.4.1 KPI system Architectures

Manufacturers ready to leverage IT strategies based on evolving integrated product architectures from one vendor, should form a partnership with the vendor to build a strategic KPI development relationship. These vendors normally specialize in certain vertical segments and have the domain expertise to embed vertical-specific KPI capability in their products.

Vendors should provide KPIs that promote demand-driven manufacturing processes. In response to the rigid architectures of many plant solutions, manufacturers have built home-grown data repositories, data marts and warehouses to support KPI monitoring. There are several problems with these approaches:

- Solutions are proprietary and long-term costs of change and IT ownership are high.
- Application and integration complexity and limited user skills inhibit effective implementation of KPI solutions. Homegrown applications seldom have the 24-h support system that leading packaged application vendors offer.
- Applications and KPIs in an integrated best-of-breed architecture must be adapted regularly to synchronize with new software revisions.
- Integration is used to move data between applications. A process-based metadata layer is seldom built as an overarching process template against which process-based KPIs can be defined and monitored.
- Home-grown data repositories are difficult to adapt to new business processes, KPI requirements and measurements.

It is difficult to build seamlessly integrated top-down KPI monitoring capabilities in a layered, tightly integrated IT application architecture of ERP, supply chain and manufacturing systems. These tightly integrated architectures make it complicated to implement process- and event-driven notification messages to support cross-functional process-based KPIs.

Decoupled process architectures should be built using enterprise application integration (EAI) and business community integration (BCI) software with process and workflow capabilities. Manufacturers use EAI software products to link information and processes so that information and data can be transferred to monitor defined KPIs. The KPIs measuring time-to-market for new products require linking transactions and processes across research and development, manufacturing, sales and supply-chain applications.

Application and information architectures must be analyzed, planned and built to support KPIs at various levels of decision-making and performance management. In designing integrated enterprise architectures, IT departments must ensure that KPI-supporting business goals are fed accurate information and data.

Figure 10.4 shows the fragmented architecture of point solution applications in a batch-based manufacturing plant. These individual applications are generally built around departmental and functional goals and can often multiply beyond control. Islands of data with different contexts make the monitoring of cross-functional KPIs difficult and the assessment of quality inaccurate.

To get meaningful and valuable views of the supply chain, performance data such as status of an order, ATP and CTP, must be available for KPI calculation. This requires software analysis and process modeling tools, which can also be used to extract and analyze data and KPIs from diverse sources. However, they must be integrated into the architecture to access data.

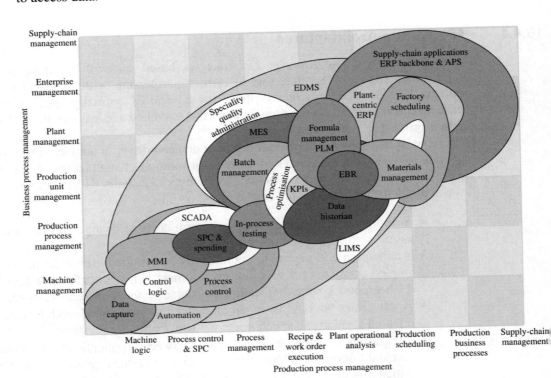

Figure 10.4
Fragmented manufacturing plant IT architecture

10.4.2   Strategic business-driven architecture concept

Companies should designate strategic business goals, combined with business process and event analysis, while identifying top-down, business-driven KPIs. A few leading manufacturers have succeeded in implementing integrated architectures with KPI capabilities as a result of formalizing enterprise architecture planning and program office concepts.

Only a few leading global manufacturers have successfully used enterprise architecture planning processes to identify and drive performance measures for changing manufacturing strategies. The success factor lies in the executive ownership of the enterprise architecture planning process and the top-down focus of KPIs. The strategies should be fully integrated into one common business strategy that provides a common framework for the identification of strategically focused top-down KPIs. This is coupled with behavioral change-management at all levels of the organization.

Figure 10.5 shows an enterprise architecture-planning model synthesized from leading manufacturers. The architecture-planning process offers the following benefits:

- Combination of business and business process change into one program to ensure alignment, common goals and focus. This combination relates KPIs to business strategies, and IT capabilities to business needs. An integrated program allows simultaneous enterprise-wide change management and KPI coordination.
- Alignment of KPIs with organizational developments. The KPIs can be used to establish platforms of capability and cross-functional problem solving in the enterprise.
- An opportunity to claim executive ownership of the one program that includes business process and IT change. This is a key point for the integrated business strategy.
- A holistic, business-driven analysis of KPIs, source data, events and triggers. This includes measurement and data sources for KPIs.

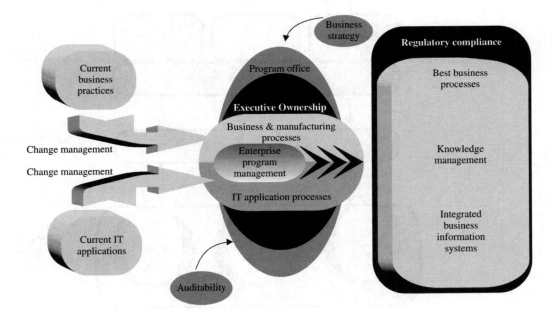

Figure 10.5
Enterprise architecture planning process

- A program office to coordinate and align business and IT priorities. The program office also supports process ownership and project management geared toward implementing KPI initiatives.
- A foundation for the integrated business and IT strategy and prioritization of initiatives.

10.4.3   Integrated architecture

Figure 10.6 shows an integrated architecture of systems from consolidated plant level to e-business applications. The architecture comprises an integrated ERP, supply chain, CRM and product life cycle management backbone, and an integrated trading exchange or portal application connected to suppliers, manufacturers and customers via the Web.

The following aspects of this architecture are required to support customer-focused, e-business collaboration processes at the supply chain level in manufacturing execution activities:

- Transparent information access across layers of the architecture must be seamless to support cross-functional and holistic KPIs. Such information includes new product development and introduction, delivery of conforming product to plan, and inventory and order status KPIs that align manufacturing with customer-facing processes.

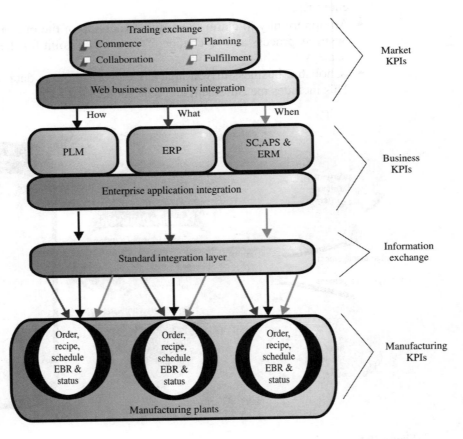

Figure 10.6
Integrated manufacturing e-business architecture

- Application layers are connected by three core, bidirectional information exchange process threads. These are a product life cycle management (PLM) information thread covering product specifications, an ERP information thread dealing with the business aspects of the order and an APS information thread. Information threads support execution instructions and feedback on actual indicators such as usage and units produced. These three threads are new priority areas for KPIs.
- KPIs supporting strategic goals should be translated from top-down market drivers into manufacturing performance. Goals and focus areas should align with common strategic business priorities, especially in the case where multiple manufacturing sites work with one supply chain. The KPIs are useful for establishing coordination and synchronization capabilities of many plants, yet they should keep plant differences and priorities in context.

Supply chain KPIs should focus on manufacturing performance across the integrated architecture of business processes, detecting problems as they occur and enabling immediate response to conformance problems. Product-based KPIs can be expected to become a priority in PLM as many manufacturers have lost sight of overall product performance in the manufacturing process. This is evident in the fragmented storage of product quality and status information.

Manufacturers will likely incorporate the following types of product-based KPIs as they attempt to meet e-business requirements: time-to-market for new products, product variability and consistency, and quality records based on the product structure.

10.4.4   KPIs in manufacturing applications

The AMR Research REPAC model shown in Figure 10.7 shows that most manufacturing application products are weak in their provision of performance analysis and KPI capabilities. The KPI application in most software packages is database- and product-centric rather than process-centric and allows limited use with dissimilar applications.

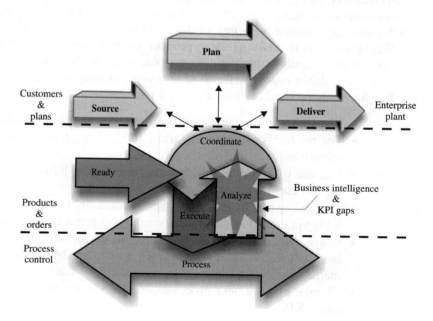

Figure 10.7
AMR Research REPAC model

The following are some areas of KPI reporting capability available in manufacturing applications:

- Most existing packaged applications have standard operations KPI reporting, but analysis is limited to operational data and measurements.
- Standard KPI reporting from existing packaged applications includes access only to defined external data in other applications.
- Configurable KPI data analysis on data extracted from different application sources are available in some packages.
- Various configurable user front-end KPI applications using data extracted from different application sources.
- Specialized tools from vendors containing embedded methodologies and tools to analyze and monitor profit flow processes and KPIs across manufacturing processes.

Return on investment (ROI) from existing IT infrastructures can be improved by leveraging KPI capabilities from packages and by partnering with vendors to further develop KPI reporting and data acquisition capabilities.

10.5   Performance analysis concepts

The collection, definition and contextualization of KPIs for business and manufacturing improvement are a core factor in business competence and change management. This is especially true in the case where manufacturing KPIs move from an internal to an external customer service focus. Only a few leading manufacturers have achieved this readiness status.

10.5.1   Organizational readiness

Readiness processes drive and implement changed practices to improve business effectiveness. In this case, the appropriate KPIs are required to ensure success. Business ownership also characterizes readiness as the process of analysing and defining top-down KPIs and performance measures aligned to business strategies and strategic priorities.

Examples include defining new product development and introduction performance measurement criteria or highlighting criteria, such as the reduction of time and touch to products as a driver toward make-to-order manufacturing strategies.

Readiness also comprises further characteristics:

- The ability of the business to define and prioritize problems and identify KPIs.
- The readiness of the IT infrastructure and applications to support dynamic KPI initiatives.
- The use of IT-based information for tracking KPIs vs the use of manual- and paper-based goals and systems.
- The ability of IT and the business community to jointly manage and implement KPIs via formal system development and project management processes.
- Users trained and skilled in the use of IT-based KPI tools – the KPI measure lies in the use of tools to increase speed of detection and resolution of manufacturing problems, such as non-conformance.
- Change management processes in place to modify practices and behaviors to attain KPIs.

KPIs help manufacturers focus on things that impact on customers and the extended supply chain, such as variability, waste and schedule adherence. Knowledge management at leading manufacturers is defined as the ability to share actionable information between different segments to increase overall productivity and organizational effectiveness.

10.5.2 Organizational change

KPIs are used to drive effectiveness and change, which is based on the following core competencies:

- Integrated business strategies and defined goals.
- Information resource management processes.
- Integrated enterprise architectural planning processes.
- Integrated IT development and business change.
- KPIs must drive specific change where needed to improve overall business performance. The ability to link changes to practices, goals and information requirements imply knowledge management competence.
- Business leadership team defines the business goals required to meet the objectives.
- Consolidate product, process knowledge and analyze specific implications to meet the goals; for example, optimize compliance management that allows the business to effectively operate while staying cGMP-compliant.
- Prioritize activities and initiatives to meet this critical business goal; for example, consolidating bottom-up-driven IT projects in manufacturing to those that will enable higher productivity means stopping any projects not helping to achieve the goals.
- Define KPIs that help lead to business change and higher productivity.
- Use IT to increase decision-support capability, enable resources to work smarter and streamline business processes for higher levels of efficiency.
- Eliminate organizational inefficiencies by using information flows and cross-functional processes to break down traditional barriers and share information.
- Implement change management processes that leverage KPIs to drive and measure change.
- Using KPIs to change practices is a quick way to focus efforts. However, knowledge management is a core capability that allows the business to focus on the right KPIs and priorities. The trick, as synthesized from comments from successful manufacturers, is to use top-down change leadership to drive the process.

10.5.3 Performance measure benefits

If people throughout an organization are well informed about their current and likely future levels of performance, and the factors that have contributed to those results, they can make more confident and more effective decisions. Measurement-based management that is focused on objectives, issues and decision-making leads to success, as employees and management are informed and allowed to take action when required. The major benefits of performance management include:

- Better achievement of objectives (Often objectives are exceeded.)
- Better and quicker decision-making (Action is encouraged.)
- All staff are aligned to common goals
- Managers and staff have greater confidence and motivation.

Performance management can be implemented through use of the balanced scorecard and other similar frameworks. The balanced scorecard is a revolutionary tool that can be used to implement corporate performance management. It motivates staff to make the organization's strategy happen and does more than just measure performance.

It is a management system that focuses the efforts of people, throughout the organization toward achieving objectives. It gives feedback on current performance and targets future performance.

10.5.4   Balanced scorecard

The balanced scorecard is a management system that enables organizations to clarify their vision and strategy, and translate them into action. The balanced scorecard approach provides a clear prescription as to what companies should measure in order to 'balance' the financial perspective. It provides feedback around both the internal business processes and external outcomes in order to continuously improve strategic performance and results. When fully deployed, the balanced scorecard transforms strategic planning into the nerve center of an enterprise.

Kaplan and Norton describe the innovation of the balanced scorecard as follows:
'The balanced scorecard retains traditional financial measures. But financial measures tell the story of past events, an adequate story for industrial age companies for which investments in long-term capabilities and customer relationships were not critical for success. These financial measures are inadequate, however, for guiding and evaluating the journey that information age companies must make to create future value through investment in customers, suppliers, employees, processes, technology, and innovation.'
The balanced scorecard views the organization from four perspectives (refer Figure 10.8):

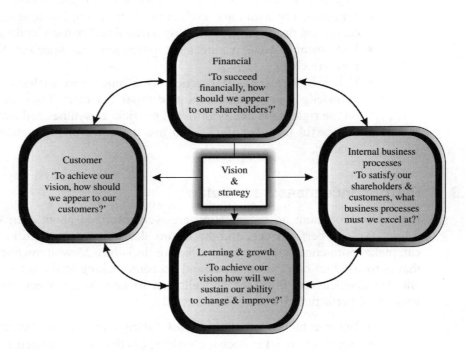

Figure 10.8
Balanced scorecard

The learning and growth perspective

This perspective includes employee training and corporate cultural attitudes related to both individual and corporate self-improvement. In the current technological change, it is becoming necessary for knowledge workers (people – the only repository of knowledge are the main resource) to be in a continuous learning mode.

The business process perspective

This perspective refers to internal business processes. Metrics based on this perspective reflect the present condition of the business and its products and services (the mission).

The customer perspective

Poor performance from this perspective is a leading indicator of future decline, even though the current financial picture may look good. In developing metrics for satisfaction, customers should be analyzed in terms of kinds of customers and the kinds of processes for which we are providing a product or service to those customer groups.

The financial perspective

Timely and accurate funding data will always be a priority, and with the implementation of a corporate database more of the processing can be centralized and automated, but an over-emphasis on finances leads to the 'unbalanced' situation with regard to other perspectives. There is a need to also include risk assessment and cost-benefit data.

10.5.5   Double-loop feedback

The balanced scorecard methodology builds on some key concepts of management ideas such as total quality management (TQM), including customer-defined quality, continuous improvement, employee empowerment, and measurement-based management and feedback. Traditionally, 'quality control' and 'zero defects' were introduced to ensure that the customer received only good quality products, and aggressive efforts were focused on inspection and testing at the end of the production line.

The problem with this approach is that the true causes of defects could never be identified, and there would always be inefficiencies due to the rejects and defects. Deming realized that variation is created at every step in a production process, and the causes of variation need to be identified and fixed. If this is done, it would be possible to reduce the defects and improve product quality indefinitely.

To establish such a process, Deming emphasized that all business processes should be part of a system with feedback loops. The feedback data should be examined by managers to determine the causes of variation, identify the processes with significant problems and then focus their attention on fixing that subset of processes.

The balanced scorecard incorporates feedback around internal business process outputs, as in TQM, but also adds a feedback loop around the outcomes of business strategies. This creates a 'double-loop feedback' process in the balanced scorecard.

10.6   Outcome metrics

Metrics should be developed based on the priorities of the strategic plan, which provides the key business drivers and criteria. Processes are then designed to collect information relevant to these metrics and reduce it to numerical form for storage, display and analysis.

The outcome of various measured processes and strategies guide the company and provide feedback. Metrics provides the following value:

- Strategic feedback to show the present status of the organization from many perspectives for decision-makers.
- Diagnostic feedback into various processes to guide improvements on a continuous basis.
- Trends in performance over time as the metrics are tracked.
- Feedback around the measurement methods themselves, and which metrics should be tracked.

10.6.1 Overall equipment effectiveness

True overall equipment effectiveness (OEE) correlates with factory output and provides a methodology to link OEE with net profits that can be used to build a solid business case for improvement projects. There are several ways one can optimize the process to improve profitability, but it is often difficult to understand where to make improvements in the overall complex operation. Overall equipment effectiveness use structured metric to evaluate the health and reliability of any process and equipment. The OEE is used as the datum to decide where to start and subsequently to monitor the improvement process (refer Figure 10.9).

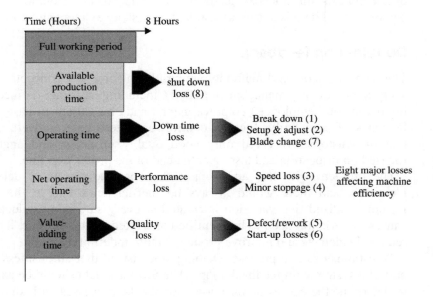

Figure 10.9
OEE

The nature of loss is categorized to aid the eventual elimination of the root cause (using a variety of problem-solving techniques). The OEE is a function of availability, speed and quality. A percentage score for each is determined and these scores are then multiplied together to give a figure for OEE (refer Figure 10.10). For consistency, it is important to agree how planned losses such as breaks, clean down and planned maintenance should feature in the calculation. Generally it would be appropriate to remove breaks, because it is useful to track clean up and planned maintenance.

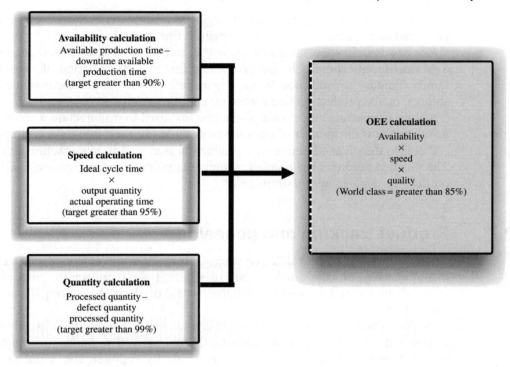

Figure 10.10
OEE calculation

10.6.2 The manufacturing profitability index

The manufacturing profitability index (MPI) is an easy-to-use tool that measures the profitability of a production line or the manufacturing assets for a given family of products. The MPI is a unique, easy-to-calculate instrument that measures the impact of manufacturing investments on profitability. Utilizing generally accepted accounting principles, it helps to analyze the performance of manufacturing assets, the profitability of products made using those assets and the revenue generated by those products.

The MPI provides valuable information to make the right business decisions with the potential to derive maximum profits from existing manufacturing operation. It is a proven instrument that communicates, in a common language, the financial impact of incremental investments in a factory to key departments in an enterprise, including administration, finance, cost accounting and manufacturing. The MPI is calculated using the following equation:

$$\text{Throughput efficiency} \times \text{gross margin \%} \times \text{revenue} = \text{MPI or}$$
$$\text{MPI} = (A/P) \times \text{gross margin \%} \times (\text{selling price} \times \text{volume})$$

Where throughput efficiency is the velocity at which raw material or components are transformed into goods to be shipped for revenue. It is defined as the 'percentage of time a production line is producing good parts that can be shipped for revenue within a given time period'. It represents the overall efficiency of all the equipment on the line, not just the efficiency of an individual work cell or piece of equipment.

Gross margin percentage is the average gross margin percentage (%) of the product family produced by the production line in a given period of time. If the product family includes several models, the gross margin percentage reflects an average of all models produced on the production line for the given time period.

Revenue is defined as the dollar value (i.e. selling price times actual number of units) of a product family shipped by the production line in a given period of time. If the product family includes several models, this figure reflects the sum of the revenue of all models shipped on the production line for the given time period.

A equals the actual output of a production line used to manufacture a product family. The actual output is the number of units the production line produced for a given period of time.

P equals the planned output of a production line used to manufacture a product family. The planned output is the number of units the production line was originally designed to produce for a given period of time.

10.7 Product tracking and genealogy

Product tracking, tracing and genealogy follows the movement of product through the manufacturing process, makes use of existing data collection processes, creates a comprehensive product genealogy in the context of all manufacturing events and provides data for analysis.

Product tracking provides the visibility to where the work is at all times and the disposition of materials and products (including WIP). Status information includes: resources working on it; component materials by supplier, lot, serial number, current production conditions, and any alarms, rework or other exceptions related to the product. The online tracking function also creates a historical record. This record allows traceability of components and usage of each end product.

Being able to identify the genealogy of products can save some companies a lot of money. Consider the following example of a motor manufacturer. An inherent defect is identified in the field after various customer and service-center complaints. The fault is identified as originating in a faulty electrical harness. Using the product genealogy function, the faulty harnesses are traced back to a specific batch of harnesses. Using the same system, the batch of harnesses is traced forward to identify the specific vehicles that were built using the defect harnesses. Only these identified vehicles are then recalled to be fitted with the new harnesses.

Without the ability to track components and products, it may have been necessary for the manufacturer to recall *all* the vehicles of that specific model made for the whole year. This alone could result in an exponential increase in cost for the manufacturer.

10.7.1 Objective/business needs

1. Regulatory compliance
2. Reduce or limit risk

   - Recall
   - Defective product liability and customer claims
   - Claims against suppliers

3. Process improvement

   - Product life cycle management
   - Supplier quality programs
   - Internal quality management

- Age/deterioration considerations
- Consistency and mixing considerations

4. Paperless manufacturing initiatives
5. Lean inventory history.

10.7.2   Tracking perspectives

Product tracking is usually one of the functions encompassed by an MES. In home-grown legacy systems or older off-the-shelf applications, it evolved from one of the following three perspectives (refer Figure 10.11).

Physical

Driven by the need to manage large moving and changing physical entities on the shop floor and understand their genealogy e.g. steel and paper.

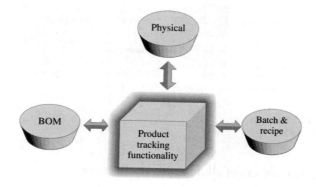

Figure 10.11
Evolution of product tracking

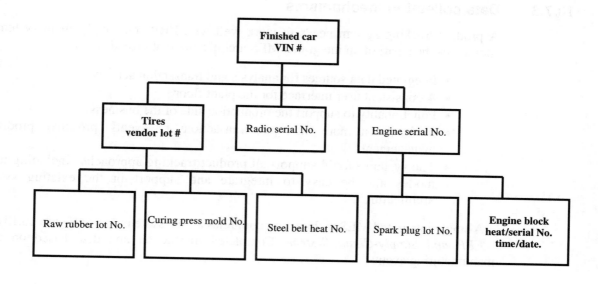

BOM approach

The concept of BOM was explained in more detail in the previous chapter. The figure below depicts a typical simplified BOM for a motor vehicle.

Batch/formula

Is similar to BOM in many ways, as in both 'many to one' and 'one to many' relationships can be present. It is different from a BOM in that the creation and consumption of whole or part entities, including byproducts are possible e.g. chemicals, pharmaceuticals, food (refer Figure 10.12).

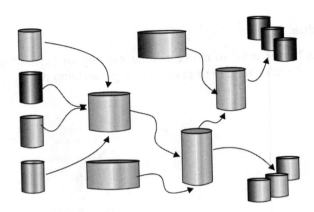

Figure 10.12
Batch/formula

Common factors of all three perspectives are:

- Unique unit identity
- Genealogy/inheritance principle.

10.7.3   Data collection mechanisms

A product tracking system needs to work well with ERP and supply-chain systems. It should also be a part of an integrated MES architecture that provides:

- Integrated data sources for analysis and transaction activity
- A consistent user interface for the plant floor
- Functionality to support the production side of the business
- A focus on more than just transactions or events: proactive production management
- The system should support all product tracking approaches including mixed modes and be easy to integrate and support in the existing systems architecture.

A single integrated MES architecture combines all the systems (refer Figure 10.13).
ERP and Supply-Chain System Completed transaction and data based on actual manufacturing events.

Production operations Drive production workflow and provide comprehensive information to influence decisions.

Customer service Comprehensive product data including specifications, 'as-built' quality and production status.

An integrated genealogy supports all product racking modes.

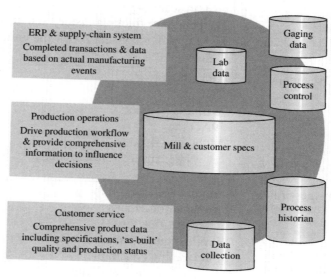

Figure 10.13
A single integrated MES architecture ties it altogether

10.7.4   Tracking implementation challenges

Complex paper processes Paper processes are configuration controlled outside the tracking solution.

Lack of infrastructure PCs and scanners on the floor require network and other hardware infrastructure.

Data integrity Tracking history and genealogy require greater accuracy in BOM and routing information. Integrity in these areas needs to reach a new level of discipline.

Implementation approach

Bill of material	Flat vs multi-level
	Tie to route enforcement
	Download from ERP
	Clean up and/or enhance ERP BOM
	Develop processes to keep BOM accurate
Bar-coded inventory	Material receiving functionality needed
	Add material receiving process if the vendor does not barcode materials
	Work on supply-chain issues to get suppliers to barcode inventory
	Perform bar-coding in MES if ERP is not capable

Routing	Add necessary routing detail if not in ERP Tie BOM validation and consumption to route steps Determine route step characteristics Download routes from ERP if possible
Electronic traveler	Defines a work order or unit S/N router Used for more manual operations to facilitate route step and BOM item scanning Backbone for tracking history and genealogy Used to provide visual ID for the unit, lot, batch or process step
PC/scanners	Tethered into PLC Tethered into PC Wireless scanner iPAQ (new) Mobile work pads Note: ensure visual feedback to operator and the ability to undo
Business rules	Material consumption Route enforcement User certifications Product quality management FG labeling constraints Note: customer defined

10.7.5   Tracking requirements

Forward and backward tracking of materials are required, i.e. when a complaint is received for a specific product, one would possibly want to know the specific batches of raw material it was made from, and if a faulty raw material is found, one may want to see all the other products the raw material was used in, to possibly recall the product or prevent dispatch or further dispatch.

It can be done only by efficient tracking, which requires processes, networks and hardware. Product tracking and genealogy software should provide visibility of manufacturing information on the factory floor and beyond. This software directs and tracks product flow, collects and recalls product genealogy data, and provides management with the visibility of the factory and process performance. It may use Web browsers throughout the factory to direct and monitor flow of both batch and serialized product along routings designed using a graphical environment.

Tools for creating flow logic enables infinite process flexibility, as well as the control of routing based on variable events. In addition to history, routing and genealogy recall, it also allows for parametric data to be input during the process, such as tuning settings, for recall later in the product life cycle for problem-solving. The system should also provide global web-based analysis and diagnostic capabilities to assist management in improving production performance (refer Figure 10.14).

A completely integrated solution spans all the architectural levels, as tracking and genealogy information is both collected and required at all these levels. The context and use of the tracking information change at each level, as each level has a specific role and objective, and collects information for that level and for the level above. These solutions typically integrate tightly with resource allocation and production order dispatch solutions (described in the previous chapter).

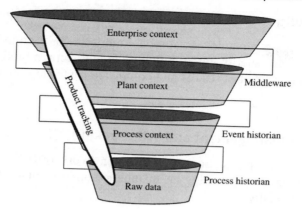

Figure 10.14
Data-centric product tracking

10.7.6 Typical solution component overview

The components of a typical tracking solution may be as follows (see Figure 10.15 below):

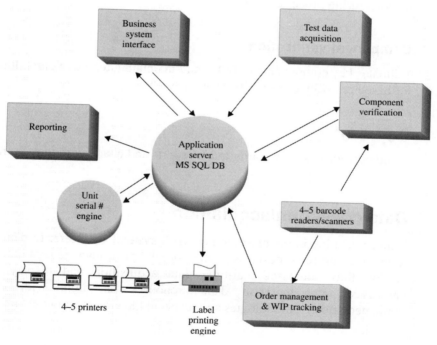

Figure 10.15
Solution component overview

Business systems interface

The interface accepts download of orders, BOM's, routes, etc. from ERP and uploads material consumption and unit production.

Serial number engine

The engine generates serial numbers for normal as well as exceptional cases (manually or automatically).

Printing engine

The engine generates serial/model numbers for each individual unit or lot in the form of labels with energy guide, rating plate and shipping information. It guides the process of collecting QC test results and completed route step information. It also includes software licenses.

Order management and WIP tracking

Supervisor (production) and operator (work order) track the released order and WIP. Manufacturing genealogy is maintained on lot, unit or other predefined basis (manually or automatically).

Test data acquisition

Integrated to process control and testing platform, which enables automatic download of test recipe from BOM or configuration database. It enables utility usage tracking. Reporting/analysis tied to 1st pass yield and historic correlation of failed units to manufacturing process.

Component verification

Validates key components to BOM and maintain alternative parts. It also enables unit and lot-based pre-packaging verification of products.

Reporting

Standardized, seamless configurable reports and charts with yield, scrap and genealogy.

10.8   Data collection/acquisition

Data acquisition in the definition of MES systems only refer to plant process data, but in terms of enterprises, business process or enterprise management data is also needed. This section thus addresses real-time data acquisition and display, real-time historical databases and enterprise data-warehouses as tools with different objectives, implementation methodologies and levels in the systems architecture.

10.8.1   Evolution of real-time plant databases

Since the early days of computers, engineers have worked to tie computers to real-time field information and use the flexibility of software to better monitor or control the manufacturing process. Over time, these efforts have produced a wide range of software products (DCS, PC-based MMI and SCADA packages) focused on process management.

The overlap is seen in the process management software industry between DCS and PLC-based controls solutions both in function and target industry. The process management software category includes any software that directly interacts with controllers or process I/O.

Examples of process management software include:

- Man–Machine Interface
- Supervisory control and data acquisition
- Data historians
- Advanced process control
- Batch process execution and management
- Cell control
- Real-time statistical process control (SPC)
- Distributed numerical control (DNC) and other program library software
- Emerging software-based control products that replace PLCs and regulatory controls.

The DCS consoles replaced large panel boards of individual instruments in central control rooms of continuous process plants. The SCADA systems were developed to remotely supervize large utilities and transmission facilities via telecommunications. The MMI packages emerged as a cheaper alternative to hard-wired push buttons and lights on discrete manufacturing machinery controlled by PLCs.

These were initially hardware products, but quickly became pure software as the PC gained popularity. Logging, reporting and alarm management functions were added as the products evolved into complete monitoring and diagnostic systems for the machine operators. In parallel with the growth of MMIs, discrete manufacturers searched for ways to make factories more productive.

Plant-wide monitoring systems provided central alarms of machine faults, so maintenance could be dispatched faster. In addition, they monitored machine productivity (e.g. uptime, downtime, production rate, etc.) and provided central libraries of PLC and DNC programs. The emergence of the batch controls market forced all these categories to expand toward the middle ground where they overlapped with each other. Small vendors with several common attributes developed new products:

- They have a method of defining and naming plant data from instrumentation or PLC memory. These are often referred to as 'tags' after the identification labels used on instrumentation in process plants. Some products are even built around the concept of a plant-wide database of this real-time data. Animated graphics were used to display process status information on the screen.
- They have a straightforward scripting language for implementing simple control algorithms. These products not only overlap in their target market, but in their underlying technology as well. Products developed by the small vendors make it difficult to create plant-wide applications, such as centralized alarming or productivity monitoring. Process management vendors quickly adopted Windows NT as their core platform for new development. Windows NT brings with it OLE component technology which allow new combinations of vendors and room for smaller, specialized component suppliers. Many plants have bought unit and cell level systems over many years and often from different vendors. Plant-wide systems are purchased to coordinate the overall facility and were integrated with a wide variety of other process management systems. Every system needs to access process data, which are potentially available in different places.
- They cannot usually share the definition information, which is redundantly defined and maintained because each system was supplied by a different vendor.

These multiple maintenance issues have frustrated manufacturers and system integrators and their preference shifted toward a more integrated approach where programing or configuring the controls and process management were part of the same system.

10.8.2   Real-time data acquisition and display

Users prefer the integrated approach of DCS, but find it too expensive for batch and inappropriate for discrete processes whereas PLCs are more cost-effective, but harder to integrate. The MMI and SCADA vendors usually have the most functional and cost-effective solutions for higher-level process management software, and these tools become the integration platform for the factory.

Their basic products include the three ingredients: data access, graphics and control scripts and all other applications can theoretically be layered on this automation platform. Using an MMI as a core platform on the factory backbone means that data need only be defined twice, once in the controls and once in the MMI. All other applications use the MMI's definition. *Microsoft's* OLE technology allows properly developed software components to be assembled into a system. Most vendors are now using this technology internally to build their applications.

Some vendors have developed tools where data only needs to be defined once, in the controls, and are then available for all other systems operating on the same platform, whether they are control, MMI or MES tools. Most other vendors are also in the process of developing similar tools to reduce work duplication and ease solution maintenance.

Realizing efficient information flow is a prime focus for business groups the world-over. In modern manufacturing, business performance can be directly linked to the successful flow of information across the enterprise. A recent report by the US industrial research group, AMR Research, suggested communication and coordination between corporate systems and the plant floor is the key to competitive advantage.

To properly manage production resources requires immense amounts of data collection and distribution. It requires efficient and reliable data acquisition and distribution strategies through process historian, Web-based publishing, data transformation, report generation and data warehousing.

A real-time data server, creating an information exchange backbone in manufacturing plants, is used to link data from a number of production sites to a central corporate data center. The function is to manage delivery of process work orders to the plants and communicate the manufacturing results back to centralized business systems and the supply chain. A typical data acquisition architecture, from instruments/field devices to the integration with the ERP is depicted in Figure 10.16.

Following are some general types of instrumentation needed for data acquisition systems:

- Temperature measurement – thermocouples, thermostats, non-contact devices
- Pressure measurement – pressure cells and differential pressure cells
- Flow – differential pressure cells, Venturi tubes, magnetic flow meters
- Displacement – resistive sensors, non-contact inductive and capacitive sensors
- Stress and strain – strain gages
- Gas detection
- Level – ultrasonic, resistive and optical

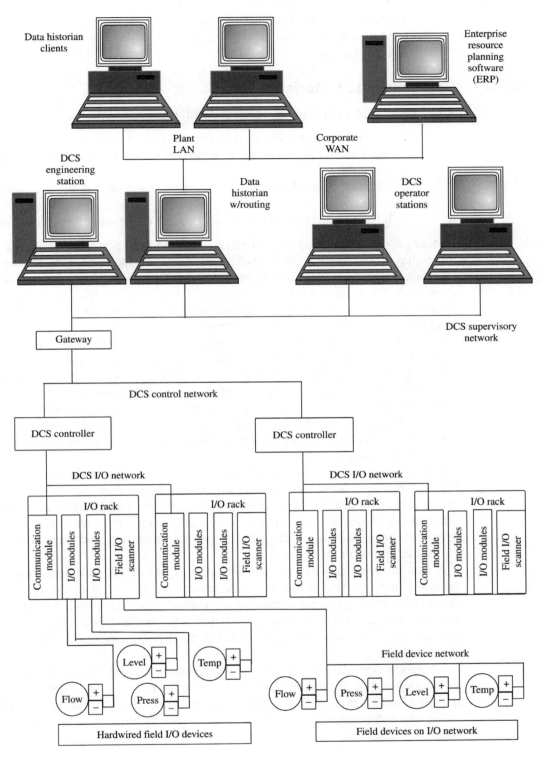

Figure 10.16
Data acquisition

- Chemistry – pH, Conductivity, etc.
- Meteorology
- Many others.

10.8.3   Real-time data historian

A successful historical data server has to fulfill three basic criteria:

1. It should enable prioritized business processes such as new product development and close the loop between enterprise planning and plant scheduling.
2. It should be used as a part of a complete solution and not as a stand-alone toolset.
3. It has to satisfy the entire IT and process community through its functionality.

Most popular production data historians provide high performance with a low cost of ownership and business decision-makers the access to appropriate information. Historians empower production personnel at every level to improve efficiency and product quality throughout the manufacturing environment. Historians are used to provide a common point of access for all production information, a single platform for the development of production applications and a single interface to business systems.

A data historian normally provides complete flexibility in defining the sources for the data to be stored and data may be automatically polled in real time from any I/O server. All data, regardless of its source or time of entry, is fully integrated (or encapsulated) into a unified storage location, giving immediate access to chronologically referenced (time-stamped) data. It contains the complete production information and acquires and stores high-resolution process history directly from industry-standard servers. These servers provides incredible flexibility by accommodating inserts, updates and data imports from non-realtime sources.

It fully integrates this imported and non-real-time data with the real-time acquired history, thus providing seamless, easy retrieval from any client. It integrates this information with configuration, alarm, event, summary and lab data, as well as data from other server products. Data becomes useful information only when it can be accessed to user advantage, so most packages also provide reporting and data-view tools for the generation of useful information.

A data historian acquires production data in real time, at full resolution, and provides seamless incorporation of non-real-time, manual or off-line data. Conventional relational database technology is not suited to high-speed acquisition and storage of plant data. Historians acquire plant data hundreds of times faster than relational databases, and stores the data in a fraction of the space using various data compression techniques (refer Figure 10.17).

All production data is then fully integrated into plant data with event, summary, production and configuration information, bringing the power of the relational database to the industrial environment. Fault-tolerant functionality provides robust system architectures with no single point of failure, virtually eliminating the need for expensive redundant hardware.

Objective

- To gather, store and beneficially use the real-time data from 'electronic' data sources such as RTU, SCADA and DCS systems.
- To provide secured, centralized and high-resolution historical data store.

- To provide a definitive, time-stamped record of plant performance and process information such as pressures, flows, temperatures, set points, on/offs, etc. for incident analysis and predictive calculation.
- To provides users with an access level using common desktop applications, Microsoft compliant software and Intranet browsing.

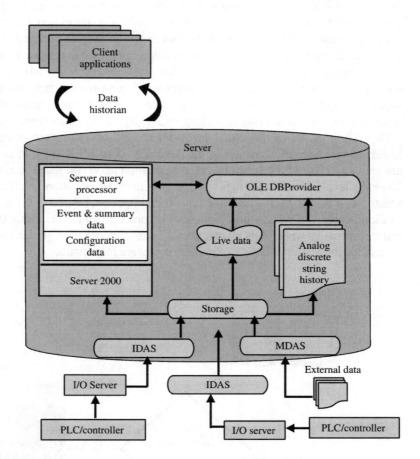

Figure 10.17
Data historian configuration

Advantages

The data historian will capture/store data over a substantial length of time and make it available to those who need to know (with convenient desktop access to data) in the form of reports, graphical presentations and trend plots. The data historian will make data available electronically to application programs for added business benefits e.g. rotating plant stops/starts and downtime for reliability-centric maintenance.

Significant benefit will accrue from use of the data historian as a platform to investigate incident cause/effect and to analyze and learn from 'near misses'. The Data Historian can be used to provide real-time information of the process such as leak detection, corrosion monitoring, chemical injection monitoring, etc.

Apart from problem analysis and faultfinding, data historians can also be used to identify 'golden batch' or 'golden campaign' situations where yields and efficiencies were particularly high. The specific process conditions can then be analyzed and set points, etc.

and can be changed to replicate the conditions in day-to-day operations so as to duplicate 'golden batch' results.

10.8.4   Data structuring

In its raw form, real-time data may not be useful for reporting purposes. It may be very well suited to show trends, but as it is time-based, unrelated to any specific product or order, it is not useful for reporting without some manipulation and structuring. Data needs to be summated or averaged (or otherwise calculated), related to a product, ordered or shifted before it can be used in other systems such as MES, ERP or SCM. Data structuring technology converts unstructured data into structured data to enable useful reporting and analysis. Relationships can be established with data originating from unlinked sources to provide reporting and analysis of events (refer Figure 10.18).

Depending on the architectural level where data historians are placed by an organization, historians can be viewed as part of the control level or MES level. In cases where the historian are viewed as part of the control level, data manipulation and result storage are often done at the historian level using 'super-tags'. Other structured information (such as LIMS data, product identity and tracking information, etc.) is then imported, related to these 'super-tags' and stored in the historian, providing one repository of process data. This method increase maintenance effort, especially where tags change that affects 'super-tags'. Queries also tend to be slower as no relationships are available to eliminate irrelevant information.

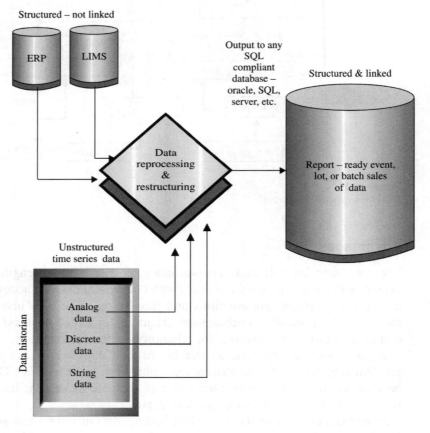

Figure 10.18
Data structuring technology

In cases where the historian is viewed as part of the MES level, only a limited amount (if any) of data manipulation takes place in the historian. Data from the historian is queried, manipulated (summated, averaged or otherwise calculated) and the result is stored together with other related data in a structured relational database outside the historian. This approach is less maintenance and system resource intensive, as the historian is used only as a data acquisition engine with real-time process-trend views without the complication of 'super-tags', frequent management queries and data inserts from other systems. It also separates the structured and unstructured data, creating a clear boundary between the control and MES level.

Features

Collect data from dissimilar sources and associate data with differing types:

- Translate data names to meaningful terms
- Backup critical data and it can be copied or linked to output tables
- Verification flag can be set manually or automatically
- Configure and execute from multiple processing nodes.

Benefits

- Simplify reporting by eliminating complex query scripting.
- Derive information from previously unusable data.
- Use of data becomes more intuitive and meaningful.
- Can efficiently back up selective critical data.
- Report selectively on validated data.
- Maximum flexibility to fit user needs,
- Copy – enhance speed of queries.
- Linked – minimize system resources.
- Gain access to a vast array of data sources.
- Optimize performance of reporting systems.
- Adapt and evolve the system without programing.

10.8.5 Data warehouses

Throughout the history of systems development, the primary emphasis had been given to the operational systems and the data they process. It is not practical to keep data in the operational systems indefinitely, and only as an afterthought was a structure designed for archiving the data that the operational system has processed.

Many factors have influenced the quick evolution of the data warehousing discipline. The most significant set of factors have been the enormous forward movement in the hardware and software technologies. Sharply decreasing prices and the increasing power of computer hardware, coupled with ease of use of today's software, has made quick analysis of hundreds of gigabytes of information and business knowledge possible.

Another very significant influence on evolution of data warehousing science was the fundamental changes in the business organization and structure during late eighties and early nineties. The emergence of a vibrant global economy has profoundly changed the information demands made by corporations worldwide. Corporations have found markets for their products globally while competing with other companies in vastly different cultures and economic environments. The mergers and acquisition of businesses have crossed country boundaries.

The primary concept of data warehousing is that the data stored for business analysis can most effectively be accessed by separating it from the data in the operational systems. Many of the reasons for this separation have evolved over the years. In the past, legacy systems archived data onto tapes as it became inactive and many analysis reports ran from these tapes or mirror data sources to minimize the performance impact on the operational systems.

Data warehousing systems are most successful when data can be combined from more than one operational system. When the data needs to be brought together from more than one source application, it is natural that this integration be done at a place independent of the source applications. The data warehouse may very effectively combine data from multiple source applications such as sales, marketing, finance and production. The primary reason for combining data from multiple source applications is the ability to cross-reference data from these applications.

Data in a data warehouse is un-normalized data, implying that the relations between entities in a relational database (RDB) have been removed from the data, and all the data is stored as one long string. This means that the same data in a data warehouse will take more space than in an RDB, as all the data will be stored, not only primary keys pointing to more data as in an RDB. This apparent inefficiency though makes it infinitely easier to view the same data from various perspectives without complicated queries and scripts. It also makes the generation of reports and are very fast, as no relational routes need to be followed to get to the detail information required.

Nearly all data in a typical data warehouse is built around the time dimension. Time is the primary filtering criterion for a very large percentage of all activity against the data warehouse. An analyst may generate queries for a given week, month, quarter or a year.

Another popular query in many data warehousing applications is the review of year-on-year activity. For example, one may compare sales for the first quarter of this year with the sales for first quarter of the prior years. The time dimension in the data warehouse also serves as a fundamental cross-referencing attribute. For example, an analyst may attempt to access the impact of a new marketing campaign run during selected months by reviewing the sales during the same periods.

The ability to establish and understand the correlation between activities of different organizational groups within a company is often cited as the single biggest advanced feature of the data warehousing systems. Another key attribute of the data in a data warehouse system is that the data is brought to the warehouse after it has become mostly non-volatile. This means that after the data is in the data warehouse, there are no modifications made to this information.

A data warehouse is a structured extensible environment designed for the analysis of non-volatile data, logically and physically transformed from multiple source applications to align with business structure, updated and maintained for a long time period, expressed in simple business terms and summarized for quick analysis.

The data warehouse system is likely to be interfaced with other applications that use it as the source of operational system data. A data warehouse may feed data to other data warehouses or smaller data warehouses called data marts. Data marts (or data cubes) are normally preconfigured data extraction engines, providing views and reporting of specific data from various perspectives.

Data cubes are used to 'slice-and-dice' data, enabling the user to view the warehouse data in more and more detail to the lowest captured granularity. For instance, a user can view product purchases by customer, product, geographical area or timeframe, etc. using the same data, without the need to run additional queries. Once the cube is launched, all the information is available and the user only needs to change the way he wants to view the information.

The operational systems interfacing with the data warehouse often become increasingly stable and powerful. As the data warehouse becomes a reliable source of data that has been consistently moved from the operational systems, many downstream applications find that a single interface with the data warehouse is much easier and more functional than multiple interfaces with the operational applications.

The data warehouse can be a better single and consistent source for many kinds of data than the operational systems. It is however, important to remember that much of the operational state information is not carried over to the data warehouse. Thus, data warehouse cannot be source of all operation system interfaces.

Even though data mining in the detail data may account for a very small percentage of the data warehouse activity, the most useful data analysis might be done on the detail data. The reports and queries of the summary tables are adequate to answer many 'what' questions in the business. The drill down into the detail data provides answers to 'why' and 'how' questions.

Data mining is an evolving science. A data-mining user starts with summary data and drills down into the detail data looking for arguments to prove or disprove a hypothesis. The tools for data mining are evolving rapidly to satisfy the need to understand the behavior of business units such as customers and products.

11

Project motivation and benefit quantification

Learning objectives

- To understand why organizations implement software solutions.
- To understand why a lot of software implementations fail to deliver benefits.
- To understand where and how benefits can be identified.
- To understand the process of benefit quantification.
- To understand the process of performance improvement evaluation.
- Typical examples of benefits and benefit quantification tools.

11.1 Introduction

E-business and e-manufacturing initiatives are keeping companies in business and in this competitive global market it is an undeniable truth that to survive in business today, all companies must have e-commerce initiatives in place. This statement, even if not verbalized in such strong terms, is often put forward as the undeniable truth. But is it true? It may be for some, but not all organizations. There are also more publicized e-initiative failures than successes, and it is hard to find any return on investment information on software initiatives.

To understand why this is, we need to investigate the way projects are motivated and measured and how this is applied to IT initiatives.

11.2 Project portfolio

Companies implement projects, or spend capital, for improved financial results and/or for strategic position in relation to competitors in order to improve financial results in the future. The ultimate project of course is the one, which results in high financial returns as well as the strategic positioning of the company. The generic project portfolio is classified according to this financial and strategic intent (refer Figure 11.1).

A – These projects are those, which will bring about significant financial returns by improving processes or by reducing costs.

B – These projects are done to stay in business and include equipment replacement or rebuilding.

Figure 11.1
Financial and strategic intent

C – These projects are done to keep in business ahead of the competition.

D – These projects are the preferred projects, as they support the strategic intent of the company and have a high rate of financial returns.

Most IT projects are classified as B or C type projects and companies see IT as a necessary evil (B), whereas others identify IT as a strategic advantage over their competition (C). This classification neither generates expectations of real financial returns nor allows one to track and measure the benefits and it is done this way all over the world. How can we do it differently? What can we do to change this perception?

11.3 Project motivation

To change this perception we need to go back to basics:

- What is the objective of the e-initiative?
- What will the benefits be?

The following table provides a map of how a project can be motivated before implementation and measured after implementation.

Project Motivation	Project Measurement
Identify data for potential benefits	Collect data for benefits measurement
Use historical information for calculation	Use the current data for calculation
Use risk/assumption scenarios	Use actual data and calculations
Set target achievements	Measure against target
Quantify potential achievement	Quantify actual achievement

The solution is simple: identify potential benefits, establish a base line, design and develop measurement methods, and review measured results. Sounds simple does it not?

In truth, while the concept is simple, the implementation is usually a lot more complicated. That is if it is ever attempted.

The success of IT projects is normally measured as actual time and cost vs budget time and cost, and if the project time and cost is less than the budget, the project is considered to be successful. Very few companies measure actual ROI on any IT project once approved. The solution lies on identification of potential benefits, establishing a base line, designing and developing measurement methods and reviewing measured results.

11.4   Potential benefits

In the IT environment, most of the benefits are not measurable e.g. 'readily available management reports', 'access to historical information', 'enhanced ability to plan and schedule activities', 'standardized business processes', etc. These normally make up the majority of the motivation for IT systems, as they address the emotion and operate on the principle that information is power. These benefits have their place, but should they be the heart of the business case? What is needed are benefits that can be quantified and measured.

Some benefits are quantifiable, but are difficult to translate directly into money, this does not mean that they should be excluded and not measured, as they may have strategic importance in terms of the company's vision and mission. Examples are; 'number of quality complaints', 'reduced number of calculation errors and greater certainty', 'time saved by customers ordering electronically instead of manually', etc. Financially quantifiable benefits are the most important benefits for any project.

In order to identify the financially quantifiable benefits, the level of the e-business or IT architecture components that will be implemented in the organization needs to be identified, in order for the benefits and measurement methods to become more apparent.

11.5   Benefits of IT architecture components

System components may be broken into three levels:

1. An ERP or e-business level includes components such as SCM, CRM and trade portals.
2. Manufacturing execution systems (MES) or e-manufacturing level includes components such as quality management, detailed scheduling, data collection/acquisition, process management, product tracking, performance analysis.
3. Process monitoring and control systems (PMC) level includes components such as SCADA, PLC and DCS.

In order to understand the benefits realizable due to each of these components, it is necessary to identify the purpose or reason for existence for each of these levels.

11.5.1   ERP or e-business level

The purpose of an e-business level system is to capture and manipulate data in order to increase the financial effectiveness and efficiency of a company i.e. the systems provides easy interfacing to financial information on local and global levels. It enables management at all levels to optimize expenditure and to identify poor return on capital employed (ROCE). The system provides efficiency in the overall financial system by reducing the time and labor to generate orders, invoices and payments, etc. This level also optimizes logistics and manufacturing schedules ensuring that products are delivered to the customer at the lowest possible cost.

11.5.2　e-manufacturing or operations level

The purpose of an e-manufacturing level system is to optimize the effectiveness of manufacturing operations. The system provides easy access and visibility to operations data such as plans, schedules, equipment and manpower availability, key performance indicators and abstracted process information. It provides the tools and information to the plant manager to manage, optimize and control the manufacturing process, equipment and labor. It provides tools for tracking materials, enabling activity-based costing and providing information about product and material quality, yield, shelf-life and resource capacity and use. It uses this information to identify problem areas and production excellence by comparing batch/lot/process/production order results between production campaigns, using this information to resolve problems and establish ideal operating conditions. It further optimizes the production schedule provided by the e-business level by generating detailed production schedules. The system also enables the e-business level by providing order status, order tracking and production volume data and provides visibility across business unit boundaries.

11.5.3　Process monitoring and control systems

The purpose of a PMC level system is to optimize the efficiency of manufacturing processes. It actually controls the processes, generates deviation alarms, changes recipes and sets points, and records process and usage information and is normally used by the plant supervisor or operator to control the process and by management to view progress and reports. It is used to optimize and tune production processes, increasing the speed of execution, thereby increasing the throughput. It influences process conditions directly, influencing yields and material recovery, directly impacting on product profitability. It is also used to keep complex processes under control, ensuring more consistent quality and increased process safety. An APC can also directly impact product mix in continuous processes, leading to more profitable operations.

11.6　Benefit quantification

In order to identify benefits that can be used to motivate and later measure the success of an IT initiative, one needs to first consider the relationship between benefits and investment. The Du Pont system of ratio analysis involves constructing a pyramid of inter-related ratios. This pyramid was expanded downwards indicating measurements that can be used as KPIs and elements in the process that may affect them, and the key influencing factors (KIFs).

Such ratio pyramids assist in graphically providing an overall management plan to achieve profitability, and allow the inter-relationships between ratios to be checked (refer Figure 11.2). It is impossible to evaluate benefits properly without relating them to the amount of funds (the capital) employed in achieving the benefits. An important ratio is therefore ROCE or ROI which states the profit as a percentage of the amount of capital employed.

Profit before interest and tax (PBIT) or net income is the amount of profit that the company earned before having to pay interest to the providers of loan capital. Capital employed or total assets are shareholders funds plus creditors' amounts falling due after 1 year, plus any long-term provisions for liabilities and charges.

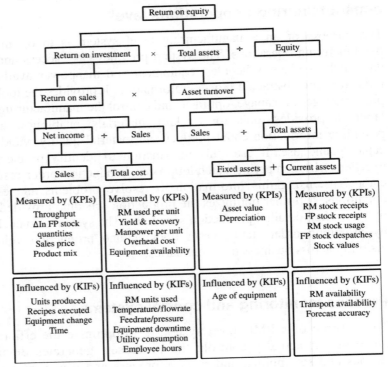

Figure 11.2
Extended Du Pont ratio pyramid

Profit margin and asset turnover together explains the ROCE, and if the ROCE is the primary profitability ratio, these two are the secondary ratios. The relationship between the three ratios is:

$$\text{Profit margin} \times \text{Asset turnover} = \text{ROCE}$$

That means: $\dfrac{\text{PBIT}}{\text{Sales}} \times \dfrac{\text{Sales}}{\text{Capital employed}} = \dfrac{\text{PBIT}}{\text{Capital employed}}$

Thus $\text{ROCE} = \dfrac{\text{Profit on ordinary activities before interest and taxation}}{\text{Capital employed}}$

Two factors contribute toward a return on capital employed, both related to turnover:

1. *Profit margin:*   A company might make a high or a low profit on its sales.
2. *Asset turnover:*   Asset turnover is a measure of how well the assets of the business are being used to generate sales.

Therefore, the increase of asset turnover or profit margin will have a positive ROI, which can be achieved by:

Higher throughput   Increase in sales through higher throughput (if throughput is a constraint).

Proper product mix   Producing more high-margin products in relation to lower-margin products.

Increased product price Claiming a higher price for better service and improved products (if possible).

Cost reduction Reducing the total cost of production by increasing the yield or recovery rate i.e. less raw material is required per unit, using less manpower, reducing overhead cost and increasing equipment availability.

Low inventory Reducing the amount of assets by keeping less raw material and final product stock.

Low raw material cost Obtaining raw material at lesser cost is having direct impact on ROI.

Any positive influence on any of the KPIs or KIFs through better control or management, will therefore result in and will be able to demonstrate a positive ROI by measuring the change in performance and equating that to a monetary value.

11.7   Benefits and architectural levels

The above two sections explained what types of benefits can be expected at each level (the purpose of the specific level), and how benefits can be calculated and identified (extended Du Pont model; Figure 11.3).

Taken together, this means that the ERP level will influence the assets, cash, stocks and sales. The MES level will influence most of the total cost and fixed asset KPIs and some of the current assets and sales KIFs. The PMC level will influence mostly KIFs. The KIFs and KPIs can be extrapolated to financial terms in order to calculate returns using the Du Pont model.

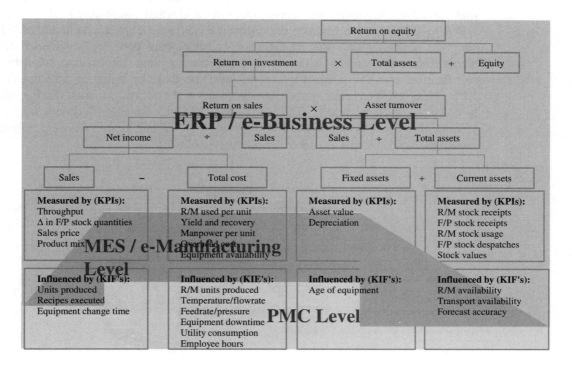

Figure 11.3
Extended Du Pont ratio analysis with benefit location

Once it is clear how to measure benefits and where to find them in relation to the IT initiative, it is required to establish a baseline of current performance and develop a business case. Hopefully the initiative will be approved by management, and the project will go ahead.

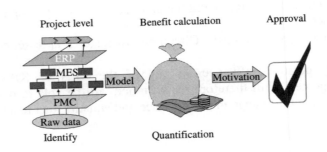

11.8   Extended benefit analysis

Once approval has been obtained, it is required to return to the initial assumptions and establish a baseline of current performance, at the same time developing measurement methodologies and targets. These should be planned, designed and captured into the system before implementation. It should be easy to periodically review performance improvement and obtain ROI calculations automatically without any additional calculation effort.

Business cases also make use of certain assumptions, such as selling price, stock levels and raw material cost which helps with the calculation of improvements. These assumptions should also be captured as part of the baseline, otherwise increasing raw material costs can negate the improvement in yield when calculating ROI.

Due to the interaction between the different levels and the integrated nature of the pyramid structure, it may mean that changes or enhancements need to be affected to existing architectural components in order to have the required information available for the calculations. The KIF information (PMC level) needs to be rolled-up and transformed into KPIs (MES level) and then further into cost, sales and assets information (ERP level) in order to calculate financial results. The installation of a process control system may mean that baseline data and calculation methods need to be stored and kept in the e-manufacturing and e-business level systems to enable the calculation of ROI.

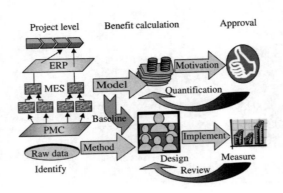

Figure 11.4
Measure for success

After implementation, if the desired improvements are not obtained, it may be necessary to return to the initial measurement design and assumptions to identify the problem (refer Figure 11.4). These may be deficiencies such as exchange- or interest rate fluctuations not made provision for/in the initial design, or sudden commodity price reduction or energy cost increases.

If the baseline, methodologies and assumptions were not designed into the system initially, identification of these shortcomings and manually recalculating and modeling the improvement would be a lot more difficult.

11.9 Benefits of an extended business case

So why go to all that trouble? If the money is approved, the initiative will go ahead, and why bother measuring after the fact? Only those who want a system or component to be installed because they believe that it will be beneficial to their department or company will ask these questions. But nobody will want to install something that will not add benefit, will they?

The base assumption of the motivation for the above example is wrong. The base assumption should not be 'Install a system because we will get benefits'. The assumption should be 'we want to achieve X in department Y, to do that we need something to help us and a system will do that'.

The difference is a matter of perspective. If the objective is to achieve X, there will be measurements implemented to prove that it was achieved after implementation. Objectives are measured and reviewed, and it should be no different for IT projects. The objective should also be focused on the project outcome, what it was supposed to achieve and why it was done in the first place.

The added benefit of this approach will be to increase project benefits, since benefits are dependent on the implementation and effective use of the system and can affect the end result. Hence, the measurements can influence the outcome 'Tell me how you will measure me . . . , and I will behave accordingly'.

Measurement drives behavior, and if a culture of measurement is prevalent for all project outcomes, IT included, it will go a long way in ensuring that the real, tangible benefits are achieved, not only the emotional ones.

11.10 Benefit examples

Manufacturers are using integrated software applications which provide a real-time look at manufacturing operations. Systems make it possible to integrate real-time data with other information systems such as production planning and business management systems. The benefits of improved shop floor operations have a positive impact on more fundamental corporate objectives such as increased market share, profitability and improved global competitiveness.

Realization of these objectives is only possible if improvements such as shorter production runs, improved quality, lower costs and improved customer response, etc. are addressed in the manufacturing phase. To achieve these goals, it is required to identify and define these benefits, quantify and finally translate them into general business benefits.

The manufacturing software industry is growing rapidly and according to advanced manufacturing research (AMR), the growth rates are projected to continue at 30% or greater annually, hence it is important to quantify the benefits from a manufacturing and financial point of view.

Benefit quantification gives answers to the following queries:

- Are manufacturers receiving the expected benefits?
- How did users approach and justify the investment?
- What are these benefits to begin with?

Two MESA surveys were conducted, one in 1993 and the other in 1996. Information was gathered using two similar questionnaires based on a list of potential and actual manufacturing, planning process and business 'benefits'.

The respondents of the survey were the user community and have at least one common characteristic – that they invested in MES and are now reaping the benefits. Seventy-five percent of the companies that provided data to MESA for this study describe themselves as 'discrete' manufacturers, 16% are 'batch/process' manufacturers and 9% are both. They represent the following industries: medical products, plastics and composites, metals manufacturing, electrical/electronics automotive, fiberglass and communications.

Specific Benefits	Survey One	Survey Two
Reduces manufacturing cycle time	60% of respondents report reduction in cycle time of 40% or better Range of reduction: 2–80% Average reduction of all respondents: 45%	Average reduction: 35% Range of reduction: 10–80%
Reduces or eliminates data entry time	60% of respondents report reduction of 75% or more Range of reduction: 25–100% Average reduction of all respondents: 75%	Average reduction: 36% Range of reduction: 0–90% Nearly half reported 50% or greater reduction.
Reduces work in progress (WIP)	57% of respondents report reduction of 25% or more Range of reduction: 25–100% Average reduction of all respondents: 17%	Average reduction: 32% Range of reduction: 0–100%
Reduces/eliminates paperwork between shifts	63% of respondents report reduction of 50% or more Range of reduction: 5–100% Average reduction of all respondents: 56%	Average reduction: 67% Range of reduction: 0–200%
Reduces lead-times	50% of respondents report reduction of 30% or more Range of reduction: 2–60% Average reduction of all respondents: 32%	Average reduction: 22% Range of reduction: 0–80%
Improves product quality (reduces defects)	Average reductions in defects: 15% Range of reductions in defects: 5–25%	Average reduction: 22% Range of reduction: 0–65%
Eliminates lost paperwork/blue prints	Average reduction: 57% Range of reduction: 10–100%	Average reduction: 55% Range of reduction: 0–100% 62% reported reduction of 75% or greater

Benefits to the planning process

The benefits to the planning process are critical but not as easy to quantify. The five common 'planning benefits' experienced by the manufacturers are given below:

- Agile manufacturing and flexibility to respond to customer demand through easier change/update of schedules
- Meeting higher sales volumes without increased cost through better planning of material availability and optimized schedules
- Empowered plant personnel can make decisions in real time using system information
- Adherence of strict delivery schedule
- Fulfillment of regulatory and compliance requirements.

Business benefits

Payback periods range from 6 months to 2 years, with an average of 14 months.

- Improvement in financial performance
- Reduction in cycle time reduces WIP and product quality problems
- The paperless approach reduces overhead, i.e. eliminates non-value-added paper handling and management
- Scheduling and order visibility reduced people expediting
- Reduction in rework and inspection
- Reduction in post-manufacturing rectification
- Lower scrap/defect = higher profitability.

Improved customer service

The improvement in customer service benefits are extremely important and are given below:

- Decreased cycle time improves service to the customer
- Improved on-time performance results in increased market share
- 100% adherence to customer order and shipment schedules
- Real-time status update of all orders.

Manufacturing and its direct impact on supply-chain efficiency

Today manufacturers are under intense pressure to reduce costs while achieving world class performance in quality and customer service. It has forced companies to concentrate on performance. Therefore optimizing supply-chain performance has become a passion for most companies. Supply-chain optimization is the right goal, but many companies still fail to recognize that the system aspects of supply-chain improvements and projects are typically evaluated solely on the basis of direct benefits and localized ROI.

Recognizing the impact of manufacturing on supply-chain performance can drastically improve ROIs and make manufacturing a key candidate for the limited available investment capital.

Optimizing supply chain operations outside the factory has become the focus for manufacturers. Inventories in distribution centers can be many times larger than the WIP, and they demand tight control. Inbound and outbound logistics can be costly and must be optimized for efficiency and smooth operation.

Collaboration with suppliers and customers can be the key to enabling strategies such as lean manufacturing and demand close relationships. Many companies are still not considering supply-chain improvement programs from a system perspective and source, make and deliver operations are optimized individually to generate the best individual performance.

Under this strategy, improvement programs are measured only on ROI for the localized optimized specific activity. Producing quality products at the lowest cost is the primary responsibility of manufacturing operations, but how this is done has an impact on other supply-chain activities. Long cycle times enables a more efficient order batching and planning strategy that focuses on load balancing that is good for the factory's KPIs, but can cause additional costs outside the factory. Quality programs that balance customer satisfaction only against manufacturing costs ignore the impact that quality has on logistics costs.

All parts of the company should recognize the systems impact of manufacturing, and manufacturing personnel should also understand how their operations affect overall supply-chain performance and develop programs to address any deficiencies.

Lead-times and variability and supply-chain inventories

Inventory reduction is the focus of most supply-chain initiatives and strategic placement of distribution centers, make-to-order manufacturing, common-part product design that facilitates assemble-to-order, and collaborative demand forecasting are strategies to address these issues. Supply-chain inventories directly reflect lead-times in the process. Manufacturing is often the major component of supply-chain lead-time, which includes the basic time to make the product as well as any scheduling delays introduced by batching orders to support the most efficient operation. Therefore, a reduction of manufacturing lead-time has a direct impact on supply-chain costs. These savings should be considered in ROI calculations for all manufacturing improvements.

Production planning affects procurement and logistics performance

Production planning solutions focus on minimizing WIP inventories and make more efficient use of manufacturing assets. A narrow focus on only manufacturing efficiency can have a negative impact on the rest of the supply chain. Optimal production planning demands constant, real-time revisions to production schedules to keep the plant running at peak performance, but frequent changes to manufacturing plans or scheduling disrupts the overall supply-chain efficiency. Procurement groups are forced to change orders, miss discount opportunities, and use more costly materials and suppliers. Logistics groups are forced to use less than full truckloads, re-negotiate favorable shipping rates, and revise shipping documents. These costs can far outweigh the apparent savings from manufacturing efficiency and are unavoidable in an organization with a silo perspective.

Poor quality and customer service

Customer satisfaction is a key to retaining market share and manufacturing has the primary responsibility for achieving good quality, as the impact of poor quality ripples throughout the entire supply chain. In the case of quality or configuration problems, the supply chain should manage the return and replacement of parts. Actually, they are effectively handling the same order three times (original, return, replacement) and incur three times the cost. These costs should be considered in the evaluation of any manufacturing improvement programs. Factory floor information and production management systems are key to achieving good quality and should be evaluated in light of these external costs.

It is therefore preferable to consider supply-chain benefits in the evaluation of manufacturing improvement programs. Manufacturing ROIs can change drastically when the full impact of supply-chain-induced manufacturing changes are considered. It is important to include supply chain, IT, automation, manufacturing engineering and factory-operating personnel in manufacturing improvement processes. Each of these groups is a key contributor in achieving optimal manufacturing performance, which is extended to resolve supply-chain problems beyond the factory walls.

11.11   Measurement examples

The following is a checklist of typical areas of focus when identifying benefits of an IT initiative:

- Identifying KPIs, to be affected by the IT initiative. These need to be concentrated on.
- IT installation should reduce time – cycle, manufacturing, decision, processing, invoicing, ordering – and these times should be measured.
- Installation should reduce the transport cost through a more accurate and effective scheduling and product placement.
- Overall, the profit margin should be increased through better planning of product mix.
- Installation should reduce manpower through automation or online ordering.
- It should increase product yield or recovery through better process control.
- Reduce rework or process waste.
- The IT installation should reduce inventory – raw material, work-in-progress, consumables, final product, placed consignment stock.
- Reduce stock turn time through smaller and more frequent deliveries.
- Immediate fault detection to reduce the impact on the process.
- Installation increases production volumes.
- Reduce breakdowns or unplanned services.
- Reduce clerical and typing errors.
- Improve customer quality or reduce customer complaints.
- Reduce power/energy consumption.
- The installation can potentially affect any of the following positively and they should be measurable.

    - Bad debt value
    - Number of billing, data processing and design errors
    - Average time spent to correct per error
    - Complaint handling – rebates, reduced price
    - Number of late deliveries of product
    - Value of accounts receivable
    - Premium freight costs due to late order completion or stock-out situations
    - Number of printing errors and time spent in handling complaints and correcting errors
    - Reconciliation time and effort
    - Quality control and sorting good from bad time and effort
    - Number of stock outs in stores
    - Value of unplanned overtime

– Unscheduled downtime hours
– Value of warranty payments.

- Number of hours per day/week/month will be saved in relation to manpower requirements.
- Value of X percentage increase in yield or recovery on an annual basis.
- Reduction of rework percentage or wastage percentage after installation.
- Current level of inventory and the value of the reduction at cost of capital percentage per annum.
- Number of instances of process defects, product loss or waste prevented through early fault detection.
- Annualized value of increased throughput.
- Value of potential throughput/sales lost due to unplanned downtime.

Inventory

Item No.	How Will Benefit be achieved	Current Value	Cycle Time	Lead Time	Potential Value	Potential Benefit	Certainity	Expected Benefit	Annual Benefit
RM	Safety stock	2000	90 d	2 w	5000	15 000	80%	12 000	1 860
WI	Remove Constr								
RM	Order quantity								
RM	Price negotiation								
FP	Sales forecast								

Efficiency

Product or Process	Current yield	Possible Yield	Yield Increase	Feed Volume	Feed Conc	Unit Price	Expected Benefit	Certainity	Annual Benefit
Product 1	93.5%	94.5%	1.0%	576 000	25.5	40.00	58 752	60%	35 251
Product 2									
Product 1									
Product 2									

Labor

Process or activity automated	Activity time	Activity Frequency	Year Hours	Equivalent Employ's	Employees cost	Expected Benefit	Certainity	Annual Benefit
Month end	24 h	Monthly	288	0.15	90 000	13 500	80%	4 050
Reciepe simultation	2 h	Daily	730	0.38	234 000	88 969	30%	26 690
R/M loading	6 h	Daily	2010	1.05	42 000	43 969	90%	39 571
Plan and Schedule	2 h	Daily × 3	2190	1.14	90 000	102 656	60%	61 593
Equipment Tracking								

Others

Current Negative Consequence	Cost Per Occurance	Frequency	Yearly Cost	Estimated Reduction	Expected Benefit	Certainity	Annual Benefit
Batch research	120	3/month	4 320	30%	1 296	30%	388
Overtime	240	3/month	8 640	30%	2 592	30%	778
Rebates	5600	1month	67.2K	40%	26 880	60%	16.1K
Lost Sales	N/A	1month	167K	75%	125K	60%	75.2K
Spares frieght							
Peoduct loss							

Measurement Design

KPI/Benefit	Person Responsible	Baseline Measure	Measure Frequency	Baseline Value	Current Value	Predicted Benefit	Actual Benefit
Safety Stock	Joe Soap	Stockvalu	Weekly	20 000	10 000	1 860	1 550
Efficiency	Batman	Yield%	Daily	93.5%	94.0%	35 251	29 376
R/M load	Lone Ranger	FTE	Monthly	1.05	0.05	39 571	42 000
Overtime	Tonto	Frequency	Monthly	3/month	1/month	788	2 880

12

System integration models and concepts

Learning objectives

- To understand the need for integration between systems.
- To understand the issues surrounding integration between different hierarchical levels.
- To understand the eight different integration styles.
- To become familiar with different integration methods.

12.1   Purpose of integration and interfacing

The main purpose of interfacing is to improve the synchronization between the 'real world' and the administrative world. This interfacing speeds up the information flow between planning and process control systems. It allows tighter control of process timing and improvement of the quality of the production process. By feeding back progress and capacity information to the planning system, several benefits can be obtained such as early detection, and resolution of process disturbances into the schedule can reduce late deliveries.

An up-to-date available capacity overview helps the sales department to improve available to promise (ATP) accuracy. Better visibility of progress information leads to improved sales order control and follow up for the customer service department. Improved inventory accuracy and production timing allows for lower stock levels, thus reducing inventory costs. Defining interfaces require information analysis to sort out reliable and accurate process data for the ERP system, and to define proper production campaign and recipe information for the manufacturing execution system (MES).

The MESs consist of plant-wide information systems, providing information to effectively execute operations to meet business goals. Most manufacturers need some functionality from six categories to succeed in their field. The scope and detailed functionality needed from each application category may vary widely based on process mode i.e. continuous, batch, discrete, assembly or mixed mode. Functionality also depends on the business offering style i.e. make-to-stock, repetitive, make-to–order, assemble-to-order, engineer-to-order (refer Figure 12.1).

The MES touches all of these categories of information systems.

SCM   – Supply-chain management
SSM   – Sales and service management
ERP   – Enterprise resource management
MES   – Manufacturing execution system
P/PE   – Plant and process engineering
Control – Real-time process control

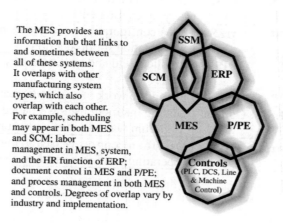

The MES provides an information hub that links to and sometimes between all of these systems. It overlaps with other manufacturing system types, which also overlap with each other. For example, scheduling may appear in both MES and SCM; labor management in MES, system, and the HR function of ERP; document control in MES and P/PE; and process management in both MES and controls. Degrees of overlap vary by industry and implementation.

Figure 12.1
MES context model

This makes MES very important in the overall information infrastructure. Integration between MES and the other five major types of systems is a key to gaining full benefits, not only of MES but also of the other information systems. The MES provides a link between the various systems. It generally links them to the actual production status and capabilities. The context of MES is one of information flow from, to and through the system.

12.1.1   Overlapping with other systems

Manufacturing execution system provides an information hub that acts as a connection between all information systems. It overlaps with other overlapped manufacturing systems, such as:

- ERP and MES can dispatch work to the shop floor.
- SCM and MES include finite scheduling.
- PPE and MES provide process plan and documents.
- Control and MES may include data collection functions.

12.1.2   MES and other systems – information sharing

From	To	Information Transfer
MES	ERP	Costs, cycle times, throughput, and other production performance data
ERP	MES	ERP's plans feed the MES work dispatch
MES	SCM	Actual order status, production capacities and capabilities, shift-to-shift constraints
SCM	MES	Supply Chain's master plans and schedules drive the timing of activities in the plant
MES	SSM	Quoting and delivering depends on the status in the facilities at any given moment
SSM	MES	SSM configurations and quotes provide the baseline of order information for production
MES	PPE	Product yield and quality is very important for fine-tuning of PPE
PPE	MES	PPE drive work instructions, recipes, and operational parameters
MES	Control	Getting instructions downloaded which reflects the facility-wide optimum way to run at a given moment
Control	MES	Data from controls is used to measure actual performance and operating conditions as they change in automated processes

12.2   The gap between ERP and PMC

The different time scales, subjects and goals cause differences between an ERP system and a process monitoring and control system (PMC) which are not readily compatible, but it is quite possible to reconcile data from both into meaningful information for each system.

The ERP systems mainly focus on controlling material flows and have transactional actions such as issue purchase orders, or produce a batch to regulate stock levels after 'disturbances' due to unexpected sales orders. The transactional character of this control requires planning of these actions and time-scales are usually expressed in terms of hours.

A PMC focus on controlling all sorts of process parameters, such as throughput, feed-rate, pressure and temperature. It has continuous controllers at its disposal to regulate these variables subjected to disturbances from the environment. The real-time continuous nature of the process requires immediate action, and the information time-scales are usually expressed in terms of milliseconds.

The incompatibility of these worlds is caused by the transactional vs continuous character. Closing the gap implies being able to reconcile continuous process data with transactional status data. For example, continuously produced material reported by the PMC must be recorded as batch output in the ERP system. In general, timeframes, routings, recipes and statuses can be used for reconciling process data into the ERP system.

To bridge this gap, an MES layer is required. The MES, as defined by AMR, is 'information systems that reside on the plant floor, between the planning systems in offices and direct industrial controls at the process itself'.

There are three layers within AMRs MES-integrated enterprise model (refer Figure 12.2):

- *Planning:* The front office/accounting/financial systems
- *Execution:* The factory level coordinating/trending/ tracking systems
- *Control:* The factory floor/process control systems.

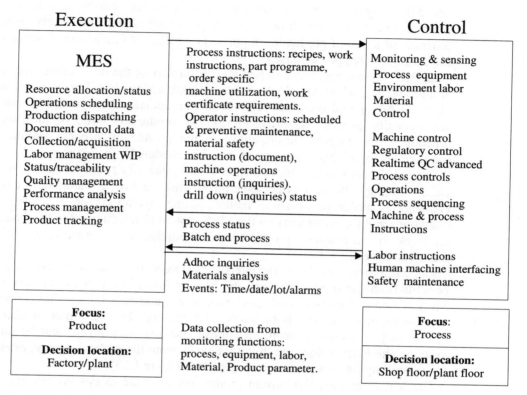

Figure 12.2
The controls layer: controls definition and MES to controls dataflow possibilities

The execution layer is represented by those functions that are in charge of the overall flow of the product and/or process, such as those found in 'SCADA' (supervisory control and data acquisition) systems. A SCADA system would control the execution of a specific recipe using a specification passed to it by the execution system. Central repositories of data are collected from various locations within the factory, rather than being localized within a particular area.

The specific performance of a piece of equipment, or an operator, may not be as important to the execution layer as it would to a control operator. The results from these discrete operations would be blended into the process control data, and then those results would be communicated to the execution systems. The execution layer downloads instructions (for a specific recipe) to the control layer. These instructions supply direction to the people and machines required to carry out the manufacturing operation. It is the function of the controls people and machines to then monitor and control their own operations, to ensure that the outputs are in compliance with the requirements dispatched from the execution layer.

Drill-down inquiries, or status indicators, from the execution layer can spontaneously access information created on an as-needed basis for process control.

There are also bilateral inquiries that can emerge from either layer: these inquiries can be used to measure progress-to-plan; to communicate unscheduled changes; or to announce alarms, events or changes that have occurred.

12.3   ERP–MES integration

The MES interfaces to process control systems for storing and retrieving data are straightforward. The real-time data can be accumulated into a relational database to generate all kinds of production reports. The MES also maintains a production schedule and manages a recipe database.

An ERP system contains an open loop controller of the production process in the form of a production schedule, which proposes future actions based on a model of the process. The scheduling tool does not account for the actual current situation, nor is it able to automatically process disturbances to the production process into the production schedule. Any progress information on the production process should be manually acquired and explicitly incorporated into the schedule by the planner.

The MES needs information, which is not generally found in office systems. In this case, planning systems provide some information, and MES should define additional data elements in order to organize manufacturing operations on a realistic basis. Such deviations and improvements by the MES system also mean that there should be constant feedback to the planning program so that purchasing is aware of any changes in the required materials.

The MES provides the basic interface between planning and execution systems that benefits both the front office and the factory floor. The benefits are even bigger when MES is oriented toward 'real time' production and scheduling. The expanded interface has information flowing both ways with factory floor information aiding the office system in job costing, payroll, lot control and inventory levels, and by interfacing both systems, the ERP system can directly benefit from the process data reported by the MES for generating its own progress reports (refer Figure 12.3).

The MES can copy the current production schedule to synchronize its own schedule. Better visibility of information gives the customer services department a more up-to-date overview of the current situation and a more accurate insight into the progress of orders, allowing it to give richer and more accurate information to customers and improve order follow up. Process data reported by the MES is used to update the proper material flows, inventory levels, and batch statuses in the ERP system enabling more accurate materials administration. This results in better inventory control, allowing for lower safety stock levels, thus reducing the cost of inventory.

It will also lead to reduced costs due to actions taken to cope with unexpected shortages caused by less reliable information. It is even possible and very beneficial to feed progress information back into the production schedule. Usually, problems are only noticed on the planning level once the production schedule is affected, i.e. only by the time a batch should have been finished is it detected that it has been delayed. By then, valuable time has been lost due to a lack of feedback, time which could have been spent on dealing with the problem at a much earlier stage, leading to less severe and less painful corrective actions.

By rescheduling earlier, the impact of process disturbances to customers, i.e., late or postponed deliveries can be reduced. This will reduce costs due to expensive last-minute corrective actions. Besides dealing with problems earlier, closing the loop by feeding progress and capacity information back into the planning and scheduling process has advantages during normal plant operation. The production schedule is normally more realistic, i.e., no actions are proposed unless they can actually be accomplished. No unnecessary pressure is put on the production department.

This will improve production – process timing and quality and will reduce unnecessary delay and production cost. Integrating ERP and MES for automatic information exchange in order to synchronize the administrative system with the real world can be accomplished, but as process control systems and ERP operate in different worlds, the MES should be considered to close the gap between them with respect to ERP's data requirements.

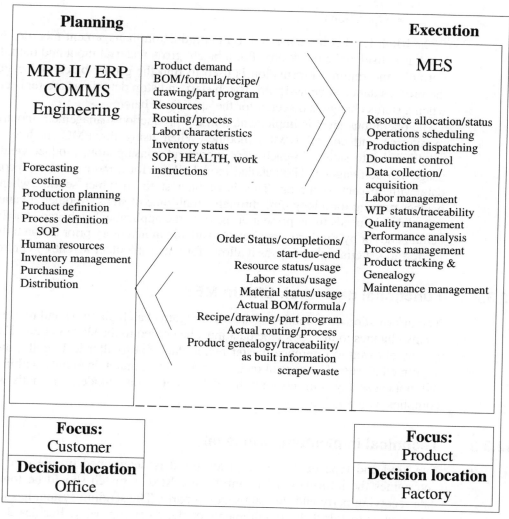

Figure 12.3
ERP to MES data flow possibilities

The ERP–MES interface conveys transactional data only. Its goal is to record dynamic data and to share static data. Examples of dynamic data are events (start, stop) and material flows. Examples of static data are the production schedule and recipes.

The interface should be addressed on two levels:

1. Functional data level what data should be exchanged in what format.
2. Technical implementation levels how should data be exchanged between systems.

12.3.1 Functional data level: MES to ERP

The set of data items offered by an MES differs from implementation to implementation, but, as a rule, this set will cover all production progress data requirements of an ERP system. Progress information on material and resource usage, reconciled with the proper batch global data, is sufficient to enable synchronization with the ERP system.

At the ERP end of the interface, there are two levels for process data collection and both levels are supported by API calls. The levels are:

- Batch
- Operations.

The batch level is implemented in the production management module. At this level, a batch is considered a black box. For a batch, gross material input and output is registered. Formula and routing information is used to find the proper batch and material to which the process data can be assigned. Batch start and stop date/times are registered and stored when a status change is detected for the applicable batch.

The operations level is implemented in the production management module and process monitoring and control (PMC) module. When using the PMC module, each batch is subdivided into steps. To each step, ingredient consumption, product yield and resource usage can be assigned. The detailed model allows for a more accurate recording of batch data with respect to timing. This is of interest to customer service when dealing with batches running for a long time through a multitude of steps. Another advantage is that raw material only needs to be present at the time the applicable batch step is activated, and that certain end- or by-products are already available in inventory prior to batch finish. This is of interest to materials planning, as it allows for much smoother inventory control.

12.3.2 Functional data level: ERP to MES

The information downloaded to the MES regards batch planning and master recipe data.

Any changes to master recipe data must be copied to the slave system. The slave system is only allowed to reference master recipe data, not to alter it. Usually, the MES will be appointed master. The control recipe is created and maintained only within the MES. This will not cause any conflicts with the ERP system's data model, especially when the model complies with S95.

12.3.3 Technical implementation level

The kind and type of data to be transferred is determined by the receiving system. It determines the informational requirements. Next, a match should be found between the data receiving party and the data sending party. The sending system has to be updated or extended as to fulfill the requirements of the receiving party. Besides determining what data needs to be transferred, the reliability and accuracy of data are equally important and also need identification.

The quality of the data exchanged will considerably affect the validity of the synchronization between the ERP system and the MES. This issue concerns the real-time process data to be uploaded to the ERP system. This issue is not as critical for static data and status information as it will always be available and does not depend on the accuracy of measuring equipment. Real-time data validity issues should be solved in the MES system through the use of data reconciliation.

S95 aims to standardize the interfaces from process control systems toward other manufacturing systems, such as MES, LIMS, HS&E and ERP.

The physical transfer of data can be accomplished in two ways:

1. Message file transfer
2. Database table replication.

When employing message file transfer, data files are exchanged across applications, most likely across platforms. Message file transfer can be performed using FTP or some specialized communication tool. Such a tool usually offers features such as persistent store and forward (guaranteed delivery) and automatic recovery after link breakdown. Some even offer remote program execution. When using FTP, additional scripts which implement such fault tolerance explicitly should be programed.

Database table replication is used between tightly integrated systems, where one (slave) system monitors a table in the other (master) system, and updates its own duplicate table when the table changes in the master system. This method is used when there is complete trust between two systems. Data is not transferred in batches on specified frequencies, but in real time as they occur, making the information available immediately.

12.4   MES within an enterprise – data flow diagram

From the left (refer Figure 12.4) we can see the MRP system, which normally does not work in 'real time', but rather in a batching mode. The MRP system notes product usages, customer orders and materials requirements, and sends requests to the execution (MES) layer to build more products to fulfill these needs. The MES systems are responsible for carrying out the product manufacture, and all operations associated with the creation of those products. 'Recipes' can be stored at the MES layer that detail the 'How to build' instructions for the control layer. These recipes are used for both labor and physical devices. The MES systems work in short time spans, but normally not in the speeds associated with control systems.

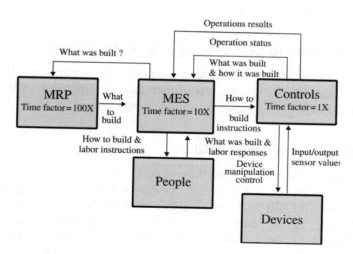

Figure 12.4
MES in an enterprise dataflow diagram

Once the instructions, programs, documents, software and other manufacturing requirements for support systems are transmitted, the controls layer is responsible for carrying out the process. Controls work in real time; this means that there are always operations occurring to refine or correct the process in order to maintain desired tolerances (or outputs). The physical devices and people carry out the finite instructions for process output. Comparative ratios between ERP, MES and PMC in terms of user numbers, focus, amount of data, time between events and amount of interface points are shown below in tabular form.

Boundaries

Level	Users	Focus	Data	Time	Interface
ERP/SCM	High	Enterprise	1 M	100 T	1 K
MES	Combined	Operation	10 M	10 T	2 K
PMC	Low	Process	100 M	1 T	4 K

12.5 Integration architectures evolution

Integration architectures and information flows that evolved between the plant and business came from tactical and strategic business directions, such as the degree of centralized control over the plant and manufacturing strategies, including continuous improvement initiatives, supply-chain management and capacity planning strategies. Integration problems should be solved within the confines of existing business priorities, plant systems, and the current status of ERP and supply chain implementations.

Plant and enterprise systems have their own objectives and characteristics. The lack of standardization across manufacturing sites requires custom integration from the business to each site. Business systems coordinate resource and planning transactions around the customer order, whereas plant systems coordinate and control execution events around available production capacity. To integrate plant and business applications, two or more systems should be connected so that information can flow across system, business function, departmental and application boundaries. This increases the responsiveness and reduces costs in the business.

12.5.1 Planning and processing customer orders

The ERP systems focus on resource planning, procurement, financials and customer order management processes to coordinate demand forecasts, order processing, manufacturing plans and shipping finished goods to customers.

12.5.2 Managing order execution

Plant systems translate business orders into coordinated shop floor schedules such as for work orders or batches and provide results for costing and customer management processes in ERP, so that available production capacity, production resources and work schedules can be coordinated to fulfill customer orders at the lowest cost.

The integration involves technical and business issues that are critical to the success of the project and ultimately affect performance of the business. It is even necessary with packaged applications, as each system is built in its own time with its own context and technology standards.

There are a number of issues that impact the flow of information.

Linking business and plant by information flow

The importance of the link to business performance, such as business planning and plant scheduling, determines how many resources to allocate to the project and who is to be accountable for its execution. For example, integrating quality management information flow is a priority in process manufacturing.

Identifying the business or plant event triggering the flow of information

The trigger may be information flow as a result of a specific business event (such as requesting the status of a customer order) or a technical event, rule, or trigger (such as a batch download of data or a machine stoppage). Three important issues which are driven by the business need for information flow, should be analyzed and planned in the technical and business architecture: technical connectivity, application semantics and data ownership.

12.6   Eight systems architecture alternatives

Manufacturers use eight different approaches or styles of integration, supporting information flows between business and plant systems (refer Figure 12.5).

12.6.1   Style 1: shop floor MRP terminals

The terminals capture data for material reconciliation and labor tracking. It is usually combined with manual and paper-based, shop floor processes, and lacks flexibility to coordinate business changes and support performance improvements on the shop floor.

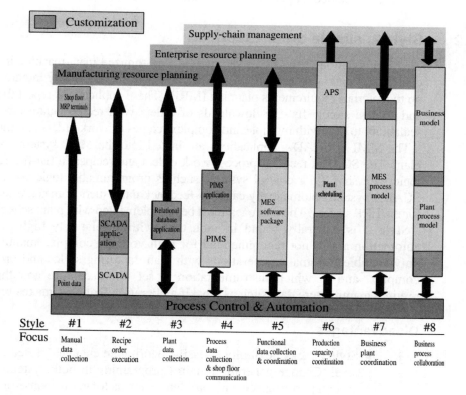

Figure 12.5
Eight systems architecture alternatives link the plant and the enterprise

IT terminals on the shop floor display work orders including schedule, recipe, order specifications, bill-of-material (BOM) and routing information. Operators use production requests with documented shop floor procedures and work instructions to execute production tasks and manually key result data into the system for costing and material reconciliation processes.

A simple top-down transaction-orientated approach, Style 1 is only feasible in smaller manufacturing operations. This style is found across all types of manufacturing in simple assembly or consumer-packaged goods (CPG) operations that do not have complex operations or products. The MRP system provides the connection between business, production and supply-chain processes, and is dependent on accurate shop floor feedback for coordination. This integration style is not recommended for high-volume, complex manufacturing environments.

Disadvantages

- Errors are introduced in the disconnection between manually receiving instructions, executing them and feeding the results back to the business.
- Style 1 does not drive and monitor standard manufacturing execution processes or schedule adherence on the shop floor. Paperwork is still needed.
- The reactive feedback is transactional and not based on time, and it requires planners to drill deeper directly into the plant processes to establish the exact status of order execution. This model is unsuitable for a supply-chain scenario, in which capacity must be coordinated at higher frequencies across multiple sites.
- The style does not support continuous improvement processes on the shop floor. Change must be introduced at either the shop floor procedure level, for instance, paper manuals, or the IT application level.

12.6.2   Style 2: simple order execution

This is a variation of the first style, in which customized man–machine interface/supervisory control and data acquisition (MMI/SCADA) applications directly integrate from the bottom up to materials requirements planning (MRP). These applications report the same information and receive simple flat-file downloads of orders with recipe parameters. The option lacks scalability to deal with multiple and complex recipes or work orders that must be executed.

The MMI or SCADA applications are linked into the MRP systems bottom-up from the plant. The SCADA receives process work orders and recipes in flat-file format and executes logic with shop floor control systems such as programmable logic controllers (PLCs). The SCADA system automatically captures feedback data, such as material usage and quantities, for the ERP system. The approach must be complemented with plant scheduling support.

Style 2 is generally found in industries with a relatively high level of shop floor automation and is not recommended for high-volume, complex manufacturers. It is also only suitable for smaller operations with manufacturing styles and products that are not complex, and in which instrumentation is set up to capture and then feed back key product manufacture data such as a baking operation that only makes bread rolls.

Disadvantages

- Style 2 lacks flexibility to easily incorporate changes caused at customer or plant level. Change generally requires programing in both systems and impacts plant system performance with problems such as longer response times.
- It can only deal with a limited number of product variations such as recipes, BOM and routings.

- The MRP system lacks production and shop floor coordination. Therefore, synchronization with supply-chain planning is impractical, and viewing information across system boundaries is difficult because the two systems, SCADA and ERP, are technically different (transactional vs real time) and are supported by different parts of the organization (engineering vs IT).
- It tends to push more real-time data up into production management and is engineering driven.

12.6.3   Style 3: plant data collection and decision support

A plant database, containing data like material usage and quality, is used for plant-level reporting and feeding the ERP system. It is often custom-built with relational databases as no consolidated application exists across systems in the plant. Instances are data-centric and site-specific and therefore proprietary, difficult to scale and costly to own.

A plant relational database, or a data historian, collects shop floor data from stand-alone plant systems. The approach is found in most types of manufacturing operations. Generally, an internal IT department develops it and, as such, emotional ties lie rooted in agendas of internal, pure IT development groups. The plant database stores data for use by the ERP system. The benefit is a consolidated, standard view of plant data that can be used for analyses with standard reporting tools such as crystal reports and business objects.

The approach is not recommended as a first step to integrating plant and ERP systems. Manufacturers with data-centric approaches should be planning to move to more packaged process-based applications, especially where integration is planned between a number of production sites and the supply chain. The philosophy behind the approach says: We cannot manage it if we do not collect the data! The real issue, however, is that it is difficult to coordinate between business and plant activities if the business process linkage and time context are missing in the data. Leading manufacturers realize the approach is not effective, and as IT groups are downsized and refocused, they are moving to packaged applications.

Disadvantages

- The approach is technical and unique to the company. The costs of change and ownership are high. Database designs in manufacturing take long to build and deploy, and may take such a long time to deploy that the original needs of the company change and the original scope cannot fully meet the new business requirements. The IT groups tend to hang on to the developments to survive in the changing IT organization.
- The approach does not concentrate on production execution and is seen as a tool instead of an application.
- The data lacks business context and does not contain business rules to transform data for the ERP connection or real-time process context in the plant data for use on the shop floor.
- User access to data in the database and technical support is specialized. Sometimes users do not use their system because access to data is complex, a fact exacerbated by a lack of training in internally developed systems.
- Style 3 is not suitable for fast and dynamically changing plant environments, in which constraints tend to move depending on the performance of the process, for instance, a packaging line where the characteristics of the technology are unsuitable for the application.
- Data must be extracted and transformed from real-time shop floor systems.

12.6.4   Style 4: shop floor coordination

This is more robust than a data repository. It also adds a production information management system (PIMS) application to the database as the heart of an MES, and manages the interaction with ERP and adds business context to the raw plant data such as yield accounting, production quantities and quality results.

Style 4 is basically the same as Style 3 with few added applications. Functionality is added to support plant management functions such as, providing time and production context to data, and business rules to add context to data for the ERP system. Style 4 is widely used in the chemical, high-tech and discrete assembly industries and is emerging in the food and beverage industries, in which few packaged application exists to support total plant management. This style is widely used in industries in which information needs to be consolidated for decision support systems (DSSs) and transferred to ERP. It also provides an entry into the MES application domain and requires configuration models to map data to the plant.

The shop floor coordination style is recommended for data-intensive plants that need to be linked to ERP systems and where no packaged application can be used. However, vendor systems still lack functionality to deal with aspects such as scheduling and user-defined business analyses but are evolving. The emergence of products that can extract and display data from a variety of systems onto a common screen, is extending the lifespan of these applications.

Disadvantages

- The system requires special integration to the shop floor and ERP systems.
- The approach is data-collection-focused and not geared toward providing balanced execution, coordination, and analysis capability to plant management. For example, it contains little to no business transactions unless they are extracted from ERP. These solutions tend to be seen as engineering solutions.
- The applications must be custom-configured for a particular site, and replication across sites is difficult.
- A collaboration role with business and supply-chain systems is unsuitable because the approach lacks business context and tends to be data-centric.
- Vendors with developed products in this space are ideal candidates for acquisition by companies that are building suites of larger software product architectures. They provide a means for business-transaction-orientated systems to access plant data via a standard data structure.

12.6.5   Style 5: functional coordination

Packaged MES applications drive functional execution of production orders from ERP according to standard operating procedures (SOPs) or batch recipes. They collect plant data about quality and material, and automatically report it to ERP.

This approach was the first to use packaged modules of functionality in an application. The packaged modules are responsible for coordinating production activities such as configurable workflow processes, dealing with manufacturing standards, work instructions, defect tracking, recipe management and materials management. They are tied directly to ERP systems via process threads like quality, materials and recipe management. Completed work orders provide consolidated information of lot records to the ERP system. They tend to be vertically orientated around solving specific problems for specific industries.

The MES is the best example of applications managing variability or driving hazard analyses and critical control point (HACCP) procedures into shop floor such as in food operations. The advantage of these systems is that the applications are packaged. Manufacturers integrating to ERP for the first time should consider model-based applications using object-orientated modeling techniques that map plant operations and processes to the system configuration.

Disadvantages

- Integration experience with ERP is limited in process industries such as CPG. In industries, such as discrete manufacturing (make-to-order) and high-tech, in which tracking of fast moving production lines is key to customer relationships and sales processes, implementation experience is more common. Integrated applications tend to be built for specific projects.
- The approach still requires data collection and integration to other plant systems such as scheduling, laboratory information management systems (LIMSs) and materials management systems.
- The applications must be custom-configured to fit a particular site. Efficient replication can only be achieved if standardized processes and configurations are used in implementation projects across different sites.
- The applications are moving toward a business process focus but still lack inventory management capability.

12.6.6   Style 6: coordinated production capacity

This style adds scheduling to enhance the coordination of planning and finite capacity scheduling between the business and sites. This style is a logical addition to Style 5.

Style 6 uses the functional coordination role of MES, which includes batch management, work-order tracking and quality management to make plant events like the status of work orders and recipes as they are being executed, available to finite capacity scheduling applications. The resulting set of information flows provides a practical approach to optimize production capacity, and close the loop between supply-chain planning, ERP resource coordination, plant scheduling and execution.

The approach works for all manufacturers with ERP, scheduling and MES applications, and is required to optimize capacity scheduling. It is particularly applicable to manufacturers that intend to leverage MES and ERP investments. These application products intercept process events in the plant for real-time plant scheduling and then uses the real-time events, with MES tracking the order, to integrate to enterprise planning systems.

Disadvantages

- Integration between ERP, supply chain and plant scheduling must still be constructed.
- The applications must be custom-configured to fit a particular site and can only be efficiently deployed across a number of manufacturing sites if effort is put into standardized processes and configurations.

12.6.7   Style 7: business and plant coordination

In this style, MES is integrated with ERP production requests for tighter real-time connections between the plant and the business. The plant may directly execute customer orders based on recipe and specification information.

This is an emerging trend formed out of still-developing model-based MES applications. The difference is that the object technology-based model, configured to fit the production site, is more business process orientated and can support logical information flows between the plant and ERP system. The configurable applications are also easier to replicate across plants within a supply chain. However, they are still new architectures and software products.

Not many users and technology suppliers are experienced at integrating with ERP systems. Manufacturers interested in the approach should start with a simple level of integration to the ERP system. The model-centric approach and emerging standards for software object components, such as S95, make information flows and integration to ERP systems easier. The process threads available from ERP vendors, are evolving to support logical information flows between the plant and business.

The threads include the following:

- Integrated quality management from ERP through to plant execution and LIMSs.
- New product development and introduction from laboratory-based formulation to scale up into manufacturing.
- Total order fulfillment, starting with converting ERP process work orders into executable plant instructions and capturing data for an electronic batch record for ERP.

Disadvantages

- Limited reference sites exist for new technology applications and architectures implemented to support ERP integration on a company-(or even site-)wide basis.
- Functionality of the products still lack support for manufacturing such as manufacturing analyses, inventory management, and integration between planning and scheduling.

12.6.8   Style 8: business process collaboration

This is an extension of Style 7 and uses emerging applications and standards (such as S95) to link the integrated business process model to the MES model to form a common process information structure for plant and business information flows.

A variation of Style 7, business process collaboration is based on a business process model in which the enterprise collaborates with the production process model of production operations covered in Style 7. Still a developing technology, Style 8 still lacks significant references. These architectures will become more prevalent in the next 2 to 3 years as web technology opens up development of process-based information flows.

The application is ultimately aimed at supply-chain-orientated manufacturing, including CPG operations. Moving in this direction should be done cautiously, as there is not enough of a reference base to prove success.

Disadvantages

- Business process collaboration creates a new business process model for enterprise information. While the application can inherit models from the ERP system in place, some of the information flows will need to be based on legacy-application processes and workflows.
- Manufacturers will need to develop formal change management disciplines to ensure that any change in applications is addressed at the business process level and information flow level.

12.7   Integration data identification

The REPAC model (refer Figure 12.6) can be used as a high-level template for the analysis of processes in the plant that need to be connected to the business.

Integration in the EXECUTE process tends to be intrusive as data moves from planning to execution functions. For example, a process work order in the business system is converted to work instructions and routings for execution in the plant, or a general recipe in a process of manufacturing operation is converted to a control recipe to be executed on the shop floor.

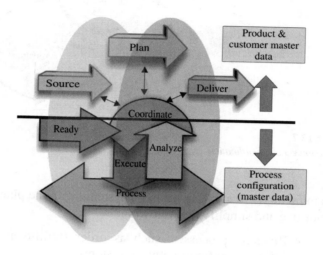

Figure 12.6
Integration style guideline

Integration across *analyze* and *coordinate* processes is generally non-intrusive as decision-support data only needs to be visible across system boundaries, for example, monitoring of cross-functional KPIs or data warehousing applications.

Business analyses should deal with the different coordination processes between the plant and the business (refer Figure 12.7). Information flows between the two processes coordinate the enterprise so that customer response is optimized, and lead-times and costs are reduced.

The two sets of systems have different characteristics and focus:

The transaction-orientated business information system focuses on coordinating, planning, and resources using automated business workflow processes, business logic, and transaction histories. The coordination focus is cost, customer interactions, and plans to fulfill an order in time.

The collection of fragmented plant systems focuses on coordinating order execution in the plant where information and business process are not consolidated. The plant architecture contains a mix of real-time systems to monitor real-time statistical process control (SPC) trends, plant transactional systems to process quality records and automation logic for work order or control recipe execution. This complexity increases the need for architectural planning and business analyses when selecting new plant applications and designing integration.

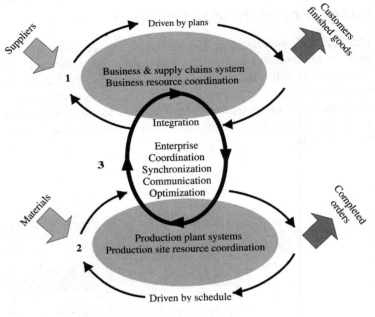

Figure 12.7
Integration enable coordination

Manufacturers integrating business systems with the plant should use business analyses to prioritize and simplify the following:

- Business processes, such as order fulfillment that define touch points and information flows between systems.
- Functionality requirements, such as inventory management, customer order management and recipe management that define the software requirements.
- Integration scope to define the IT infrastructure and technology requirements, network bandwidth and switching requirements.

12.8   Common communication protocol

One of the most prevalent protocols emerging for data transfer between systems and devices is the TCP/IP protocol. The Internet protocol (IP) is a network-layer protocol that contains addressing information and some control information that enables packets to be routed. Along with the transmission control protocol (TCP), IP represents the heart of the Internet protocols. Internet protocol has two primary responsibilities: providing connectionless, best-effort delivery of datagrams through an Internet work, and providing fragmentation and reassembly of datagrams to support data links with different maximum-transmission unit (MTU) sizes.

12.8.1   Internet routing and routing protocols

Internet routing devices have traditionally been called gateways. Interior gateways refer to devices that perform these protocol functions between machines or networks under the same administrative control or authority, such as a corporation's internal network. These are known as autonomous systems. Exterior gateways perform protocol functions between independent networks.

12.8.2   IP addressing

As with any other network-layer protocol, the IP addressing scheme is integral to the process of routing IP datagrams through an Internet work. Each IP address has specific components and follows a basic format. Each host on a TCP/IP network is assigned a unique 32-bit logical address that is divided into two main parts: the network number and the host number.

When computers communicate with one another, they exchange a series of messages. To understand and act on these messages, computers must agree on what a message means (a common language). A Protocol is a standard set of rules that governs how computers communicate with each other.

12.8.3   Transmission control protocol

The TCP is low-level software logic that makes data transfer and communication possible between system components. The TCP provides reliable transmission of data in an IP environment. It corresponds to the transport layer of the open systems interconnection (OSI) reference model.

Figure 12.8 maps the many protocols of the IP suite onto their corresponding OSI reference model layers.

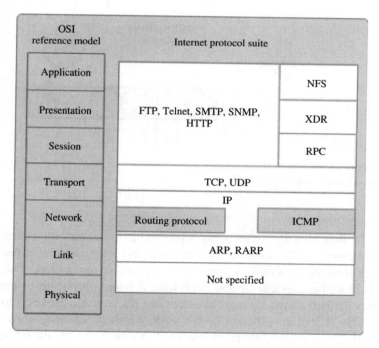

Figure 12.8
Internet protocol suite and the OSI model layers

Among the services TCP provides are stream data transfer, reliability, efficient flow control, full-duplex operation and multiplexing. With stream data transfer, TCP delivers an unstructured stream of bytes identified by sequence numbers. This service benefits applications because they do not have to chop data into blocks before handing it to TCP. Instead, TCP groups bytes into segments and passes them to IP for delivery (refer Figure 12.9). TCP offers reliability by providing connection-oriented, end-to-end reliable packet delivery through an Internet work.

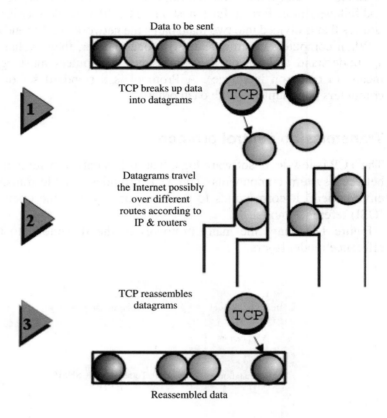

Figure 12.9
Interaction between TCP and IP

It does this by sequencing bytes with a forwarding acknowledgement number that indicates to the destination the next byte the source expects to receive. Bytes not acknowledged within a specified time are retransmitted. The reliability mechanism of TCP allows devices to deal with lost, delayed, duplicate or misread packets. A time-out mechanism allows devices to detect lost packets and request retransmission.

The TCP offers efficient flow control, which means that, when sending acknowledgements back to the source, the receiving TCP process indicates the highest sequence number it can receive without overflowing its internal buffers.

- *Simple mail transfer protocol (SMTP):* It is used to send and receive electronic mail.
- *File transfer protocol (FTP):* It is used to transfer files between different types of computers (MAC, PCs, UNIX, etc.).

- *Hypertext transfer protocol (HTTP):* It is used too transmit information on the World Wide Web.
- *Network news transfer protocol (NNTP):* It is used to transfer network news.

Information technology in today's environment is the most important cornerstone of an enterprise's ability to successfully compete in the global marketplace. The channel used by the organization to leverage the potential of information is critical.

The impact of this visibility in IT terms is that systems interaction can now be handled by a standard communication protocol. The suite of Internet protocols thus enables seamless information sharing from both a systems integration and geographic point of view. The Internet and the resulting e-business possibilities are not only a change in technology, but also a change in the way that people are communicating, working, playing and living. In the short term, e-commerce offers organizations the attractive possibility of differentiating themselves from their competitors in a commodity market by making their products available via the Internet.

There are three primary interconnectivity schemes:

1. Direct database access
2. Interprocess messaging
3. Flat files transfer.

13

Product and vendor evaluation methodology

Learning objectives

- To understand the implications of choosing a product vendor.
- To understand how vendors should be evaluated.
- To understand how vendor product should be chosen.
- To develop a vendor/package evaluation process.
- To understand the latest trends in user expectations.

13.1   Software vendor functional scope

Software applications are not uniform as every plant operates quite differently, both to accommodate uniqueness and to preserve a competitive edge (refer Figure 13.1). This diversity means the software must be configured to match the plant model and business rules and there are many types of software to choose from. Buyers of software applications must determine and evaluate their own requirements, system functions and types.

There are two major dimensions of choice in software:

1. Functional scope or breadth and associated architecture issues.
2. Match of functionality to how the plant actually operates and its business needs.

Process engineers, management and procurement professionals are increasingly being called upon to take more and more factors into account when purchasing IT solutions. While the universal system with perfect fit for every situation is often sold, it is seldom supplied. Predictions indicate that the best supplier suite is likely to become increasingly chosen over integration of best of breed products.

An organization's needs and the software products that serve them may include different combinations of functionality. In some plants, the primary need is resource allocation and status and having an ability to see and direct what each machine and person is doing, the other measurement and tracking functions may not be needed to produce competitively. Some plants need many functions; quite a few plants may opt for only a few of them in a full-fledged version and a few others at a rudimentary level.

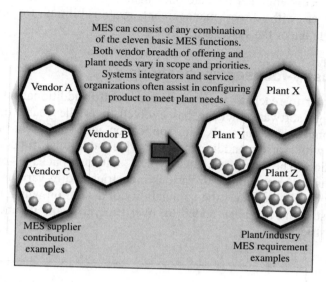

Figure 13.1
Producer and consumer model for MES

13.2 Software selection trends

Within the industrial workplace it is becoming increasingly difficult to choose software which is applicable for the users needs and provides the required value addition to the business. The gap between software writer and software user is increasing at an alarming rate.

13.2.1 Software suite evolution

The days when a process engineer developed a piece of software, using commercial tools, on a desktop platform, are all but over. The special skills required to configure, adapt or modify a piece of software to do something slightly different are no longer within the realm of knowledge of the typical process engineer. Whereas 10 years ago, a staff manager could study a bit of Basic and write a small program to, for example, transform some data into a locally used unit of measure, the expertise for doing this within XML is no longer as easily gained.

During the same 10-year period the complexity of the average industrial IT system has increased many-fold. What previously was a simple graphical user interface, with limited options for the display of information, is now a highly customizable package with the ability to read information from multiple sources on the network, integrate and manipulate the information in various ways, and display the information on, for example, an Intranet page which can be viewed anywhere around the world.

Within a chemical manufacturing environment, for example, the process engineer of 10 years ago was usually charged with the maintenance of the plant's entire control system, including data acquisition, manipulation via PLC ladder logic, tuning of PID control loops, display on SCADA HMIs, generation of printed reports, regular backup of the software and even development of new functionality for plant extensions/changes. As new developments, such as advanced control methodologies, Internet/Intranet technologies, sophisticated data storage mechanisms and others, were developed, so that the process engineer became once again, as was the case before the IT revolution, a staff manager with direct management responsibilities for control rather than intimate knowledge of the systems involved.

13.2.2   Industry-focused applications

In order to reach a larger market and fulfill additional needs at current customers, the average software package has evolved, over the years, into a multi-functional, highly customizable set of tools with application across different industries, in various purposes, and with various additions to provide additional functionality and improve market position. The result is that a typical software package, or offering, could be mapped against its applicability to various industries and in various roles within an industry. The table below illustrates the applicability of a solution which has been written for a specific industry and has been added to, over time, to make it functional in a number of related industries.

Profile of a typical solution package (industry focused)

	Food & bev	Chemical	Water & waste	Petro-chem	Pulp & paper	Metal & mining	Power
Plant control	Difficult but possible	Easy application	Customizable to purpose	Perfect fit	Easy application	Customizable to purpose	Difficult but possible
Advanced control		Customizable to purpose	Customizable to purpose	Easy application	Customizable to purpose		
Product tracking	Customizable to purpose	Customizable to purpose		Customizable to purpose		Customizable to purpose	
Design simulation				Difficult but possible			
Asset management	Customizable to purpose	Customizable to purpose	Customizable to purpose	Easy application	Customizable to purpose	Customizable to purpose	Customizable to purpose
Alarm management	Customizable to purpose	Easy application	Easy application	Perfect fit	Easy application	Customizable to purpose	Difficult but possible
Field device control		Customizable to purpose	Customizable to purpose	Easy application	Customizable to purpose		Customizable to purpose

Legend:
- ● Perfect fit (made for purpose)
- ◖ Easy application
- ◗ Customizable to purpose
- ◢ Difficult but possible

This hypothetical solution has been written for the petrochemical industry as a plant control and alarm management system. As time has passed it has been adapted within the industry to provide functionality in all the other areas except design simulation. Characteristically this would have been done at one site and then become a universal tool with various degrees of applicability elsewhere. In similar fashion when the package has been used within a different industry vertical, for example food and beverages, some of the functionality has been found to fit well but other areas may have needed extensive rewriting for good application.

13.2.3 Functionally focused applications

The second table depicts a system written as a plant control solution with applicability to a wide range of industries. Advanced control methodologies have been incorporated into the system, but the specific requirements of some individual industries (e.g. legal requirements in the pharmaceutical industry) may not have been met. The system has been applied in other horizontal areas with some success too.

Profile of a typical solution package (Function focused)							
	Food & bev	Chemical	Water & waste	Petro-chem	Pulp & paper	Metal & mining	Power
Plant control	◕	●	●	●	◕	◕	◕
Advanced control	◕	◕	◑	◕	◑	◑	◕
Product tracking	◑	◑	◑	◑	◢		◢
Design simulation							
Asset management							
Alarm management	◑	◕	◑	◑	◑	◑	◢
Field device control	◑	◑	◢	◑	◑	◑	◢

● Perfect fit (made for purpose)	◕ Easy application	◑ Customizable to purpose	◢ Difficult but possible

In deciding on the optimal solution for a specific plant, the process engineer is faced with a number of complex problems. Should a single system be chosen which addresses most of his needs well, but performs poorly in some areas, or should the best possible solution for each need be chosen and the various systems integrated to give an overall solution?

The trade-off is between lack of functionality in some areas, or additional cost to develop functionality where it is needed, and the cost and possible difficulties of integrating systems which are often perceived to be written specifically not to integrate with competing products. The next section explores the choices available to the plant manager.

13.2.4 Selection issues

Four hypothetical products are here compared as a plant manager would compare them. Vendor A's product is an overall good product which provides some functionality well and some which can be customized. Vendor B's product has excellent plant control capability but is mediocre in some other areas. The product from vendor C is excellent in most areas but will be quite difficult to apply in the design simulation and field device control areas. Vendor D's product is similar but has good characteristics in other functional areas.

Choices for a specific application

	Vendor A	Vendor B	Vendor C	Vendor D
Plant control	◕	●	●	◕
Advanced control	◕	◕	◕	●
Product tracking	●	◑	◕	◗
Design simulation	◕	◕	◴	●
Asset management	●	◕	●	◔
Alarm management	◐	◕	●	◐
Field device control	◐	◗	◴	●

The decisions the engineer may make include:

- Purchase best-of-breed modules from each of all four vendors and integrate them to work together, receiving the best possible solutions but possibly significantly increasing total cost.
- Purchase modules primarily from only one supplier and add modules from other suppliers to give the required application.
- Purchase a single solution and accept the additional cost required to customize the system to provide the required functionality.
- A final option would be to ignore the four vendors and have a solution built specifically for the application (bespoke solution). In recent times bespoke solutions for any application of significant size have become very expensive due to the expertise required and not often considered.

ARC Advisory Group published the results of a survey to determine trends in these decisions. Specifically the question asked was how systems are currently purchased and how will they be purchased in future? A number of interesting results were shown:

- Some systems, for example control systems and HMIs, are usually purchased together. Other related systems, such as production scheduling and production control systems, are generally purchased separately.
- The overall predicted trend is that systems will be bought more as a cohesive whole and less custom built integration will be required in years to come.

System integration suppliers thus face significant threats to their continued existence. Some have already responded by aligning themselves with specific products and focusing on making those products work in as many environments as possible. Others are focusing on the integration of systems and reducing the cost of integration, the perceived cause of the trend. No single optimal methodology is on the horizon yet.

In conclusion the message is this: decisions made now are going to be proved wrong or right in the long term. The success of plant IT system installations depends on many factors and this number is growing rapidly. Choose your consultants wisely, ensure they are working for your best interests, and make your decisions quickly.

13.3   Product landscape

Software vendors historically had a hierarchical focus on application development. They focused on a specific hierarchical level and attempted to develop the best products on that level. This brought about various interfacing/integration issues between the different levels in any organization attempting totally integrated solutions. The following graphic (Figure 13.2) depicts a typical enterprise systems functional architecture that are not integrated but patched together.

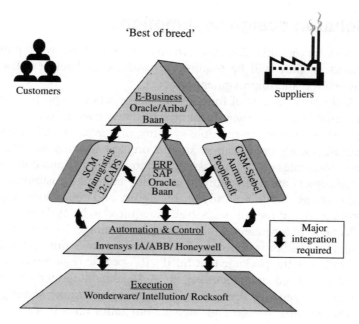

Figure 13.2
Best of breed solution

The above graphic is currently a typical reflection of the e-manufacturing landscape. This situation is changing though, and is soon going to look quite different. Most of the big, reputable global software solution providers have started acquiring smaller vendors with point solutions to integrate into their product suites to enable a totally integrated solutions platform from one vendor.

The ERP vendors have started either acquiring or developing their own SCM, CRM and e-business solutions. This strategy seems to be a good fit in terms of strategy and functional fit, as these enterprise planning systems are mostly transactional based and use similar time-based information. They have also started to move down into the MES hierarchical level in terms of functionality. The success of these initiatives are yet unproven, and will only be proven with time, especially the move into the MES domain.

PMC and pure SCADA vendors have started either acquiring or developing their own HMI, SCADA, Historian and/or MES solutions. These vendors have a vision of providing a totally integrated solution from the instrument or PLC right up to the ERP interface. This also seems to be a good strategic fit, as a lot of the data is real-time based or plant-event driven. These initiatives seem to be progressing at a faster rate than the SCM/ERP convergence, and seem to be quite successful in terms of actual product integration. Currently few vendors are able to provide total functional coverage of the PMC and MES domains, although this may change in time.

With the current convergence of systems, it is quite possible that within the next couple of years, the five hierarchical levels will be completely covered by only two types of vendors, enterprise solution providers and automation solution providers, with the enterprise-control integration standard (S95) providing the standard for the only integration interface between systems from different vendors. The likelihood of one vendor providing a total solution from instrument to SCM seems remote at this time.

13.4   Solution design assumptions

Unless an organization has already aligned itself with preferred automation solution providers, they will be required to evaluate and select software and vendors for future automation and improvement projects.

A simple checklist of functions does not serve the purpose. The selected product should be suitable for the manufacturer's business over a broader range of functions, with the ability to integrate to multiple systems, on the same and different hierarchical levels. The manufacturing organization should prepare for this new technology, and the software must integrate well with existing business planning and controls level systems.

Customers are demanding more agility and mass customization from manufacturers and the systems should not only enable agility and continuous improvement in manufacturing operations, but the software itself is required to be agile and easily changed/improved.

Vendor independence is a strategic advantage for both the system integrator and the client, as neither is committed to any one package during the design phase. The client can now choose the package that fulfills its specific requirements, after the requirements have been identified and documented. A system integrator can propose the best fit for the client, and generate revenue by configuring and implementing the package.

This sounds simple and logical, but unfortunately very few clients actually execute initiatives in this way. To understand why, two basic assumptions need to be explored.

13.4.1   Strategy of vendor-independent design

The proposed business process design and functional solution design are based on business requirements and not on any package functionality.

When designed in this way, any proposed functionality adds value to the business process due to its merits, but is not included just because it already exists in a package. The client then has the right to choose the package of his/her liking.

13.4.2   Strategy of fit-to-package design

The package is a product that needs to be sold at all costs and business processes are to be modified to accommodate package functionality. The functional design is based on package functionality and lot of non-value adding functionality may be proposed as it is already part of the package.

Clients have no option but to choose a product from the supplier's portfolio. Package functionality evaluation is biased toward supplier experience and expertise, and there is a chance that a supplier will attempt to bias the evaluation and selection process based on its track record of implemented packages.

13.4.3   Design conflict resolution

The apparent conflicts of the two strategies need to be resolved through a process that instills confidence and negates the areas of conflict. The conflict is: 'What is best for the business' vs 'What is best for the package vendor'. The proposed approach will reduce the perception of vendor bias and provide the client with the best possible solution.

13.5   Proposed approach

The MES software evaluation/selection process is intended to help manufacturers shorten the time to evaluate and select the software, whilst getting the most suitable and beneficial product to satisfy the business requirement. This process is specifically critical when an organization is in the process of considering interactive multi-functional or cross-functional systems with broad organizational impact. This process guides the manufacturer in ascertaining who to involve, what to be done, and how to communicate with the software vendors.

It is necessary for the manufacturer to make a good selection in a systematic timely manner. The overall time required to complete the process depends on the scope of the requirements, the number of vendors and the experience of the evaluation/selection team. Software evaluation and selection can take a long time (4–6 months) but can be shortened (2–4 months) if the proper time is committed to all the processes and steps recommended. The level of detail to consider in determining requirements and evaluating vendors is critical to timely selection.

The following process (Figure 13.3) is proposed for clients who indicated that they would prefer a vendor independent business design, and for the selection of vendor packages to satisfy the business requirements.

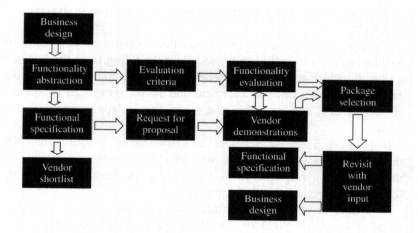

Figure 13.3
Evaluation and selection process

13.5.1   Business functionality design

The current business process, current systems architecture and functionality, and client functionality requirements are analyzed and evaluated, and a 'To Be' or 'Future state' business process and functionality design will be performed based on:

- Business objectives and measures
- Business processes supporting objectives and measures
- Business functions that need to be enabled
- Critical success factors for solution success.

In the case of MES and SCM projects, the organization needs to identify the specific requirement that will serve the business best in terms of functionality and cost. This is especially important when sourcing MES products, as many vendors claim to have MES products, but in fact only provide manufacturing information systems (MIS). The differentiating factor between these two is execution ability.

S95 does not specify any execution processes, only the data models and structures, so complying with this standard does not necessarily qualify the vendor product as MES, even if vendors may claim so.

Manufacturing information systems is a tool that collects and manipulates data into decision-making information, makes it available to management in the form of reports, and also structures the information in such a way that it can be integrated to the ERP system. Users may make decisions based on the information required, but these are not necessarily forced or captured in the system, merely reported on after the fact.

Manufacturing execution systems is a tool that not only collects and manipulates data, but also contains business rules that govern and drive the manufacturing business processes of the company. It does not merely report on historical activities, but intervenes at critical stages of the process according to business rules. Direct user interaction with the system is more prevalent with MES, as users need to make decisions and feed this into the system at the stages where choices are available. An MES may also interact directly with control equipment such as PLCs, pumps and valves, starting and stopping processes much like a SCADA, but on the business process level.

For instance, the MES may receive a signal that a specific piece of equipment has stopped unexpectedly. The MES will request from the operator a decision regarding next steps as required by the business process that want to be forced – 'Generate maintenance request?'

The operator may decide not to generate a maintenance request on the ERP system, but will be required to include a reason from deviating from the business process. This drives a specific work-method that is best for the company, but also allows flexibility in the process, so that work can carry on and is not prevented by the system in extraordinary circumstances, and provides history of these circumstances through the capture of the reasons.

If it is decided to generate a maintenance request, the MES will populate the maintenance request fields automatically from the configuration tables, and the only thing the operator is required to do is to confirm the information and possibly add notes regarding the reason for the stoppage.

It is thus critical for organizations to decide what they really need, MIS or MES. Depending on the industry and business requirement, MIS only may be enough (e.g. material tracking in a continuous chemical plant), or a full-blown MES need to be implemented to satisfy the requirements. It is also possible that different components of the total solution will be either more MIS or more MES, depending on the business need.

13.5.2   Functionality abstraction and specification

The next step will be to abstract from the 'to be' specific, critical, business functionality and generate a functional specification based on:

- Functionality that support the business processes
- Functionality and other technical specifications that are critical for the success of the project
- Functionality that supports the business objectives
- Functionality that will enable measurement of objectives and improvement after implementation.

This functionality will be used as the evaluation criteria for package selection and to create a vendor information document.

13.5.3   Vendor shortlist

As high-level requirements are being defined, the evaluation/selection team should create a list of products and vendors that may potentially meet the general, high-level requirements. Researching various products, attending product overviews, demonstrations, etc. allows the team a chance to review requirements and evaluate vendor competence.

The initial vendor list can be prepared from the following sources:

- Trade and professional shows (APICS, MESA, SME/Auto Fact, etc.)
- Trade and professional journals and organizations (APICS, MESA, SME, ISA)
- Directories of manufacturing systems software
- Specific industry consortiums (i.e., aerospace and defense, automotive, chemicals, electronics, food, metals processing, pharmaceutical, semi-conductor, etc.)
- Manufacturing consultant publications (AMR, ARC and Gartner)
- The experience of a systems integrator and/or manufacturing consultant, many of them have already participated in evaluations, selections, and imple-mentations
- Reference from manufacturers in the same or similar industries
- References from current hardware/software vendors for alliances.

The team creates and distributes a vendor survey form to solicit vendor responses to questions related to the manufacturer's industry in order to check the vendor's experience in this area. Various automation products have evolved with unique industry-related strengths that may or may not support the high level requirements the team has selected.

This list will eliminate those vendors that will not fit the business model of the customer in terms of more commercial criteria such as:

- Commercial considerations

  - Financial strength (financial reports, profitability)
  - Vendor company and product – mission and vision, % revenue into R&D

- Support approach (local and international)

  - Methodology and access (technical and functional) – training, escalation support, direct access, helpdesk, 24/7, response times, website
  - Maintenance policies and service level agreements

- Upgrade/licensing approach
    - New releases (frequency), previous releases supported (number), upgrade paths
    - Licensing cost structures
- Installed/user/resource base (local and international)
- Available functionality
    - Customization/configuration/business rules/modeling – environment and tools available
    - Skills/training required for in-house maintenance
- Solution platform and integration issues
    - Open environment, architecture, web-enabled, network protocols
    - Data extraction capabilities, data conversion tools, ease of integration (up and down)
- Security, audit ability and control
    - Built-in audit trail, access control of users/groups, change/access logs kept
    - Built-in security and authentication, synchronized with external systems.

After comparing the high-level requirements to what is available, by looking for vendor experience in specific industries, and by sticking to a preferred technological architecture, the list of possible products can be refined and reduced to three or four.

13.5.4 Request for proposal

Detailed requirements help the organization and the vendor identify and estimate the cost of customization. Future cost overruns and delays can be removed or minimized by more concise and complete documented requirements. With the ultimate goal of selecting a single integrated solution, the organization should minimally prepare a requirements document. This document is submitted to the short-listed (not more than three or four, if possible) vendors for their responses. The more formal and proposed alternative is to submit a request for proposal (RFP) that contains the functional specification. It includes more information for and seeks more in return from the vendor.

The RFP should contain three elements:

Description of the business and the areas to be addressed

This section identifies the business, the industry and how the products are produced, from the point at which orders enter the shop to the point they are ready to be shipped. A current situation analysis describes the driving factors and the objectives for the implementation. This information should be the same as that of the functional specification.

A list of functional and technical requirements

Adding detail to the high level requirements generates the detailed list of functional and technical requirements as per the functional specification document. The vendor is required to address the functional and technical requirements, point by point regarding whether or not the package complies, or requires custom work.

For example, when detailing the requirements for product tracking and genealogy, some of the detail could include such line items as:

N.0 product tracking and genealogy;
N.1 must associate material ID of material consumed with lot ID;
N.2 must keep track of serial numbers of components within lots;
N.3 must be capable of attaching comments (up to 100 at 80 characters each) to lot ID
N.x etc.

This section also provides some typical business scenarios that must be managed by the system. Some of the basic elements of these scenarios include:

- The name of the process or function and its objective
- Inputs to, and outputs from, the process
- Processing performed.

This section allows the vendor an opportunity to provide demonstration of compliance so the user can actually sense the operation of the system.

Commercial issues regarding the proposed project

This section requires the vendor to provide fixed or budgeted costs for the solution software and implementation. It may also contain information regarding the implementation approach for the proposed project. The costs should ideally be broken into:

- Software costs
- Hardware costs
- Manpower costs
- Other expenses (such as S&T, travel, accommodation, taxes, discounts, etc.).

Meetings with vendors (or drafting replies to questions) should be allowed for to clarify requirements and reach a common understanding of the scope of the functional specification and requirements. A formal correspondence process should also be implemented to ensure that all vendors have access to the same information (answers to vendor questions and vendor meeting minutes) after the clarification process has been completed. This process needs to be explained to all vendors in the RFP so as to protect them from revealing trade/technology secrets during meetings or written questions. The RFP should also indicate the need for a solution demonstration for vendors that proceed to the next stage.

Responses to the RFP and the cost estimate to implement the solutions may vary widely and some vendors may decline to bid after understanding the requirements. Vendor responses provide the opportunity to narrow the list of potential products further before proceeding into the next stage, package demonstration.

13.5.5   Evaluation criteria

In parallel with the RFP generation process, the evaluation (scoring) criteria need to be established from the functional and technical requirements. A MUST and WANT scoring matrix for functional and technical requirements are proposed. MUSTS carry a GO/NO GO decision, so if a package does not comply fully, it will not be further evaluated (NO GO) as it misses one of the fundamental value-adding functions. Not having this will defy the objectives of the project.

If however, not any of the evaluated packages meet that specific criterion, its criticality needs to be re-evaluated, and it may be demoted to a want. For instance, if the objective of the system is tracking in order to determine product genealogy, and the system can satisfy all requirements but tracking and genealogy, is it still worth evaluating? Unfortunately, some companies will answer yes to this question.

The requirements not as critical as MUSTS (will not disqualify a vendor package) are called WANTS (functional and technical requirements) and are weighted 1 to 10 where 1 is the least important WANT and 10 the most important WANT. The others are weighed as compared to the 1 and 10 wants.

Product Evaluation: General & Instrumentation

		Product 1		Product 2		Product 3		Bespoke
MUST		Comments	Go/No	Comments	Go/No	Comments	Go/No	Comments
Windows-Like GUI			Y/N		Y/N		Y/N	
NT4 / Windows 2000 Server			Y/N		Y/N		Y/N	
Configurable by non-programmer			Y/N		Y/N		Y/N	
Full Support In SA			Y/N		Y/N		Y/N	
Bar Code Support			Y/N		Y/N		Y/N	
Instrumentation Integration			Y/N		Y/N		Y/N	
CAN	Wt	Comments	Sc Wt Sc	Comments	Sc Wt Sc	Comments	Sc Wt Sc	Comments
MIMS Integration (registration & publication)	5		4 20		5 25		3 15	
MS SQL 7.0 Compatibility	5		5 25		5 25		0 0	
Fully Auditable Transaction Based System	4		5 20		5 20		4 16	
Training Database possible	2		5 10		4 8		5 10	
Archiving Possible	3		5 15		4 12		4 12	
Online help & SOP's	3		4 12		4 12		4 12	
Instrument Maintenance Management	2		3 6		3 6		4 8	
User support available	4		3 12		2 8		4 16	
	28	Total	120	Total	116	Total	89	Total
Percentage Fit			86%		83%		64%	

Product Evaluation: Reports

		Product 1		Product 2		Product 3		Bespoke
MUST		Comments	Go/No	Comments	Go/No	Comments	Go/No	Comments
User Definable Reports			Y		Y	Oracle Reporting	Y	
CAN	Wt	Comments	Sc Wt Sc	Comments	Sc Wt Sc	Comments	Sc Wt Sc	Comments
Support for Crystal Reports	4		5 20		4 16	None	0 0	
Consumables Tracking	3		4 12		4 12	Additional Software	4 12	Partial
COA's with comments	4		4 16	Code Required	3 12		5 20	
Scheduling Functions	3		4 12	Code Required	3 9	Additional Software	4 12	
Analyst Performance Tracking	4		4 16		5 20		5 20	
	18	Total	76	Total	69	Total	64	Total
Percentage Fit			84%		77%		71%	

Product Evaluation: Summary

	Product 1		Product 2		Product 3		Bespoke
Group	Comments	Go/No	Comments	Go/No	Comments	Go/No	Comments
General & Instrumentation		86%		83%		64%	
Sample							
Configuration							

13.5.6 Demonstration, evaluation and selection process

The next step would be to invite the vendors to demonstrate their competence and package functionality in order to do an evaluation. The evaluation and selection process is facilitated using the evaluation criteria score-sheet for the following reasons:

- It enables objective scoring of functional requirements.
- The evaluation criteria score sheet helps to guide the process.
- It eliminates the danger of being side-tracked by 'bells and whistles'.
- It provides structure for the evaluation process.
- It provides a checklist to ensure that all criteria have been demonstrated/evaluated.
- It removes vendor bias and provides an even chance for each participating vendor.

The criteria are evaluated and the WANTS are scored on a scale of 1–5 or 1–10 (to be decided before the meetings commence) by each participant individually. The scores are then combined and the winner identified based on the highest accumulated score.

13.5.7 Conclusion

By using the proposed vendor and package selection process the bias is reduced, as the client is responsible to do the evaluation and the system implementer only acts as a facilitator. All scoring is done by, and decisions are made by the client.

13.6 Design revisit

Once the most appropriate vendor has been selected, the design should be revisited to decide what to do with criteria with low scores, will they be removed from the requirements document, or will they be custom-developed.

If any of the MUSTS have been demoted to a WANT, the business design as well as the functional specification needs to be revisited after discussion with the vendor.

13.7 System functionality and architecture design

Once the package selection and design revisit process has been completed, the system architecture and functionality can be documented as a part of the technical design for solution delivery.

13.8 Evaluation and selection teams

After the need for the solution has been analyzed and the company leadership understands how it affects the entire organization, the software evaluation/selection team can be formed. Experienced consultants and/or systems integrators could be a part of the team. Manufacturing consultants should provide an outsider's experienced perspective in analysing business process improvements and ensure that the new system's requirements are driven by the entire organization's requirements, and not by any individual department's needs.

Systems integrators should assess prospective systems as to whether they will accomplish the performance objectives set forth by the organization. Systems integrators should assist in evaluating integration requirements, responsiveness, flexibility, and delivery and retention of critical information. Both consultants and integrators provide experience in methodologies and processes as well as knowledge of software and hardware vendors.

The evaluation/selection team can be divided into two groups:

1. The review team
2. The evaluation/selection team.

The two groups are differentiated by the number of people, areas of expertise, responsibilities and time commitment as follows:

13.8.1 Product review team

This team consist 5–6 persons from various manufacturing functional areas and 1–2 persons from manufacturing leadership.

Responsibility

- Review progress of selection team
- Ensure that the evaluation/selection process is being followed
- Provide functional area expertise as needed
- Approve the final decision.

13.8.2   Product selection team

Consist 3–4 full-time expert representatives, one each from manufacturing information systems (MIS), production, production control or scheduling, quality assurance or process engineering, and an outside consultant to facilitate the process and one full-time project leader having the following capabilities:

- Knowledge of entire plant floor operations with experience in more than one functional area
- Needs to manage the level of detail
- Excellent communications, coordination and conflict resolution skills
- Connections/contacts throughout the organization
- Responsible for arranging the team's meetings, trips, presentations
- Acts as single point of contact for vendors.

Responsibility

- Determine high-level requirements
- Create initial vendor list
- Review/evaluate products, survey vendors
- Prepare Requirements Document and/or RFP
- Meet with vendors
- Compile results and select the product (make the final decision).

13.9   Visits to reference sites (if required)

The vendor should be able to provide references in the same or a similar industry to that of the manufacturer. A visit to such a site can afford the opportunity to:

- Estimate implementation times and effort involved
- Assess vendor reliability and cooperation
- Witness user friendliness of product and responsiveness
- Assess vendor post-sale product support.

13.10   Vendor survey form example

1.	Vendor contact, title
2.	Organization background
	Date established
	Total number of employees; number supporting MES products
	Company ownership; private or public
	Percent of business in USA; list other areas and percentages also

3.	Financial Information Annual sales Annual net income Current ratio (current assets/current liabilities) Net worth Annual sales applicable to automation solutions % of solutions toward software licensing % of solutions toward hardware sales % of solutions toward consulting/implementation support % of total revenues put back into solutions research and development

4.	Alliances/partnerships
	Interfaces developed for other software (ERP, controls, data collection, reporting,
	    MES functions, others)
	Software from other vendors integrated into the solution
	Alliances with hardware/operating system vendors and systems
	    integrators

5.	Product support Help desk support? 24-h hot-line? Maintenance agreements and costs? What types of training available? What documentation provided?
6.	Product information Number of customers; number of installations; notable customers by industry Frequency and types of releases? Major product functions Applicable industries (discrete-lot, discrete-repetitive, batch process, continuous process) Development languages used International languages provided Product pricing; by user? site license? high/low/average
7.	Technical architecture Hardware platforms; current plus future (6 months to 1 year) Operating systems; current plus future (6 months to 1 year) Databases used Graphical user interface (GUI) Network protocol Client/server?
8.	Industry experience Representative customers in same or like industry Capability to configure industry-specific processes
9.	Miscellaneous: Security Ad hoc reporting capability Case tools utilized Application program interfaces (APIs) available? Is source code available? Quality assurance procedures

14

Software project management

Learning objectives

- To learn how to manage solution project execution.
- To learn how to manage project scope and budget.
- To learn what the different phases of solution development entail.
- To learn the concept of critical chain project management.

14.1   Development life cycle

A 'life cycle' is a framework of processes, where processes are sequentially organized activities, arranged in a linear or iterative fashion: waterfall, spiral and evolutionary. It focuses on both business management and systems management. The processes that deliver one or more end-results are commonly referred to as deliverables. It provides developers with guidelines, management with checklists and milestones, and the organization with indicators about how and when to incorporate non-system processes. Life cycles include methodologies of real-world concerns such as project management, the production handover and maintenance.

14.2   Risk minimization

Software project management is strictly based on project management principles to achieve the stringent results required in an effective way. These principles are generally tailored to suite the market in which it is operating in, and accordingly a project methodology is to be used to make it successful in real-life applications.

There are different phases in the life cycle of a project, from the original decision, to the final phasing out/replacement of the solution. It facilitates a phased approach to projects, where each phase can be clearly defined, estimated and even contracted separately, with specific deliverables at each point of completion. The completion of each phase reduces the risk for subsequent phases by increasing the level of information available about the required solution. The cost of identifying and removing defects increases exponentially once development starts, as changing code or configuration within a package is much more expensive than changing the words/pictures in a document (refer Figure 14.1).

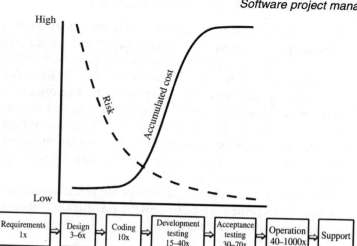

Figure 14.1
System life cycle vs cost

Additional cost spent on extra checking at the beginning stage(s) of a project is cheaper than the errors at the developed stages because testing is more complex and costly, and documentation changes are more widespread when more people are involved in correcting the error. Seventy percent of software costs go on maintenance of existing software, so it is very important to ensure that the design is easily maintainable. The design phase impacts dramatically on the development costs and the total development costs.

The later in the process, the more it costs to effect a change. This does not even take into account the effect of non-functional software on product quality, throughput and yield. This approach lessens the risk for both the customer and software provider and produces better systems at lower costs.

Some software solution providers use IPEC (Initiate, Plan, Execute, Control) as the high level approach to all projects, where the initiate component is an internal process of the provider leading to a proposal being prepared. The plan, execute and control components are directly applicable to the deliverables and client involvement.

The detail project plan is finalized in the project-planning phase and any uncertainties regarding the project scope are clarified and the quality assurance plan is finalized. All ambiguous/unclear requirements will be clarified and abbreviated requirements will be expanded on by means of request for information (RFI) forms. These RFIs will be filed in the established project note book (PNB), as they will then form part of the contractual baseline.

It is preferable to obtain a fixed price for each phase of the project from the solution provider, and to handle each phase as a separate project. This ensures that each phase is measured and controlled independently, and that the relationship with the provider can be terminated after completion of any phase, should the relationship deteriorate beyond saving.

14.3 Solution design requirements

The following are generic requirements that should be taken into consideration in the design and implementation of any proposed development and configuration project.

14.3.1 Single point of entry

The solution should have a single point of entry and real-time updating of information.

This is effectively performed to eliminate data duplication. The only way to eliminate data duplication is to effectively design the systems architecture and select the appropriate solutions to efficiently enable each business process from the boardroom to the shop-floor.

A golden rule is to only capture data once in the most appropriate place within the information architecture. Data that is more appropriate to capture in the transactional environment should be captured there and data that is generated in the control system or shop floor, should be captured there. Once the information is captured it can be summarized by another system that may require the summarized or subset of data, but should preferably not be duplicated for data integrity reasons. This principle should rigorously be applied in the design.

Since operational solution processes are not transaction driven as is mostly the case in business management systems (ERP), real-time event-driven updating of information is the norm instead of the exception.

14.3.2 Effective reporting facilities

Reports that do not support the principle of 'the right information, in the right format, on the right time, in the right place, and to the right person', do not add value to a business and are merely a record of past activities making them redundant for effective operational management.

The design of reports therefore takes high priority in any solution to ensure that the right mechanisms (standard pre-generated reports, dynamic query, graphs, etc.) and technology are selected to ensure maximum business benefit as qualified above.

14.3.3 Automation of administrative processes

Operational personnel only add value if they apply their process knowledge and experience to increase productivity and implement improvements to the process, instead of performing repetitive or administrative tasks. As many administrative or repetitive processes (production of reports, third-party information capture, systems integration, archive and recovery of data, etc.) as possible should be automated in the solution design to ensure maximum availability of operational personnel.

Where tasks cannot be automated (validation of production results, validation of production plans, approval of items and manual capture of data) the aim is to provide simple and intuitive methods as well as user interfaces to quickly and effectively accomplish these tasks.

14.3.4 Transaction traceability

Transaction traceability in the operational system should mean audit trails are available for all transactions.

According to S95 specifications, that history of each transaction should easily be generated, archived and restored for record taking, tracing and auditing purposes. Traceability should therefore be possible for all transactions since relevant information can be captured with each transaction such as user id, time of transaction, background process number, schedule, etc.

14.3.5 The system should be near-paperless

The system should be as paperless as possible with minimum human interaction required. All data transfers should be as automatic as possible. In the instance where the system is unavailable for a period of time, some data capturing on paper could however be required. If a near 100% availability of the operational solution is required, redundant servers can be set up that will ensure system availability and eliminate the eventuality of capturing data on paper in some instances.

14.3.6 Employment of generic principles

All naming conventions, coding standards and principles should be generic throughout the organization, not only those areas as addressed by S95.

The naming conventions and coding standards are normally a combination of standards as required by the customer's organization and what is generally accepted in the IT world (such as S95).

The exact formats should be clarified during detail design and should be approved by the client before implementation.

14.3.7 Policies and procedures

In general, the methodology employed by solution providers for system development should have the aspects of backup and recovery, change control, service level agreements and conformance testing integrated in such a fashion that it cannot be separated from the execution of the project.

It will be the responsibility of the client to manage these issues after the completion of the project, and the supplied comprehensive user documentation (that should form part of the deliverable) will greatly assist the client to develop effective systems to address these issues.

14.3.8 Participation of key-users

Key-users must participate in the specification of detail requirements for customization and/or development of applications. Applications should be driven by business requirements and should not be IT-driven.

Key-users/business process owners from the client should be involved during the detail design phase of the project, to ensure that business processes are properly modeled and implemented in the system, and that user ideas and inputs in mechanisms, user interfaces and required outputs are properly accommodated. The detail design should be approved by the client before the implementation can proceed to ensure that the client is familiar with the specifications of the proposed operational solution and agree with all aspects thereof.

Some providers strongly encourage the participation of suitably qualified technical professionals from the client during the development phases of the project to ensure continuity during the support phase of the system and to be able to implement future modifications and extensions.

14.3.9 Provision of management information

During the detail design, the aspects of time, format, scope and business rules should be addressed to ensure the provision of timeous, accurate, reliable and complete management information from the solution. Participation of key-users/business process owners from the client will be required during the detail design phase to ensure that these aspects are properly addressed to their satisfaction, since they should sign off these aspects before development can proceed.

14.3.10 Consistent 'look and feel' system

An integrated solution should have a consistent 'look and feel' throughout all functionality. This very factor, as well as its ability to seamlessly integrate different applications into one consistent 'look and feel' environment, influences the selection and recommendation of software solutions. This should prevent multiple 'logins' and complex navigation to load a specific application.

Open standards such as ActiveX, OLE, OPC and COM should therefore dominate the design and should also provide the vehicle to achieve such a high degree of systems integration.

14.3.11   Intranet-enabled user access via a browser interface

The solutions selected may all be web-enabled and could be implemented where specifically requested or offered as an option.

14.3.12   Emphasizing 'off-the-shelf' packages

The emphasis should be on 'off-the-shelf' packages as first choice. The basic philosophy is 'buy' first, 'buy and configure' second and only then consider building bespoke applications.

14.3.13   Preference for open systems

Open systems are preferred to proprietary systems or solutions. As discussed above, standard well-known software packages from reputable companies should be used in most cases. Where modifications or custom implementation of certain functionality is required, easily available and flexible tools from reputable companies employing open standards should be investigated.

14.3.14   Use proven, reputable and recognized technologies in the solution

All the tools used for the solution should be well proven in the industrial information technology domain, supplied by reputable suppliers employing internationally accepted standards.

14.3.15   Establish a technology infrastructure to ensure best IT practices

Use international standards supplied by AMR Research, ISA, Supply Chain Council and Microsoft to develop architecture and infrastructure. This ensures alignment with development trends such as the emerging web environment.

14.3.16   Use standardized middleware throughout the solution

Integration among applications needs careful consideration from both a business and a technical perspective. The business requirement should dictate integration from which the technical specifications for mechanisms such as middleware can be developed and selected.

14.3.17   Legality of software

Ensure the provider uses no illegal or pirate software, and the client should ensure legality audit trails of software within its organization.

14.3.18   Conformance to standards

De facto and international standards should be used for all interfaces and APIs. By conforming to existing and emerging standards, development effort is reduced, standard application components can be used, and maintenance is significantly reduced and simplified.

The design should make provision for standardization of interfaces using the relevant standards such as ActiveX, COM, DCOM, OLE, OPC, etc.

14.3.19   Scalability of systems

All systems initially installed should be scaleable to the defined size and needs of the client. It should be understood that requirements may increase after commissioning of the solution. These scalability requirements should be handled as part of the project management and quality assurance procedures.

14.3.20   Access/security control

Access/security control of all data should be controllable per user (i.e. per log-in code and password).

Most software products have some form of access control. These methods should be standardized in the solution to ensure consistent access to the applications in the integrated solution application portfolio, as well as the avoidance of multiple 'logins' into the different applications that make up the complete solution.

14.3.21   Easy modification

Easy modification to business processes, responsibility structures and information requirements should be possible.

The application portfolio and products selected should be very flexible since it provides a solid foundation with access to most of the data sources, presentation and business tools that can be tailored to the unique requirements of the client users and business processes.

Throughout the years this requirement has led to the development of bespoke solutions that gives the user total flexibility vs the limited functionality and flexibility offered many times by package solutions.

14.3.22   Minimized complexity

Complexity should be minimized through reducing the discrete number of applications. The business requirement of the application portfolio is by no means simple. It requires a high level of business, information and integration knowledge. This is one of the reasons why maximum utilization of proposed packages is considered to prevent a complex 'best-of-breed-mix-and-match' application portfolio. Where different products have to be used, integration should be done in such a fashion that it should be hidden from the end-user to create the impression of one system.

14.3.23   System's user-friendliness

The system should be user-friendly. Although it should be standardized throughout the client's business unit/s, it should be easily adaptable to individual sections' and users' needs.

The tools selected should provide a degree of flexibility on the user side where applicable, but some of the business issues are fixed in the background to prevent different business practices to evolve around a specific division's culture, which could result in significant system mismatch and organizational inefficiencies.

14.3.24   Disaster management

Disaster prevention, recovery and backup practices should be specified and complied with in accordance with the client's guidelines.

14.3.25   Single master security and authentication domain

A single master security and authentication domain (with single log-on authentication) should be deployed for all network centric services. The most suitable way should be investigated and recommended during detail design phase.

14.3.26   Modular design

An individual component (modular) design approach should be followed to assist incremental improvements and upgrades. The individual applications should be de-coupled to ensure that once the real-time and transactional databases are installed, the other applications can be 'plugged and played' without influencing each other. This should be achieved through employment of proper interfacing standards such as for instance COM and DCOM. It should further allow for alterations to individual components or the addition of new components in the future without influencing the rest of the system.

14.3.27   Reuse

An infrastructure that supports commonality through reuse and shared services should be established. The system databases can serve as a plant data repository/mart/exchange that can be used for many different data storage and exchange purposes. The applications residing on this virtual database should share devices, services and components which should ensure commonality throughout the applications and should significantly simplify future maintenance.

Development of reusable components increases the cost of development the first time the components are developed, in order to save money in later projects when the same components are to be used. The impact on the costs of reusable components is difficult to determine, as it is not always known how many times the component should be used in the future, at the time of development.

14.3.28   Standardization

Through the standardization of application, information and hardware components, complexity can be minimized. Considerable effort should be spent to maximize the use of products. The design should further focus on achieving optimal integration between the applications to ensure standardization of user interfaces. Lastly, hardware should be shared between the applications to ensure maximum utilization of the hardware components.

14.3.29   One source of core data

One source of core data should be kept and managed centrally and preferably entered into the integrated solution once only. The philosophy behind the system design should be to provide an environment where time and relational based data can be seamlessly integrated. This provides a single virtual source of core data that can be used in multiple applications without the need to re-enter this data into different systems.

14.3.30   The system support end-user computing based on raw data

The tools recommended should make provision for seamless extraction of data into end-user tools such as MS Excel and MS Query as well as the generation of ad hoc reports.

14.3.31   Distributed user access

Distributed user access should be supported and users should be able to extract information from the core database(s), for manipulation at their workstations.

14.3.32   Scheduled tasks on time or on events

As discussed before, the environment should be based on events and the design should focus on automating administrative and repetitive tasks where required. Such tasks could for instance be the scheduled production of reports.

14.3.33   Provide task configuration environment

A system should increase productivity of operational personnel, and a task configuration environment where tasks can be compiled, modified and scheduled such as system back-ups, the compilation of shift reports, a follow-up scheme or task lists are essential. Where applicable this should be provided as part of the system.

14.4   Software development life cycle

There are many methodologies for various industries and for various types of developments. Each methodology promotes certain types of tools and even specific tools to perform the prescribed steps. Two of these distinct methodologies for development life cycles are:

14.4.1   Spiral

The spiral development model (see Figure 14.2) defines four phases of the project: an idea/need analysis resulting in a vision/scope document; a design phase resulting in a functional specification; a construction phase ending in a complete piece of code; and finally a stabilization phase ending in release of the product.

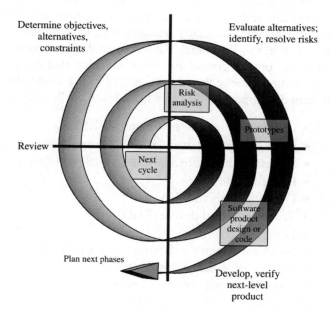

Figure 14.2
Spiral development model

14.4.2   Waterfall

Most solution providers use the waterfall life cycle approach for software solution development. The waterfall approach (refer Figure 14.3) helps to understand the extent of the residual risks and allows one to work conscientiously toward reducing those risks.

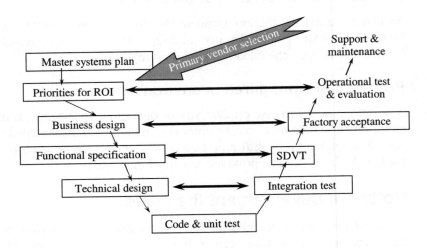

Figure 14.3
Waterfall life cycle approach

At the end of each phase, a baseline is established which becomes the platform for the next phase, and until the previous phase is officially approved no next phase should be started, hence delays on any approvals lead to project delays and have cost implications. The execution component is responsible for the final deliverable of the project and is built around pure code development, system configuration, or a combination of both.

The waterfall approach is best suited for a simplistic, yet systematic approach to meet the exact requirements of the client. The immediate benefit to the client is the constant realization of the benefits in terms of the expectations of the final deliverable. This approach also ensures that the provider can constantly measure itself to interpret the requirements of – and deliver the best solution to – the client. The tools that this methodology prescribes should have built-in quality and project control measures, ensuring that a certain quality level is maintained. These properties enhance the management of time and specifications of the project.

A system development life cycle includes the following steps:

- User requirement specification (URS)
- Business process analysis (AS-IS) and design (TO-BE)
- System architecture analysis (AS-IS) and design (TO-BE)
- Benefits case
- Function analysis (FA)
- Technical detail design (TDD)
- Development
- Software design verification testing (SDVT)
- Factory acceptance testing (FAT)
- Installation and commissioning
- Site acceptance testing (SAT)
- Training
- Support and maintenance.

User requirements specification (URS) is a general term used in industry and it comprises the masterplan and the operational detail design (ODD), while a functional specification is a high level combination of the ODD (business design) and technical detail design (TDD).

Each of the testing steps in the development/configuration phase is tested against a step in the design phase. This ensures that the objective of the system is met and that it is fit for purpose.

	Waterfall	**Spiral**
Essence	Preceding phase must be completed before the next starts	A spiral model implies learning at all stages, redefining the problem as work progress, and vigorous examination of the solution's viability
Advantages	Provide structured approach. Clearly defined phases, appropriate to a contracted project	Intermediate goals lend closure and easily traceable progress to the process
Disadvantages	Difficult to define requirements at the beginning and difficult to change at a later stage. Maintenance and reuse difficult. A fault at the top might filter down to all the levels	Makes project feel out of control, tends toward chaos
Use	Structured analysis and design techniques. Difficult to use in modern development, as the projects are complex and it is difficult to lay down and know all the parameters at the design time. Good for simple projects where the deliverable is well defined	Fits well in the development environment and allows for changes as development proceeds. The learning culture leads to spirals. Work well with projects that 'grow' during development

14.5 Life cycle components

14.5.1 Master system plan (MSP) or conceptual design

The master systems plan (high level business requirements and design) can be interpreted as the indirect business requirement and its purpose is to put the direct business requirements in the context of the broader requirements of the business, to ensure strategic alignment. A high level business design is proposed in the MSP. It is an important building block to eliminate possible wrong or improper assumptions in further processes in the project, which may impact on the strategic direction and intentions of the client.

During the master plan phase, the solution provider does a business analysis and defines the client's requirement using workshops, document studies and interviews. Hence it is extremely important that the client management, IT staff and operational personnel make themselves available to the design team as the architecture of the integrated system is established during this phase.

The MSP document content generally consists of the following:

- *Current business analysis:*   It enables all levels and departments to understand their role and influence on the operations and overall business success. It consists of:
    - Process flow and description, where the process flow of each operations or support process is defined and documented
    - Key performance indicators (KPIs) and key influencing factors (KIFs), where all the critical measurements and influences on each operational or support process are defined and documented
    - Information flow, where the flow of information of the specific operations or support process is modeled using an agreed model as base sin order to identify information gaps
    - Information technology used, where the current information technology used by the process is identified and documented to enable functionality mapping
    - Current initiatives, where the current systems or process expansion projects planned or in progress are identified as they may influence the design
    - Opportunities are identified from the defined and documented gaps in information with input from users and managers.

- *Functional requirements analysis:*   It abstracts and orders (using the modeled functionality) all required systems functionality and integration needs for future systems from the documented opportunities.
- *System mapping:*   It maps the current systems functionality against the required systems functionality and integration, to indicate the fit of current systems set-up and usage to operations needs. It also identifies current systems shortfalls and improvement opportunities within operational systems for future incorporation.
- *Systems architecture design:*   It provides a conceptual architecture for systems development and integration which is understandable to all levels of the organisation, ensuring maximum employee buy-in.
- *Operational benefit:*   The MSP document provides a roadmap of 3–5 years for systems development and integration implementation, prioritized using the identified KPIs, KIFs, client company vision, strategic intent and critical success factors to ensure maximum operational benefit.

The MSP may also contain a business case (value proposition, benefits analysis) and high level vendor package evaluation.

14.5.2   Business requirement specification

This step in the design process can also be called operations detail design (ODD), or functional analysis (FA) or future-state-study. The purpose of the ODD exercise is to establish and design the business processes and functionality that should be enabled by the proposed system solution. This exercise can be built on the initial high level requirements of the MSP.

During the ODD phase, consultants generally perform a business and functional analysis of the client's requirement for one or more of the identified system modules. The client's personnel should review the business and functional design document in detail as it should establish the baseline from which development/configuration will eventually take place. This document is critical, as it establishes the purpose of the solution, the processes involved and the business rules related to the solution. It should be studied in great detail to ensure all requirements have been addressed, as it is time consuming and expensive to add overlooked requirements later in the project.

An ODD document consists of the following:

Current business analysis, which includes:

- *As-is business processes*, where, by means of workshops and interviews, the current business process flow diagram is developed and described in detail.
- *As-is business functionality*, where the functionality within the identified business processes is studied with descriptions of business rules (gathering, calculations, formats, outputs, etc.), value of information and interaction with other functionality being documented.
- *As-is systems architecture*, where the architecture of the current systems used by the client is investigated to give a baseline for future development.
- *As-is systems functionality*, where, in order that it is retained throughout the transformation process, the functionality present in the systems in use is also investigated and documented in detail.
- *Summary of current business analysis*, where detailed analysis of the current business is done and deficiencies and opportunities for improvement identified.

Proposed business design, comprising:

- *Conceptual functionality requirements*, where generic models used are described and the basic philosophies underlying the design discussed.
- *Client functionality requirements*, where the requirements of the management and staff of the client are detailed for inclusion in the business and system design, thereby maximizing buy-in to the system at a later stage.
- *Design features*, where the benefit-generating features of the design are identified and discussed.
- *To-be business process*, where the proposed business process is developed using some modeling technique to identify inputs, outputs, mechanisms, controls and interaction between processes in the solution.
- *To-be business functionality*, where the functionality within each element of the proposed business process is described in detail with descriptions of business rules, value of information and interaction with other functionality documented. It should also differentiate between system and manual functions (what the system will do automatically, and the interaction of users with the system for decision-making, driving business processes, choosing alternatives, etc.).
- *To-be systems architecture*, where diagrams and text describing how the as-is systems architecture needs to be modified to realize the proposed business process and functionality are presented.
- *To-be systems functionality*, where the high level systems functionality needed to realize the proposed business process is designed.

Critical success factors (CSFs), comprising:

- *Project CSFs*, where business risks endangering the success of the total project are identified and discussed.
- *System CSFs*, where factors crucial to the success of the designed system in meeting the needs of the client are identified and discussed.

Benefits analysis, comprising:

- *Operational benefits*, where the benefits, both qualitative and quantitative, of implementation of the system are identified and discussed.
- *Quantifying the benefits*, where operational information is used to quantify the benefits of installing the system to the client in order to determine an ROI.

14.5.3 Design and specification

Design and specification is the general term used in industry, however some providers prefer to use the term technical detail design (TDD). This phase follows the ODD phase where a detailed technical analysis of the client's requirement is made.

The purpose of this phase is to establish and specify the precise user requirements with respect to business process enablement, screen layouts, report layouts, integration requirements and clarification of technical aspects such as software requirements, server requirements, interfacing amongst third-party systems and general system look and feel. The client and the project team interact in this phase to review the business design and technical design document carefully.

The activities of this process include:

- Confirmation of solution design findings with the client's technical and management personnel
- Workshops to establish a detailed system look and feel, screen layouts, reports and background functions required
- Technical evaluation of functionality
- Database and interface design
- Development of business rules and constraints
- Documentation.

The TDD document details the application, system functions and user interface, and needs to be reviewed by the client to verify its accuracy.

The TDD document consists of the following system design concepts:

Functional overview

It summarizes the business design on a high level, relating the high level system layout to the business requirement.

Database design

It includes descriptions and graphics of the following:

- Entity relationship diagrams
- Entity type description
- Data store description
- Record descriptions
- Table of code descriptions
- Background and stored procedures, logical flow diagrams.

Screen requirements and layouts

They graphically depict the screen layout with descriptions of field values and attributes, button functions and screen purpose.

Report layout and requirements

They graphically depict report layouts and the types of data (relating to database fields) that will populate the variable fields (such as averages, sums, area names, site names, tag names, etc.).

Process models

They graphically depict the following:

- Data flow diagram
- System/process flow diagram
- Decision tree diagrams
- Use case diagrams
- Transactional sequences diagram.

Interface records

They describe how interfacing will be done and the data structures that will be used. It also describes interface methods, the types of data that will be transferred and the source and destination of the data.

Standards and conventions

They describe table- and tag-naming conventions, front-end development standards, screen layout standards, standard system buttons and meaning, colors, number and text fields, screen size, toolbars, key sequences, keyboard conventions, messaging (notices and errors).

Physical network model

It graphically depicts and describes where the solution will be located on the existing network infrastructure and how it will interact with other network components.

Hardware and software requirement specification

It describes any specific hardware or software that is needed to ensure system operation, such as processing power, disk space, licenses, printers, monitors, etc.

Business rules and constraints

They describe the technical business rules relating to the rules in the business design, and the specific system rules relating to front-end use and user choices.

14.5.4 System development/configuration, integration and unit testing

During this phase the hardware and software is purchased and development/configuration and integration of the system specified in the TDD document is undertaken. Software coding and configuration begins and continues until all software modules are developed and tested under strict quality assurance and quality control.

Modules pertaining to each functional area should be developed and tested, together with module milestones, which should follow each successful test cycle. In this phase, the client and the original design team are encouraged to participate in the development and testing of the system. After unit-tests have been completed, all the components of the system are integrated, and integration testing is performed. This approach ensures that the system is designed to the expected functional requirements and that knowledge is continually transferred to the client.

14.5.5   Software demonstrations

Software demonstrations are done to reduce defects early in the development cycle, and reduce the effort of subsequent testing phases. For the purpose of these demonstrations, the development/configuration of the client application is generally divided into logical units of work, each of which is then demonstrated to key client personnel and for each project milestone, the developer and the client's project managers should agree to the content.

Cosmetic changes to the appearance of the client application and defects in the application functionality are addressed during the demonstrations, which should be recorded using QCRs (quality control reports). The client's approval is required for any additional functionality. Budget and schedule implications should be clearly indicated and may only be approved by the client project manager or his duly authorized representative. A contract variation order (CVO) handles cost implications for changes in scope or effort.

The demonstration phases are not mandatory, and should be viewed as early information gathering by the client to identify project progress. As the development and integration of these demonstration units will not necessarily be completed at the time of the demonstrations, it will normally not be to the same level of detail as the official test phases (FAT and SAT), and clients can expect to find defects/bugs of various kinds.

14.5.6   Software design verification testing (SDVT)

It is an internal test phase and involves the testing of the entire system by the solution provider and an independent quality assurance specialist and/or process specialist for conformity with the design documents (operations and technical). This approach ensures objectivity during system verification, which in turn contributes to the early identification of any defects and minimizes client exposure to basic system errors. The client is normally not involved during this phase of the project.

At the end of SDVT, the client application is to be certified in accordance with the TDD and is ready to be tested in the presence of the client. The 'as build' TDD, factory acceptance test (FAT) manual and site acceptance (SAT) manual forms part of the deliverables of this internal phase of solution configuration.

14.5.7   Factory acceptance testing (FAT)

The objective of FAT is to validate that all the software modules are functional and that the application can be regenerated from the original source code without loss of functionality. If necessary, simulation should be used to provide the client application with data that should be supplied by integration systems. During the FAT, QCRs should be generated for defects in the application or where the application does not comply with the design baseline.

In the course of the FAT, the solution provider corrects the QCRs raised and those sections of the FAT manual, which are directly affected by the QCR, are to be re-tested. The FAT milestone is achieved once the test procedures described in the FAT manual have been completed successfully. The FAT milestone is completed once the FAT has been accepted/signed off. The client may choose at his own discretion to accept the FAT milestone with a small number of QCRs outstanding which can be tested during SAT.

Test results are to be monitored and documented and after the successful completion of the FAT, the client application is ready for delivery to the client site and for system population.

FAT manual

The solution provider compiles the FAT manual by performing a requirements analysis on the ODD and TDD. The FAT manual details the exact procedure to be used to test the functionality of the client application and the client's project manager should review and accept the test methodology as specified in the manual before the FAT process starts. This manual should be studied in detail to ensure that all the necessary business functionality will be covered during the testing of the system. This is important, as once approved, the testing will take place and any additional testing or other requirements will have cost implications. One printed copy and a software copy are generally given to the client as project deliverables.

The FAT manual describes the tests to be performed during factory acceptance testing and the following are the main headings:

- Applicable documentation
- Required factory acceptance test configuration

    - Hardware equipment list
    - Software equipment list

- Test script

    - Acceptance form
    - Quality control report

- Test to be performed:

    - Hardware installation
    - Hardware configuration
    - Software configuration
    - Functional areas of application
    - Test cases (the most important but usually the most neglected).

14.5.8   Installation and start-up

Installation and start-up involves the delivery to site of all necessary computer hardware, software and peripheral equipment, installed into its final position and started up. The following steps are executed during start-up and commissioning:

- Physical installation of computer hardware and related peripherals
- Network connections with the existing plant network(s)
- Software installation
- Software configuration of the actual peripherals
- Integration testing.

All the software required for the operation of the system is generally installed on the computer systems by the developer prior to shipment to site. It includes packaged software as well as bespoke software developed for the project by the project team. Source code and development/configuration tools are removed from the computer systems before commissioning, and delivered to the client on removable magnetic media. Configuration management of the completed software is very important at this stage of the project, and all changes required are to be made under formal configuration control procedures. Formal configuration control procedures should also be applied to the 'as-built' system documentation.

14.5.9 Site installation and acceptance testing

In this phase the system is delivered to site, installed into its final position, connected to the relevant plant equipment and networks, and started up. Training is also performed during this phase and would typically include applicable technical training on the software packages used to develop the application, as well as application-specific training on the solution components as documented in the various design and user manuals.

This phase further involves the operational verification procedures, which should ensure that the system that was verified in the FAT phase is functioning correctly on site with the actual plant equipment and network connected to it. At the end of this phase, the user should have the full benefit of the new systems.

14.5.10 System population

System population involves the capturing of technical data in order to populate the client application data tables for complete operational functionality. It includes obtaining complete data of product streams, products, product groupings, tag names, production units, etc. The client should be responsible for populating the master data tables of the client application. The solution provider facilitates the process by issuing RFIs detailing the data required and the associated integrity constraints. The provider should check the completeness and the correctness of the data and the client project manager should be responsible for coordinating the data capture personnel and the integrity of the data.

14.5.11 Site acceptance testing (SAT)

The objective of SAT is to test functionality not tested during FAT and to verify that functionality dependent on integration to legacy and control systems is compliant with the procedures laid out in the SAT manual.

During the SAT, QCRs should be generated for SAT procedures that are not completed successfully and are to be corrected and retested. The SAT milestone is achieved once all the test procedures described in the SAT manual have been completed successfully. The SAT milestone is completed once the SAT had been accepted. SAT procedures are the means by which the client declares that the systems are in compliance with the approved design requirements (ODD&TDD). SAT ensures that the client application, which was verified in the FAT, works correctly on site with the actual plant equipment and neighboring systems connected to it and integrated.

The following documents may form the basis by which compliance of the systems can be established:

- The Operational and technical detail design documents
- Any other relevant correspondence between the project manager and the client's personnel such as contract variation orders (CVOs), minutes of meetings, etc.

Site acceptance testing should be conducted by the solution provider in conjunction with the client's technical personnel and then handed over to the client.

At the end of this phase, the user should have full beneficial use of the client application.

SAT manual

The SAT manual details the exact procedure that should be used to test the integration of the new client application to existing client systems/applications. The SAT manual should be given to the client project manager for review and acceptance well in advance of moving the system to site.

The SAT manual may contain specific requirements on site that may be the responsibility of the client project manager, so early receipt of the requirements are essential (such as network points, electrical connections, physical space in server rooms, PCs and system administration changes – new groups, etc.). One printed copy and a software copy should be given to the client as project deliverable.

The manual describes the tests to be performed during site installation acceptance testing and the main headings are shown below:

- Applicable documentation
- Required site acceptance tests configuration:

    - Hardware equipment list
    - Software equipment list

- Test script:

    - Acceptance form
    - Quality control report

- Tests to be performed:

    - Hardware installation
    - Hardware configuration
    - Software configuration
    - Functional areas of application

- Testing of interfaces:

    - Pre-approved sets of data should be generated for each type of interface
    - Transferring data between the different systems, using the interface files to verify that the correct data has been exchanged and the interface controls are adhered to

- Outstanding FAT QCRs
- Test cases.

14.5.12 Commissioning with parallel run

During the parallel run (a period as agreed contractually), the various components of the client application are to be used in parallel with the existing systems. The objective of the parallel run is to verify that the calculations used to generate key reports are correct and comply with business needs. The parallel run is also used to test specific procedural aspects, step sequences and business rules built into the solution in a real-life operational plant.

The solution provider works with the client to establish a procedure that is used to test the method in which the client application generates these reports. This procedure is used during the parallel run to enable an iterative comparison of the two techniques and to highlight any transactional discrepancies. Investigations are to be done for any discrepancy and corrections are to be made, if necessary, to the calculations used by the client application in order to reflect the correct business procedures.

At the end of the parallel run, the client must accept the system if discrepancies between the two methods can be accounted for and none can be attributed to defects in the solution provider application in terms of the procedure established. The solution provider provides on-the-job training and client application support to the client's personnel throughout the parallel run where required.

The duration of this phase is dependent on the following factors:

- Other systems availability
- Super-user availability
- End-user availability proficiency and skill level
- Functionality changes required.

14.6　Critical chain project management

Critical chain is a project management technique used to schedule a set of interdependent tasks. The completion of each of them at the right time is important for the completion of the project as per the schedule. To protect against uncertainty within a project schedule it is required to identify tasks that if delayed, would make the project duration longer. Those tasks should be considered the most important and be protected.

An example of critical chain tasks is shown in the top part of the Figure 14.4. The boxes represent tasks, and within each box is the number of weeks to complete the task and the name of the resource needed.

A delay of any of the bold tasks should delay the project. This is different from the traditional critical path in two ways: resource conflict is taken into account, and tasks are placed at their late start times. Unlike the critical path, the critical chain can hop from one path to another as a result of resource conflict.

Resources are fixed, so one way of adding protection is by adding time. Traditionally, time was added to each individual task to protect the schedule. In the critical chain approach, the project completion date is protected rather than the individual tasks.

This is done by adding scheduled blocks of time, called buffers, at the appropriate places in the project plan. Buffers are inserted in the schedule so that the impact of disruptions is not severe. This might result in a longer schedule, but is more realistic. An important point that has to be taken care of is 'move risk earlier and slack later'. The critical chain determines the project duration; the critical chain itself needs to be protected. If work is not ready for the critical chain tasks to start, the critical chain should be delayed, thus likely delaying project completion. This means there must be some protection every time a non-critical chain task feeds the critical chain. This type of protection is called a 'feeding buffer'.

With the feeding buffer, the critical chain is protected from fluctuations. To protect the commitment date from fluctuation, a 'project buffer' is placed after the last scheduled task. Figure 14.4 shows the fully buffered schedule.

The feeding buffer protects the critical chain task 3:HW from uncertainty in the task 6:Prog. The project completion date is also protected by a project buffer of five weeks. This means that every task has at least five weeks of protection. By specifying average task durations and by removing the necessity to keep everyone busy, the need for people to take on multiple tasks to keep busy is reduced. This, in turn, helps reduce the normal chaos associated with fighting fires across multiple projects.

The critical chain approach helps projects to complete more quickly by encouraging tasks to start early by prioritizing them. Furthermore, the buffers allow this to happen.

Let us consider in above figure the task 5:HW completes in four weeks. Unless 6:Prog is very late, the feeding buffer ensures that 3:HW can start a week early, thus speeding the project along.

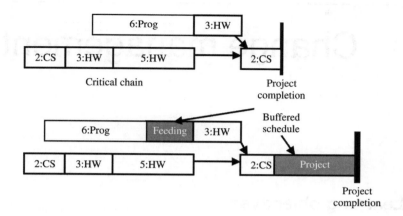

Figure 14.4
The buffered schedule

There is a useful refinement that can be added to the schedule. Consider what should happen if the engineer is working on another project before the 3:Eng task starts, and that other work is delayed. That delay can delay the entire critical chain of this project, and potentially use up some of the project buffer. To avoid that, a 'wake-up call' (also known as a 'resource buffer') is scheduled some time before resources are due to start their critical chain tasks. The resources are told in advance when they should be needed to perform these key tasks. This lets them know that they need to be ready for high-priority work, and adds further reliability to the critical chain schedule.

In creating a critical chain schedule, the five-step improvement process has in fact been carried out. The leverage points (constraints) were identified by resolving resource contention and identifying the critical chain.

The leverage points should be exploited by focusing on critical chain tasks. Everyone else is allowed to subordinate to the leverage point by inserting buffers. Increasing the resources available to work on critical chain tasks can elevate the leverage point. If late tasks are not tracked, the project status can be monitored by monitoring how much of the buffers have been used up, compared with how much work remains on the path feeding it. For example, suppose delays have pushed completion of the final project task into the project buffer, so that only 30% of the buffer remains. If the project is 90% complete, the project is probably in good shape. If it is only 50% complete, there may be a serious problem.

15

Change management

Learning objectives

- To understand the issues surrounding the behavior of people in the face of changing working conditions.
- To understand the relationship between measurement and behavior.
- To learn a process with which change can be managed during system implementation.
- To understand the three phases people go through before new behavior is fully ingrained.

15.1 Organizational readiness

Today, manufacturers are battling with the challenges of falling revenues, increasing costs and decreasing levels of customer service. Improved quality service has become a major marketing tool.

The key challenges facing manufacturers facing today are:

- Lowering operational costs
- Reduction of personnel
- Improved operational efficiency
- Increasing demand for ROI on infrastructure spend
- Managing increased network complexity
- Automating processes to enable the rapid introduction of new services
- Improved customer satisfaction
- Proactively monitoring and increasing customer satisfaction.

Technology integration, together with process and organizational readiness deliver a complete business solution, which adds real value. Integrated, flexible and scalable operation system architectures, which automate end-to-end business processes, improve operational efficiency and supports the rapid growth of the business (refer Figure 15.1).

E-business technology is enabling fundamentally new business models in manufacturing enterprises. Collaboration processes are only possible if manufacturing is an agile component of the overall extended supply chain. A product must be made to a demand signal or forecast to fulfill the order and ensure a satisfied customer.

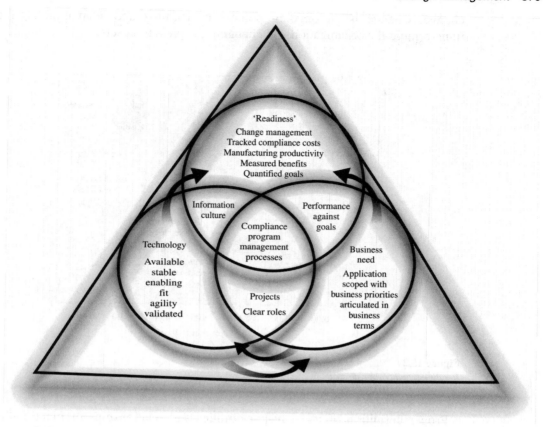

Figure 15.1
Organizational readiness

Trading exchanges will move their focus from procurement to sell side processes and more inventories will be eliminated from the supply chain. Reliable delivery from manufacturing will become even more critically important. Manufacturers must focus on consistency and delivery to demand, integration of plant information processes, and product specification management. S88 and SP95 have laid a valuable framework to support interlinking process threads between manufacturing plants and the extended supply chain to enable responsiveness. Manufacturers are challenged to translate business strategies into prioritized IT and business initiatives. Research has indicated that organizational readiness is one of the key areas to consider in developing a successful e-business environment. Manufacturers must build organizational readiness and governance, manage integration and integrate information across the extended supply chain against priorities set by the business strategy (refer Figure 15.2).

15.1.1   Readiness interdependencies

Organizational readiness indicates the relationship between people, processes, systems and performance measurement. It requires synchronization and coordination without which no implementation will be successful. The organization should therefore have processes and people in place to coordinate the efforts and communicate changes. The organization (people and management) will need to be ready to accept but rather embrace change. Change is an ongoing process of learning and adaptation. The goal is on transforming the organization and changing people's mindsets.

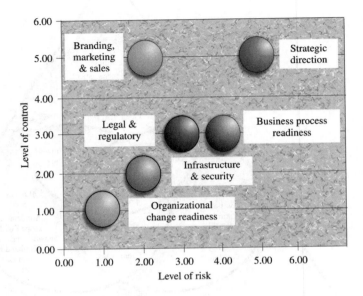

Figure 15.2
Readiness and corporate strategy

Proper definition of roles, responsibilities and relationships of functions and specific positions before or shortly after the new structure reduces confusion, anxiety and resistance, which is critical for successful implementation. For greater commitment and success, more people should be involved in the change process. Organizational readiness and culture are key criteria in KPI initiatives. Only a few leading manufacturers have achieved a readiness status.

Readiness processes drive and implement changed practices to improve business effectiveness such as: the ability of the business to define and prioritize problems and identify KPIs, the readiness of the IT infrastructure and applications to support dynamic KPI initiatives, change management processes in place to modify practices and behaviors to attain KPIs.

Integrated business strategies and defined goals ensure breakthrough level of performance.

15.1.2   Process and organizational readiness

A customer-oriented, service-driven process management methodology gives manufacturers flexibility and a clear understanding of operational activities. This enables cost-effective management and ensures quality delivery of customer services. The following questions need to be answered to establish the readiness of the organization for the intended change or system implementation:

In terms of organizational readiness

- What is the skill base?
- Can the organizational structure change?
- What funding has been set-aside for projects and training?
- What change management plans have been developed?

In terms of business process readiness:

- Has the impact been assessed?
- What is the effect on business processes?
- What is the effect on products, suppliers and customers?
- Will new competitors enter the market – from where?
- How will the efficiency of business processes be affected?
- May not always be a positive effect
- Has demand for services been assessed?
- Who is driving the change – customers, partners or competitors?

Once the corporate direction has been set, there is a need to look into the existing processes and redesign them to maximize resource utilization. To simplify the change in an organization the three Cs of change management – complexity, conflict and consensus – has to be addressed.

Complexity – one of the main inhibitors of change. As organizations evolve, complexity increases as informal routines and practices are added to the formal procedures, creating 'once-off' bespoke processes.

Conflict is created by a failure to manage consistent goals and objectives within an organization.

Consensus is one of the keys to managing change successfully.

A strategic plan is like a route plotted onto a map. It identifies major areas of opportunity, obstacles that cannot be overcome and the path that will lead to the successful realization of the strategic goals. In business the landscape is continually changing, with new opportunities and obstacles appearing from nowhere.

As the strategy is implemented it is important for senior management to monitor how effectively the strategy has been implemented and to take account of its impact on the competitive environment. It has been found that when large organizations implement solutions affecting multiple departments, a methodology is needed that can bring together complex inter-relationships and processes into one melting pot.

15.2   Reason for change

Reasons for change can be expressed as a personal/departmental/organizational pain or desire, a business outcome, or a strategic focus or change in focus. Fundamental changes aimed at improved process effectiveness and improved product or service potential are essential in today's marketplace.

In today's economy, building capacity to change is no longer an option; it is a strategic imperative for most organizations. In an environment of constant and unpredictable change, the capacity to change becomes a key source of sustainable competitive advantage and an environment that enables a culture of continuous learning and adaptation to change needs to be created.

In the 1980s and 1990s, companies pursued several strategies to gain improvements in business performance (refer Figure 15.3).

These strategies have various shortcomings:

- Business improvements gained through downsizing are often not sustainable.
- Re-engineering results alone are often disappointing or ineffective.
- Despite their popularity, mergers, acquisitions and alliances often fail to achieve their goals due to conflicting cultures.
- Technology-led change is costly and often results in short-term competitive advantage, but not necessarily any return on investment.

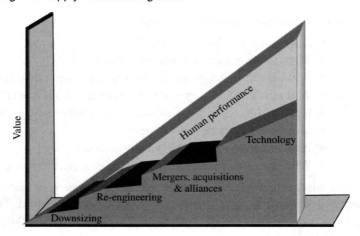

Figure 15.3
Improvement initiative value

The missing, untapped dimension is explicit attention to human performance. It provides a business's signature and capability that is unique and not easily copied. Given identical equipment and resources, the company with the best people will win the race.

The following tables indicate the reason, measure and approach for effecting change within an organization, department or strategic direction of a company.

Personal/departmental reasons for change

Reason: Current Pain's / Desires	Measures of Success	Recommended Initial Approach
Tired of fire fighting Maintenance costs are too high Need greater predictability Tired of losing contracts to the competition Looking for a new market, requiring improved capabilities	Personal: Promotion, rewards and recognition, more challenging work with less overtime Department: Less chaos, meeting objectives, less staff turnover, improved credibility	Focus efforts – apply the 80/20 rule Increase knowledge about driving change Document core processes Start tracking product/service defects

The organizational or business reason for change

Reason: Desired Outcome	Measures of Success	Recommended Initial Approach
Competitive advantage	Short-term: cost, revenue and profit Long-term: market size and market penetration	Define a business goal As a pre-requisite, market research and/or applying benchmarking techniques can be considered. The next action should be strategic planning followed by a change agent to facilitate the achievement of the objectives

Strategic reason for change (can be any combination of)

Reason: Strategic Focus	Measures of Success	Recommended Initial Approach
Quality and/or business capability	Product quality, service quality, work predictability, sale success, etc.	Start quality and business capability improvements by using a standard or model to document the core process and its key supporting functions
Work efficiency and/or time to market	Productivity Cycle time	Conduct a self-assessment to determine the key areas of opportunity and the best course of action, including short-term wins, considering the current situation and needs
Customer loyalty	Customer referrals Repeat business Reliable/predictable products and services	Tailor and shape products/services to fit the needs of customers Change efforts can be started by measuring products/services and by gathering/analyzing the customer needs
Business alignment	Operational excellence Strategic effectiveness	Start by determining business vision and related goals. Assess the current situation against these goals, determine key areas for improvement, and use a change agent (internal or external) to facilitate the implementation of the agreed action

15.3   Strategies for change

Various change processes and change leadership strategies are put forward by change consultants. There is no standard solution for enabling change as the circumstances surrounding any change program are unique to an organization and a customized change roadmap needs to be designed. The roadmap should revolve around best practices for enabling change in an organization such as:

- Determine organizational readiness and business case for change
- Articulate the organization's vision for change
- Design a tailored change strategy
- Build leadership capacity and stakeholder commitment
- Deploy a two-way, multi-audience communication strategy
- Align organizational design and performance management systems
- Build individual and team capacity to change
- Align culture with the new strategy and vision.

A facilitative approach ensures the involvement of employees and sustains the change process well after the initial energy is spent. The following table depicts how resistance to change can be overcome and what should be avoided as it can promote resistance.

Success	Resistance
Compelling case	Inertia
Vision	Confusion
Values	Corruption
Strategy	Diffusion
Information and communication	Uncertainty and fear
Leadership	Misdirection
Resources	Frustration
Capability	Exhaustion
Motivation	Resentment

15.3.1 Problems in major change programs

Various things in an organization can prevent change from being adopted and embraced by employees.

- There is no clear reason for change.
- The leadership is not aligned behind the change.
- There is a lack of resources to support the change program.
- There may be significant mistrust between departments.
- Management underestimated the complexity of the change program.

15.3.2 Initiating and sustaining organization change

Change requires a number of change agents within the organization and senior managers should be involved and lead the change process (refer Figure 15.4).

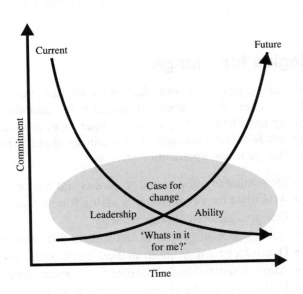

Figure 15.4
Building commitment to change

External resources could be leveraged to bring new perspectives on the change. Management's objectives should be very clearly communicated to the staff, and employees will accept change easier if they realize that the organization is going to grow, not downsize, or that the system will not lead to staff reduction. A change champion should be appointed to ensure action plans are implemented and to support the initiative throughout all phases of the change process (refer Figure 15.5).

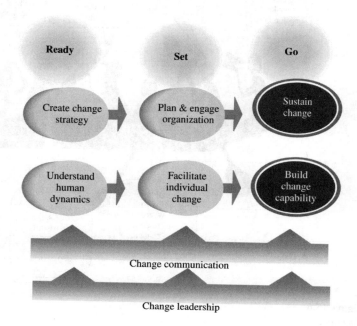

Figure 15.5
Sustainable change framework

Effective organizational change requires that a large mass of people be committed to the change effort. It cannot be driven by only a handful of people in the organization. This can only be enabled through good change leadership and company-wide communication. A clear strategy and good understanding of human nature must be developed by management to ensure that the change is effectively implemented.

Without a common vision and objective, people do not form teams and rally behind a common cause. Teamwork is not possible when people are confused and do not agree with the direction the organization is going. When the organizational goal is unrealistic some people give up and start working in a different direction (refer Figures 15.6 and 15.7).

Only when an attainable objective is set do teams come together and work toward a common goal.

15.3.3   Change leadership

The following areas need to be considered when leading change in any organization:

- Organizational change is an ongoing process of learning and adaptation. The goal is to transform the organization and changing people's mindsets. The proper definition of roles, responsibilities and relationships of functions before or shortly after system implementation reduces confusion, anxiety and resistance.
- People change the way they think only when they choose to do so voluntarily and the key is to get people to want to change. Engaging people early in the change process is one of the best ways to achieve this.

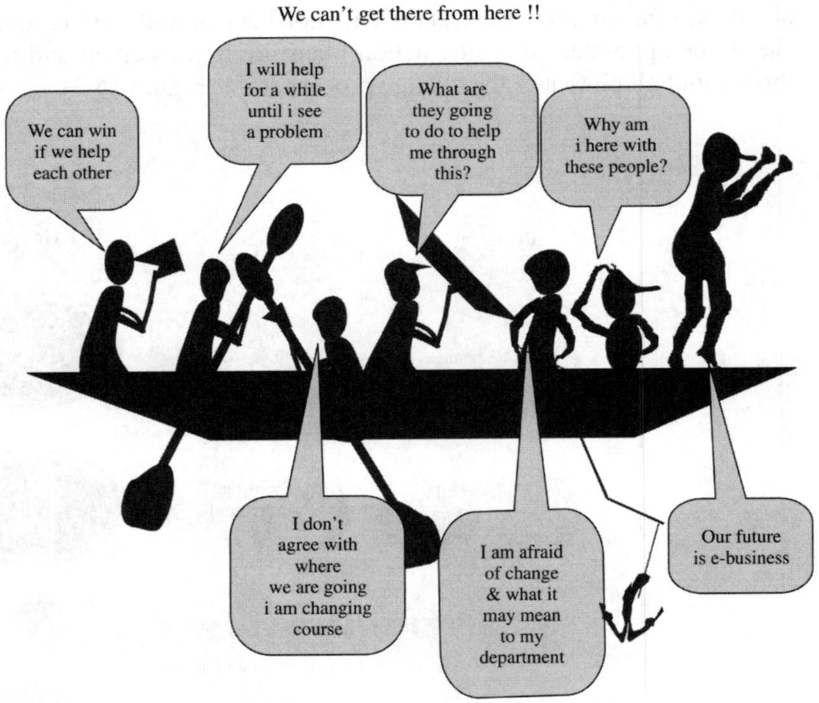

Figure 15.6
Uncertainty

Figure 15.7
Teamwork

- The organization should not treat employee resistance as something to be feared, as resistance is natural, can be managed, and the organization can learn from it.
- The more people involved in the planning and the wider the communication, the greater the commitment and success rate of change implementation. Employees may comply with top-down changes, but this will not change their minds or hearts.
- Change leadership is required at all levels of the organization. For less complex changes or system implementations, a senior management champion will be sufficient, but for more complex changes crossing organizational boundaries, executive sponsoring is necessary to successfully implement the system or fully effect the change (refer Figure 15.8).
- Sustainable change almost always requires multiple, integrated approaches. These may include process redesign, individual transformation and large-scale organizational design just to name a few. The key to success is knowing when and how to apply each approach.

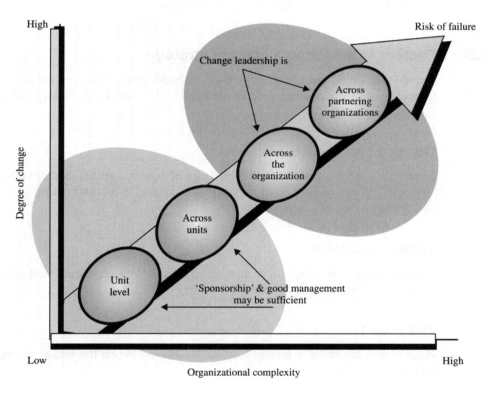

Figure 15.8
Degree of change and organizational complexity

15.3.4   Change management strategies

Rational-empirical

People are rational and will pursue their self-interest – once it is revealed to them. Individuals will accept change easier if how the change benefits them individually is communicated to them and incentives are offered.

Normative-re-educative

People are social beings and will adhere to commonly accepted cultural norms and values. Change is effected by redefining and re-interpreting existing norms and values, developing commitment to new ones and obtaining buy-in from a critical mass of employees as soon as possible.

Power-coercive

People are basically compliant and will generally do what they are told or can be made to do. Change is effected by exercising authority and imposing sanctions to drive the change.

Environmental-adaptive

People oppose loss and disruption but they adapt readily to new circumstances. Change is driven by building a new organizational environment and values, and gradually transferring people from the old one to the new.

15.3.5   Factors in selecting a change strategy

There is no single change strategy that will work every time, but for any given initiative it is preferable to use some mix of strategies. This decision is affected by a number of factors.

Degree of resistance

Strong resistance argues for a combination of power-coercive and environmental-adaptive strategies. Weak resistance or concurrence argues for a combination of rational-empirical and normative-re-educative strategies.

Target population

Large populations argue for a mix of all four strategies, identifying individual behavior patterns and applying the appropriate strategy.

The stakes

High stakes argue for a mix of all four strategies. When the stakes are high, nothing can be left to chance.

The time frame

Short time frames as for system implementations calls for a power-coercive strategy. With longer time frames, a mix of rational-empirical, normative-re-educative, and environmental-adaptive strategies can be used.

Expertise

If the organization have adequate available expertise and experience for making change, some mix of the strategies outlined above can be used. In most companies, the expertise and experience are not available, and the company will have to rely on the power-coercive strategy, or alternatively employ outside expertise.

Dependency

If the success of the organization is dependent on the skills of its people, and replacement skills outside the company is limited, management's ability to command or demand is limited, lest the employees become disenchanted and resign. Conversely, if people are dependent on the organization for work, their ability to oppose or resist is limited.

15.4 Requirements for effective change

15.4.1 Capability exploitation

Only people can deliver superior, sustainable and predictable business results (Figure 15.9). Individual capability needs to be exploited to the fullest extent in order for the organization to increase performance and capability. The individual capabilities and energy should be aligned to support company goals and strategies, and in some cases this means that individual focus and behavior need to change.

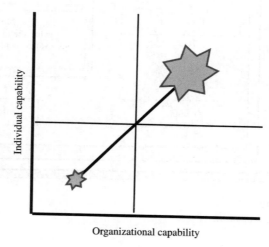

Figure 15.9
Individual vs organizational capability

Individuals perform to their maximum ability in the best interest of the organization if they understand how the organization is structured to re-inforce its strategic goals and how their individual performance contributes to these goals.

15.4.2 Human performance framework

Individual performance is the result of a number of related layers (Figure 15.10). These include the general environment (economic, social, etc.), the corporate strategy, the operation in which the individual find themselves, the organizational structures used to manage the individual and the individual's unique capabilities and personal motivation.

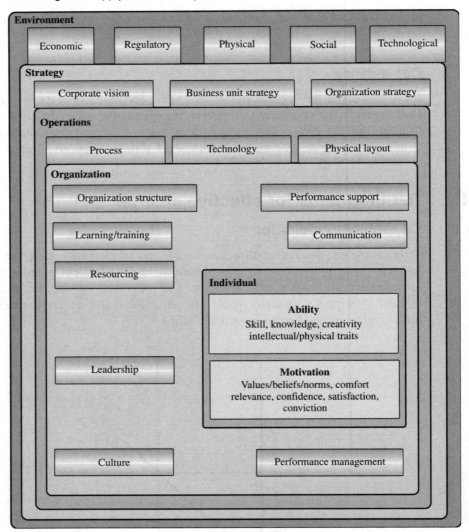

Figure 15.10
Performance framework

15.4.3   Skills development

The human performance is directly related to education and training, and it is a most important factor for a successful organization wide implementation. Training needs to be scheduled at the appropriate times during system implementation. The intensity and practicality of training should increase toward the time of implementation and thereafter (refer Figure 15.11). Too much practical system training a long time before the system is implemented will be wasted, as people will forget what they learned before they get the chance to practice and use the system. Training that takes place too late may also be ineffective, as people have already started working with the system, already figured out the basics and may find the training too basic to be of use.

The demand and supply of education and training should therefore be aligned (Figure 15.12) and could be something like the following:

- Communication regarding the impending implementation
- Information regarding the system objectives

- Education regarding the solution principles and procedures
- Practical off-line system training
- On-the-job practical system training
- Individual mentoring
- System configuration (super-user) training
- Advanced practical system training.

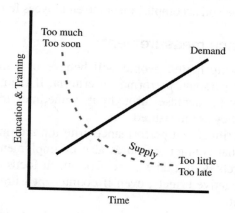

Figure 15.11
Education and training demand

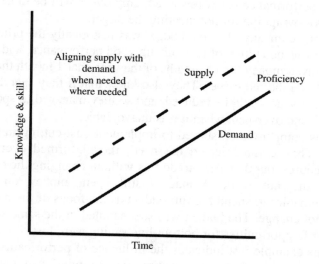

Figure 15.12
Aligned education and training

15.4.4   Reward for change

Behavioral change is required for successful implementation of change and being tough about performance requirements may be the most compassionate tactic. It is preferable to modify systems such as training, communication and employee appraisal before changing the incentive system, even for motivational purposes. While driving change, organizations should not undermine their efforts by modifying pay systems first, as this can have a negative effect. After new performance objectives and measures have been implemented, incentives can be used to motivate improved performance toward desired business results.

Any planned reward or incentive should be carefully thought through before implementation, even if it is being implemented with good intention. In one forestry company, every employee willingly and without payment gave up their free time to fight fires when it was necessary. With good intention, after such an incident management decided to pay the employees that volunteered, over-time pay for the time that they fought the fire. The next time a fire broke out, nobody volunteered their free time, as they did not need the over-time pay. Management effectively changed the community-service into normal work, taking away the pride and accomplishment the employees felt for their achievement.

15.4.5   Performance measurement

During working hours, people will behave in a manner that will contribute positively toward their personal performance criteria. If their primary measurement is throughput, they will try to increase throughput, sometimes to the detriment of quality. They will perform as they are measured.

It is thus critical that performance criterion are implemented that supports the company goals, and that counter-measures also be implemented to ensure that unwanted behavior to meet targets do not take place. Too much focus on one particular performance criteria can have negative effects, even if counter-measures were implemented but do not carry equal weight.

Equal attention should also be given to good and bad performance. People would try much harder to perform well if good performance is recognized, than they would try avoiding bad performance. If only bad performance is addressed on a frequent basis and good performance is not rewarded, employees will be more willing to take the occasional knuckle-wrapping for not meeting the target.

In one company, the cost budget was historically the primary focus of each department. There was no culture of recognizing good performance, and the manager knew that even if they saved money for 11 months of the year, that 1 month that they over-spend would mean a walk on the red carpet. They also knew that if they spend less this year, their budget for the next year would be reduced, and so they managed to spend the full budgeted cost each month and overspent when it was unavoidable.

This company then started to implement cost-cutting initiatives to improve the bottom line. They communicated the need, they informed everyone, they implemented new procedures, but they tried to do this without changing the culture or measurement system. This did not work. Managers still spent money on unnecessary equipment and consumables to spend the full budget as the focus of the measurement and reward system did not change. The budget was still handled in the same way by senior management and a lot of opportunities for bottom-line saving were lost.

This example just indicates the influence of performance measurement on the behavior of people, and how not keeping this in mind can negatively affect a project. The measurement system counteracted all other initiatives.

15.5   Change during system implementation

Installing a software solution and getting people to use it effectively are two different things as some behavioral problems are involved. Organizations often do not get the full benefit from their IT investment, if benefits are obtained at all. This is not necessarily due to a badly designed or configured system, but often due to non-use or non-optimum use.

In order to maximize system benefits and minimize time to full system usage, change management is required before, during and after system implementation (refer Figure 15.13).

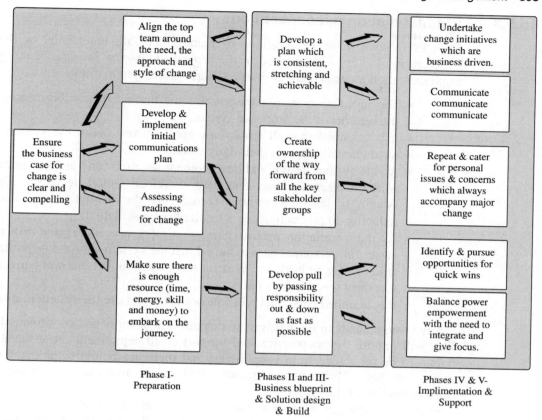

Figure 15.13
Process of managing change

15.5.1   Tools and techniques

The following tools and techniques are used to drive change during a system installation exercise:

- Individual mentoring
- Group facilitation
- Activity flow model analysis
- Software design analysis
- Compliance audits
- Compliance reports
- Review meetings
- Flash reports
- Project management methodologies.

15.5.2   Strategy

The focus of the implementation change management project is to reduce the time between system installation and effective usage of the system in order to maximize the system and business process benefits. The purpose is to provide the client with the confidence and proof that the implementation was effective and that all levels of management are comfortable with the usage of the new business process and system.

15.5.3   Pre-installation misconceptions

Systems fail when the expectations of management are unrealistic, or they have no expectations and are disinterested in the project. Following is some of the general pre-implementation misconceptions observed during new system implementation:

- The installation team/managers assume that all their subordinates understand the benefits of the proposed system.
- It is assumed that all employees are informed about the system and how it would benefit them in their daily work.
- As the system design and development required long and exhausting hours of work, managers expect that their workload would decrease as soon as the system is installed without considering the disruption and accompanying reduction in productivity as a result of the installation.
- As the installation means changes in work procedures and will make things easier for personnel, it is assumed that the system would do everything. People tend to then sit back and wait for the system to sort out problems, and are then surprised when it does not.
- It is assumed personnel will follow orders and use the system immediately.

Any change needs to be managed to enhance the benefits and coordinate the activities. This will require the cooperation and support of all departments at all levels. There are bound to be resistance from some areas, and there are bound to be areas that want to move too fast. These two extremes need to be managed to ensure coordinated implementation of the project.

Some of the changes may involve an increased workload for all levels until the system is installed at all levels. The only way to avoid system collapse will be the communication of the ultimate benefits to all employees. The project goals need to be communicated as well as the potential benefits for the employees. If they understand the goal and see how it will make their job easier they will be more willing to accept changes.

With the implementation of any new system or work-method one can expect an initial reduction in productivity. The key is to minimize this reduction as far as possible and to return to normal or higher productivity as soon as possible. The only way to do this is through the increased effort of management (refer Figure 15.14). A proper approach toward implementation change management reduces the need for additional management effort.

15.5.4   Implementation problems

In implementing any new system, we must anticipate and be prepared for potential problems that may occur. We must be aware that we are asking people to do things differently and that unless we involve them early in the process they may not understand what is required of them. Even if they know what is required of them and have been trained, they will still be unsure of the new environment, and this leads to a reduction in productivity.

When installing a new system, three types of problems occur – technical, currently existing/previously unknown and behavioral problems. Of these, technical problems are easier to recognize and solve.

If management takes responsibility for the behavior of people and if they start to effectively steer the behavior into constructive courses, the technical problems will be identified sooner and resolved more easily.

General behavior on the part of management, supervisory and non-supervisory personnel will determine the length of time needed to work on the technical bugs normally found within a new system.

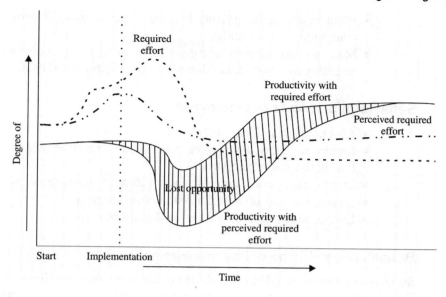

Figure 15.14
Need for change management

Installations uncover problems that previously existed but were thought to have been resolved, or problems that were not discovered before the installation.

Technical problems

- Missing or incomplete data, interface problems or incorrect calculations on reports.
- Supervisor/manager not familiar with the details of activities or priorities and therefore assigning work incorrectly.
- Inadequate training or practice to use new system effectively.
- How to handle excessive absenteeism, unusual peak loads and emergency situations.
- Not receiving work input from other divisions on time and not taking the necessary action to ensure that it is obtained.
- Failure to understand the mechanics and the principles behind the mechanics.
- Actions implemented as a result of work buildup due to initial low productivity such as overtime.
- Confusion caused by integrating new work assignments in compliance with the system, with normal problems and regular procedures.

Existing or unknown problems

- Large backlogs of work that always existed but were never known now cause concern.
- Failure to meet deadlines – input or output.
- Individual productivity differences that were never measured now become apparent.
- Equipment efficiencies and availabilities always thought to be world-class, suddenly show substandard performance.

- Weaknesses in supervisors become apparent as shift performance can now be compared between shifts.
- Manual data-capture accuracy and quality suddenly receive considerable attention as it is compared to electronically recorded data.

Behavioral problems of employees

- Absenteeism suddenly increases.
- Employees slow down work process (as a reaction, not on an organized basis, failing to meet objectives).
- Employees become careless and produce at substandard quality levels.
- Opposition and verbal expression of resentment.
- Low productivity – less than previously produced.

Behavioral problems of line management

Supervisors/managers fail to follow or adhere to normal deadlines, procedures and work sequencing, and fail to react to unusual circumstances as they previously did. Supervisors/managers may temporarily forget their normal responsibilities.

Failure to fulfill their responsibilities can be expressed in the following ways:

- Criticizing rather than actively participating in the correction of problems.
- Not attempting to keep morale at a high level but doing just the opposite, contributing to low morale.
- Failing to follow up on employees or problems properly.
- Unwillingness to participate or support the program/system.
- Unwillingness to actively supervise, talk to, or discuss problems with employees.
- Not adopting the system and data as their own, rather referring to it as 'new system data'.
- Failing to analyze or determine the facts before jumping to conclusions, reacting or verbally expressing themselves negatively. This is often done deliberately to show employees that he or she is on their side.
- Expressing their own weakness by fighting the system, its principles and basic data, rather than doing what the system calls for – taking what appears to be the least distasteful way out.
- 'Too busy' or 'Can't do it right now', procrastinating, rather than consciously or obviously resisting, exhibiting passive resistance, hoping that the problem will go away.
- Blaming anything that goes wrong on the system. Using the system as a scapegoat for current problems and even things that occurred in the past. Making non-productive statements such as, 'we can't do it anymore, the system does not allow that', or 'We're working to a schedule now'.
- Not discussing problems within the department, but voicing objections to other divisions where the system has not been installed yet, negatively influencing their perception of the system value.
- Weakness and compromising behavior shown to negative group reactions.
- Failure to communicate with divisions outside of the installed areas in asking their help in supplying information on time.

These problems should be anticipated, and action plans developed to identify and solve them when they occur. These actions should be taken during the planning stage before system implementation, as it can drastically reduce the time taken to get back to normal productivity levels.

The most effective way to prevent or resolve these problems is a positive management attitude toward the system. It is thus of critical importance that managers at all levels 'buy in' to the system objectives and functionality early in the process. Managers must display a positive attitude, verbally and non-verbally. They should not allow negative comments without investigating the root of the problem and resolving it. If they remain positive and work toward solving problems, clearing misconceptions and assisting struggling employees and supervisors, their subordinates will start following their lead.

This requires a lot of additional management effort and dedication, and change management specialists (or company employees dedicated to the project on a full-time basis for this purpose) can help relieve some of this workload, leaving the manager free to continue with his other duties. This will also give the manager more time to study and learn the system functionality so that it will be easier to solve problems and identify apparent misconceptions regarding system deficiencies.

15.5.5   Change management process

An implementation change management project is normally done in three phases. First the business process functional requirements are identified. After all functionality has been identified responsibilities are differentiated between system, user and management for each business process functional requirement. Specific compliance audit checklists are then developed for each individual or group of individuals interacting with the system. Workshops are held to explain audit requirements and individual responsibilities. Audits are then carried out on a regular frequency and areas or individuals requiring improvement are identified. Individuals are then mentored individually to increase business process and system knowledge and understanding. Upon final completion of the project, a report will be generated indicating system use, understanding and recommendations for enhancement to system and or business processes.

To facilitate this interaction with an economical commitment of time from both parties, the following procedure is recommended:

- General briefing involving all employees involved
- Detailed workshops with the different teams and/or individuals, responsible for different systems and or processes
- Preparation and acceptance of the initial draft compliance audit criterion describing how specific functional compliance will be measured
- Conducting of audits on an agreed frequency
- Individual mentoring in areas of concern
- Publication of audit results on an agreed frequency
- Preparation and presentation of the implementation change management report document
- Implementation of the recommendations in the implementation change management report document.

Clearly the determination of the compliance audit content and measurement philosophy will require considerable input from the client, since many decisions relating to the functional requirements and measurement methods will be value-judgments. The consultants will need to contribute meaningfully to this activity, in terms of practical consequences of alternatives, making recommendations based on experience in other installations. The following figure (refer Figure 15.15) explains the implementation change management process steps in more detail.

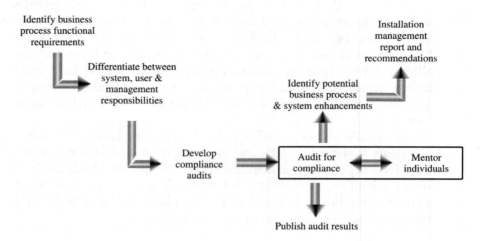

Figure 15.15
Change management process

Business process analyses

The objective of this phase is to understand the business process functionality, identify specific responsibilities that will support the business processes, and to differentiate the responsibilities of individuals from those of the systems.

The existing documentation should be evaluated to fully understand the complete business requirements. Specific functionality responsibilities should be identified and defined. The defined responsibilities should be allocated to individuals, or groups of individuals.

During this phase, the business drivers and measurements (key performance areas (KPAs) and KPIs) and how they support the company critical success factors (CSFs) should be reviewed to ensure that the defined responsibilities support the drivers and indicators. The physical and workflow processes should also be analyzed to give an understanding regarding the complete business- and management systems and work methods. Without a proper understanding of the total business, what it is influenced by, how it is managed and what tools are used, it will be difficult to design a concept for measuring business process compliance.

The basic guidelines for the identification and definition of the functional responsibilities are as follows:

- Description of the business drivers
- Description of the current and required management functionality
- Description of the current and required business process functionality
- Description of how the business process functionality support the business drivers (if required)
- Allocation of the functionality to system, user or management responsibility.

Business process functionality table

NO	Current Functionality	To Be Functionality	Tools to Be Used	KPA Influenced	Responsibility(M/U/S)

Management practices functionality table

NO	Current Functionality	To Be Functionality	Tools to Be Used	KPA Influenced	Responsibility(M/U/S)

Develop compliance audits

During the second phase the defined business functionality should be used to further allocate responsibility to specific individuals or groups of individuals. This in turn can be used to generate compliance audit checklists for measurement of actual compliance. The guidelines below should be followed.

- List functionality per individual or group to correct level of abstraction
- Define objective measurement criteria to measure compliance
- Define objective measurement criteria to measure understanding
- Define objective measurement criteria to measure usage
- Generate compliance audit checklists.

Management level 1

Management level 1 compliance measurement

Function No.	Required Behavior	Compliance Measure	Understanding Measure	Usage Measure

Supervisory level 1

Supervisory level 1 compliance measurement

Function No.	Required Behavior	Compliance Measure	Understanding Measure	Usage Measure

User group 1

User group 1 compliance measurement

Function No.	Required Behavior	Compliance Measure	Understanding Measure	Usage Measure

Audit compliance and mentor at all levels

The final step will be to conduct audits and mentor individuals or groups that require improvement. Much time should be spent with individuals, guiding them during normal working conditions to understand their responsibilities, how non-compliance affect other sections and what specific skills they need to improve. They should also be guided in system use, especially on management level, where the information supplied by the system can indicate potential problems and or solutions for operational problems.

During this phase it is possible that some business process enhancements are identified that do not require system enhancements. These should be implemented and the compliance audits adapted accordingly. Possible system enhancements may also be identified during mentoring and audit activities.

Audit results should be published at the agreed frequency and at the agreed level of abstraction. This can promote healthy competition between, shifts, department or areas. The audit results should indicate mechanical compliance, understanding and usage of the system.

The audit criterion can also be used as personal performance indicators and be built into job descriptions. They can also be used to audit system use frequently, and for training and assessment of new employees during probation periods.

Deliverables

The implementation change management report should provide proof to management that the business process and system are utilized effectively in all areas. In compiling the report, the content below should provide the proof.

- Description of the business drivers
- Description of the current and required management functionality
- Description of the required business process functionality
- Description of how the business process functionality support the business drivers (if required)
- Description of functionality per individual or group to correct level of abstraction
- Definition of measurement criteria to measure compliance
- Definition of measurement criteria to measure understanding
- Definition of measurement criteria to measure usage
- Compliance audit checklists per individual or group to agreed level of abstraction
- Compliance audit results indicating progress or improvement to agreed level of abstraction
- Description of potential business process or system enhancements
- Description of benefits that can be derived from the added functionality.

15.6   The three phases of change adoption

During a system installation, employees involved in change will have to be taken through three phases of change to ensure complete adoption of the new way of working and complete system usage.

15.6.1   Compliance

During this phase it is required to instill the discipline to use the new system and business process, totally and completely. A considerable amount of re-explaining is required as to the reason for and method of doing things. Regular activity checking and auditing is critical during this stage, as is the support of the system and positive attitude toward it by management.

During this stage, employees are usually merely complying, often mechanically, and as quickly as possible in order to get the job done. They do not necessarily understand the reason for what they are doing, what the benefits are or the business process it supports. It is thus critical that the three types of problems are identified early and actions taken to resolve them quickly, as these problems (especially technical problems) can easily derail the project at this stage.

If we take the example of driving a car, one first needs to be able to release the clutch, and depress the gas pedal at the right rate in order to drive. The method needs to be ingrained first, before variations in rate of release are attempted. Compliance with the basic rules is thus required before moving to the next phase 'Understanding'.

15.6.2 Understanding

Only when system use has become the norm, employees will start feeling comfortable using the system. It is then necessary to teach them to understand the value of the information by mentoring and explaining once more the value of their actions and the information collected. Then they will start checking their own work, especially if their supervisors/managers did and continue to do this during the compliance and understanding phase. If they start to understand that their use of the system is being monitored and their mistakes pointed out to them, they will attempt to identify these before their supervisor.

The employees will also start to take action to correct deviations, without the prompting of supervisors. They will start reporting any mistakes they made and cannot fix themselves to their supervisor, before it is identified by the supervisor. They will start using the available reporting tools to monitor production and identify problems. In the example of the car, once the basics are understood, going faster and slower will be tried, changing gears will become easier and after a while, the driver will start to understand how the gears, pedals and steering wheel interact.

Once the supervisor or employee understands the system and what it does, they are able to progress to the third and final phase.

15.6.3 Usage

Only when the system is used consistently and comfortably, will real, effective use of the system occur and its use start adding real value. At this stage, the employee will be recording meaningful and comprehensive information, and will start making use of it. The supervisor/employee will start identifying and reacting to problems immediately. They will also start exploring the system in an attempt to find any additional, useful tools that will be able to assist in making process improvements.

They will also start identifying trends, looking for ways to prevent potential non-conformances before they occur. Usage involves preventing problems that cause lost time and process inefficiencies. During this phase, employees will start anticipating production problems and request schedule changes. As they are now intimately familiar with the system and its intended use, they will start requesting enhancements and additional reports.

Using the example of the car, once the driver understands how everything works, he/she will really start using it to the fullest extent, handling the vehicle in wet conditions and testing its limitations.

15.6.4   Moving too fast

It is important to ensure that employees are only allowed to request system enhancements and additional reports once they have reached the usage phase. Allowing these requests before that time is like allowing an inexperienced driver (still in the compliance phase) to drive in wet and dangerous conditions and heavy traffic.

Management often start requesting changes to the system even before complete installation, and mostly before they even start understanding how the system works or what it does (or attempts to do).

Systems often end up with hundreds of reports within a few months after installation, of which only 20 or 30 are used after the first year, while the others just use up valuable resources. This can be avoided by following a structured process of implementation change management during the installation.

16

Conclusion

Learning objectives

- To provide a short summary of the contents.
- To identify critical issues that will ensure continued benefits from systems implementation.

16.1   Manufacturing future

In order to make the enterprises 'Internet ready' in the future e-business era, manufacturers have started focusing on people, plants and business processes.

The Internet has made the manufacturing enterprises more agile and more responsive. To respond to this changing requirement, process manufacturers are required to deploy integrated information systems, which allow them to view the actual capabilities of their plants in real time. Manufacturers need to have business processes in place to facilitate rapid decision-making and optimize their extended supply chain, integrating their internal business processes with their partners' business processes (refer Figure 16.1).

The Internet offers all process manufacturers the attractive possibility of differentiating themselves from their competitors in what has become a commodity market. Over the long term, e-business will impact the various individual segments of the process industries differently. Downstream segments such as polymers, specialty chemicals and pharmaceuticals will see more demand for customized products and will have more opportunities to sell and market their products directly to end users. Upstream segments, such as refining, will feel the most impact from the close collaboration with their business partners, which e-business demands. The implications for all process manufacturers are the same; however, where they sit in the value chain will determine where the manufacturer should focus its energies.

16.2   Establishing leadership

The e-business leaders in the manufacturing industries will not be the companies that launch the attractive website or deploy the most sophisticated Internet-enabled manufacturing technology. The companies that align their people, plants and internal business processes can gain the maximum leverage from Internet technology and tap into the vast potential of e-business. The future of e-business in the process industries is coming into focus and information is created and distributed in real time.

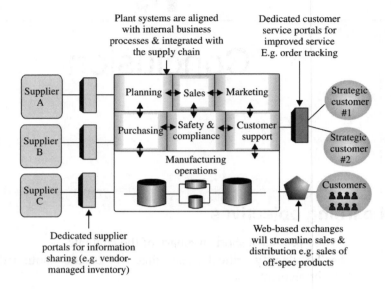

Figure 16.1
Integrated supply chain and plant systems

The gap between customers and suppliers effectively shrinks and expectations rise. The main characteristic, which sets the process industries apart from other industries, is their asset-intensiveness that is huge and costly. Due to the continuous nature of most processes, product changeovers are more complex and less frequent than in discrete manufacturing facilities.

Therefore product customization and improved customer service is less likely to come from make-to-order production than it is from improved business processes, a better and more accurate understanding of the plant's capabilities, and more accurate production planning via collaborative forecasting. This will all help process manufacturers in meeting their customers' rising expectations and serve their individual needs. It requires consistent and interoperable business processes and software and web-based services to support those improved internal business processes. Leading manufacturers will answer their customers' demands for improved customer service by providing portals into their operations, which will help integrate shared business processes such as improved production planning and collaborative forecasting.

Finally, Internet-based exchanges will streamline pricing and distribution. Commodity products will move more quickly and efficiently through the distribution network while specialized, value-added products command premium prices and become increasingly customized to end-user needs. In most industries, e-business entails a transfer of power from vendors to their customers. Vendors who understand and embrace this shift stand to reap significant benefits.

16.3   Success dependencies

Using the CMMS model in a different context, looking at processes rather than systems, we can map processes along the same three axes as we did with systems. These processes support the business goals and drivers, held together within the globe by the company employees and performance metrics. The business and physical processes provide a framework wherein the employees operate, and are optimized to contribute toward the goals of the company. The performance metrics measure the effectiveness of the business and physical processes, and drive employee behavior.

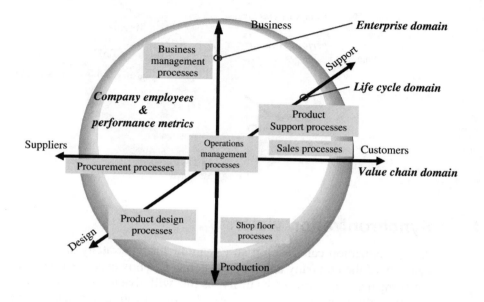

The business and physical processes are in turn supported by systems, and are also used by the company employees in order to increase performance. These systems need to be as flexible as the business and physical processes, as they are there in main to support the processes. Their effectiveness is not measured as such, but rather that of the supported business processes designed into the systems. The following figure depicts this concept, business and physical processes supported by systems, and held together by company employees and performance metrics.

Changes to any of the four critical factors (people, process, system, metrics) will influence the others. A change to the business process can potentially influence the metrics (positively or negatively), the system may not be able to support the change, requiring changes in the system, and it will have an effect on the behavior of the people, as they now have to work differently.

Similarly, a change in the system will need to be communicated to the employees so they can use the system differently, but if they are measured based on the old business process (not supported by the changed system), they will revert back to the old way of working, supporting the old business process to increase their measured performance and letting the system fall into disuse.

Changing performance metrics (or the emphasis of metrics) directly influence behavior, and this in turn can influence business processes, as people start taking short-cuts to increase their performance. The behavior may be in direct contradiction to the acceptable business processes as supported by the system, and if system compliance measure emphasis is dropped, the system would not be used again.

The above illustrates that planned improvement in any one of the above factors may have negative effects for the organization as a whole. The envisioned improvement in a system may be totally negated by the negative effects in behavior or business process.

It is thus critical that improvement and change actions be synchronized within a company to address all four factors at the same time in order to identify possible negative impacts.

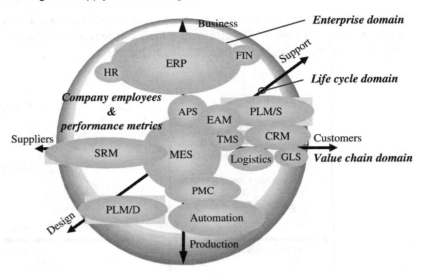

16.4   Synchronization vision

Any organization can develop an e-manufacturing strategy and vision, regardless of the maturity of the company and its IT systems. It is however, necessary to understand where the company is in terms of maturity, as this will give an indication of the predictability of supply within the organization. As the organization maturity increases in terms of systems integration and continuous improvement, the more predictable will be the quality and supply of process outputs and products.

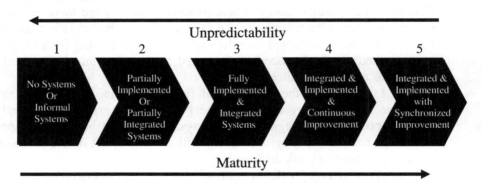

The major objective of any system is to support an underlying business or physical process. A system should not be there only because someone thought it would be a good idea, it should have a purpose, and that purpose is normally to automate some process in order to increase its efficiency and effectiveness.

Partially integrated systems only partially support business processes, as there are obvious breaks in communication between systems. These breaks in communication reduce the efficiency of the complete business process, often requiring manual user intervention, making it no more effective than a paper-based system. Once full integration has been achieved, entire business processes are automated and data can flow uninterrupted from the point of collection (e.g. instrument), through the various functions making use of the information to base decisions on (supervisor/manager), right up to the board of directors of the organization. The same information is also available to other users such as suppliers and customers where required.

Unfortunately this is not enough in today's fast changing environment, customers are demanding frequent improvement and new products. There is thus a need for change and improvement on a continuous basis. Continuous improvement can have many benefits for an organization, internally and externally. Products that stay stable and companies that do not innovate are soon replaced by better products or companies. This requirement for change and improvement has become even more pronounced during the last decade, and will continue to rise in importance.

This change needs to be managed, and the effects on the critical factors needs to be assessed and plans made to reduce their impact. Change is specifically difficult in a systems environment, as systems are often rigid in application and not flexible enough to accommodate frequent changes. The organization should therefore have a synchronized improvement vision in order to succeed in the new marketplace.

When embarking on the organization synchronization journey, it is important to know that the end vision is an organization where all the systems, people, processes and metrics support the business goals and drivers, and where methods are in place to coordinate and synchronize the change.

The correct engineering change procedures, taking into account all four critical factors can go a long way toward ensuring synchronized continuous improvement, as each of the factors influence at least one, but often more of the other factors.

Appendix A
Practical exercises

Learning objectives
- Practice the application of the techniques and concepts learned.
- Understand how the concepts can be used to identify potential problems and opportunities.

A.1   Practical exercise 1

Using the information supplied below, identify the tools used in the various levels of the systems hierarchy and identify gaps and opportunities for improvement.

Company X sells precious metals internationally from its refining facility. The figure below depicts the high level business model.

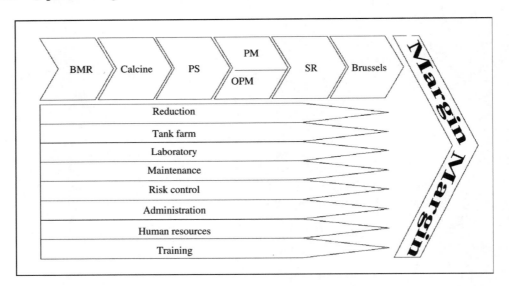

High level business model

The refining process begins when concentrate (the platinum group metals bearing material) is received from the base metal refinery (BMR). All the materials are blended together in the calcine plant to form homogeneous batches of material. The primary metals and minerals are then dissolved and separated in the primary separation (PS) plant. Further separation of precious metals is done at either the precious metal (PM) or other precious metal (OPM) plants where the different metals are extracted via chemical processes. Any rework or spillage is routed to the reduction plant where it is concentrated and then rerouted back into the process. The status (quality/make-up) of the material at that stage determines to which plant and process it will be sent. All the material from the PM and OPM, whether in liquid or solid state, is stored in the appropriate strong room. All materials are is sent to the effluent plant after refining, where it is stored, treated and disposed of.

A.1.1   PS plant

In the PS plant, the metals from the concentrate and caustic fused, recycled materials are separated. This is done in a manual batch process using a series of dissolution, precipitation, filtration and distillation steps that produce the different metals in crude form. These crude metals (Pt, Pd and Au) are then manually weighed and booked into a strong room from where it is taken to be purified in either the PM or OPM plants.

Before being used in the next process, the material is sampled and the analysis results must be available before the material can be released from the strong room to the next process. Due to the high value of the product, samples are weighed and labeled by the operator before sending to security. Security then reweighs the sample to ensure the correct weight has been recorded before sending the sample off to the lab. The sample is reweighed at each security checkpoint between the plant and the lab and the results recorded in a logbook. At the lab the sample is weighed again and the results captured in the Unix system (a bespoke or custom developed system on a Unix box).

Initially PGMs are dissolved from the concentrate/caustic-fused recycle, crude gold is precipitated, followed by the distillation of ruthenium and then the precipitation of rhodium and iridium. Finally the crude platinum is precipitated followed by the precipitation of palladium with the final filtrate going on to the reduction plant for recovery. Samples are taken on a frequent basis at each process step to test the progress of the chemical reaction. These samples are analyzed in the quick turnaround labs (QTL) located on the plant floor. Batches may not move on to the next step of the process unless the reaction has gone to completion. Based on the QTL results, processes are 'tweaked' and more chemicals may be added. These intermediate results are captured manually on logsheets.

Within this whole process, residual metal solids are generated at various steps and these are recycled within the plant.

When maintenance is required on equipment, jobcards are filled in by the plant and given to the maintenance foreman or fitter. The foreman is responsible for the requisitioning of spares for all maintenance jobs. After completion of the job the jobcard is signed-off by the plant and maintenance foreman.

A.1.2   Key performance indicators and influencing factors

The main objective of the PS plant is to separate the metals from each other with the highest possible recovery and the lowest possible contamination at each step of the process and at the lowest possible cost. The key performance indicators (KPIs) used to measure this are:

- Metals transfer weights
- Batch efficiency %
- Metals process time
- Operational cost of the section vs operational budget.

Certain factors in the process influence the attainment of the set objectives. These differ from step to step but the key influencing factors (KIFs) identified can be broken down into the following:

- The ability to control process parameters and the accuracy and availability of real-time process indicators during each step of the process. This includes chemical addition rates, solution temperatures, distillation/precipitation/reaction time, pH, specific gravity (sg), milliVolt (mV) potential and visual color checks.
- The accuracy of process material quantities whether weighed or estimated. This includes volume estimations in glass-lined vessels (GLVs), tanks and addition vessels.
- The accuracy of main lab or QTL lab results for each process step.
- The turnaround time for lab results.
- The quality and availability of utilities provided by service departments. This includes steam, cooling water, vacuum, electricity and draught.
- The quality and availability of make-up chemicals.
- The tracking and/or identification of metals quantities (metals deportment) throughout the plant.

A.1.3   Information technology tools used

The Unix system is used to manually capture material transfer transactions and the registering of samples for analyses. This results in much duplication of data capture between departments as each transfer and sample-registering transaction involves the recapture of the same batch data.

The Unix system is also used to generate time and attendance reports and capture leave. Hard copies are then generated and checked by the manager for accuracy before being approved.

The VIP payroll system is used to manage employee details and salaries. The VIP system is used in conjunction with UNIX system to manage all leave-related functions.

Excel spreadsheets are used to generate material transfer schedules (estimates). These are not always accurate or available timeously for the downstream processes. The spreadsheets also require the recapture of data already available on the Unix system.

Performance measurement is partly done through a custom-developed Unix metals accounting and Unix lab system supplemented by Excel spreadsheets in the plant for day-to-day control. Metals transfers are tracked against weekly/daily forecasts and batch efficiency against set standards.

Costs are reported monthly and the protean financials module enables the tracking of consumable costs per section daily. The protean customer order management module assists with the tracking of order and payment transactions.

The stores use Avantis to keep record of stock and show how the budget has been used. Avantis is also used to generate purchase orders for materials, spares and consumables.

A.2   Practical exercise 2

Based on the REPAC model, use the IDEF0 tool to draw a dataflow diagram of the process described below and identify the opportunities for improvement.

Chrome Company is part of a multi-national group of companies extracting and refining chrome metal. The operations process of Chrome Company starts where material is delivered into the plant by contractor transport. The materials are then mixed and processed through a smelter. Ingots and slag are separated when hot melt is tapped. The ingots are then broken, crushed and screened into final product. The slag is reprocessed through a recovery plant to recover additional metal fines that are also sold as final product. Final product is stored on stockpiles and loaded onto rail trucks for delivery to the port.

The section below describes one of the operations functions (raw material handling), the others being final product, furnace operations, ladle cleaning and chrome recovery.

A.2.1   Raw material handling

The raw material (RM) department is responsible for:

- RM ordering
- RM registering
- RM receipts
- Dumping on the RM stockpile bunkers.

The RM department orders the raw material according to the forecast as contained in the annual budget's schedule. In the monthly ore management planning meeting, actual year to date (YTD) figures are recorded against annual forecast, a forecast is also fixed for the following month. The ore requirements are calculated from this forecast. A month to date (MTD) figure and a weekly forecast are generated from the weekly planning meeting. Changes will only be made to the raw material requirements in the case of out of the ordinary circumstances. The order quantity is determined on an Excel spreadsheet taking into account the moisture content, handling losses and required quantities.

The orders are created and then e-mailed, posted or faxed to the suppliers, depending on the supplier's infrastructure.

The raw materials are received from the suppliers, internal or external to the Chrome Company group, via contracted road hauling transport. The trucks arrive at the weighbridge where the commodity's receival is registered together with the supplier details and its certificate of analyses (COA) and other information regarding the transporter. The information is captured on the standalone weighbridge system that records the truck's incoming and outgoing weight from where the net load weight is registered for the commodity received.

The trucks move from the weighbridge to the tippler were they off-load the raw material. The apron feeder feeds the material onto the conveyer belt that dumps the material on the designated stockpile bunker. A sample is also taken and sent to the laboratory for full chemical analysis, including moisture content. From the tippler the material is send to either one of the 12 stockpile bunkers of West plant or one of the 14 stockpile bunkers of the East plant. The tripper car operator is radioed by the tippler receival operator to inform him what material has been received and he then positions the tripper car over that commodity's allocated bunker, were the material is dumped and stored.

The day-bins above the furnace activate the batch mix system when a low-level setpoint is reached in the bin. The batch mix system weighs out the appropriate amount of each raw material commodity from the raw material bunkers onto the same conveyor belt and transports the 'sandwich' mixed material to the bin.

Commodity samples are taken on a shift basis for the calculation of moisture content. These samples are collected for a day period and a composite sample is made up and analyzed. The batch mix recipe is adjusted using these results. The analytical results are then used in the access batch mix application to calculate a new production recipe if required. The supplier is paid for dry weight delivered.

Material usage is calculated on dry weight for material transferred through the batch mix system. Laboratory results and weights transferred are manually captured into the AS400 accounting system.

A.2.2   Process flow diagram

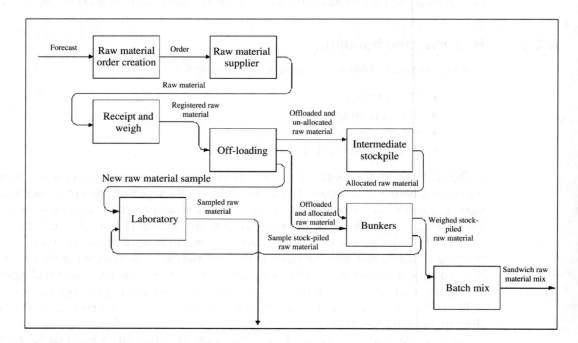

Process flow of raw material handling

A.2.3   KPIs and influencing factors

The main objective of the RM department is to make available the required, quantity and quality, material to the furnace operations. These materials must have a known quality and must be available in the requested batch quantities to enable them to create the right mix of materials. The KPIs used to measure this are:

- Tonnage received and delivered to the furnaces
- Costs, performance against budget
- The quality of raw materials delivered
- Availability of raw materials to the batch mix system.

There are factors that determine the attaining of the objective; these are called KIFs. The KIFs identified for RM are:

- Sampling methods not standard and accredited, resulting in disputable results.
- Figures used for reporting purposes.
- Availability of raw materials from the suppliers.
- The affect of climate conditions on the moisture content of the raw materials.
- Suppliers' infrastructure, lead-time and materials.
- Communication channels, methods used to communicate between various suppliers differ.

- Transport contractors. One low-tonnage supplier material arrives by rail, while the system is geared for road transporters. Some transporter trailers still have to change to hydraulic lift dump trailers for more efficient offloading.
- Bunker infrastructure. It includes the availability thereof and the ability to measure the outgoing material accurately.

A.2.4   IT tools used

Various applications are used to control and execute the RM activities.

- Excel spreadsheets are used mostly for internal planning, calculation, reporting and reconciliation purposes. This includes their tracking (capture and summarize info of materials received) data recording.
- A Word template are used to create purchase orders that are then printed out to be faxed, posted or sent via e-mail to the suppliers.
- The weighbridge system records the material received.
- The AS400's accounting module is used to record the raw material received information. This is entered manually from the daily weighbridge reports.
- Outlook 98 is used to e-mail orders to suppliers and to distribute internal management and information documents and memos.

A.3   Practical exercise 3

Using S95, draw the equipment hierarchy model of the enterprise below down to the lowest possible level, identify the functions of interest for the operational levels and identify opportunities for improvement.

A.3.1   High-level business process overview

MultiPlats produces platinum group metals (PGM) through various processes and in various locations as can be seen below.

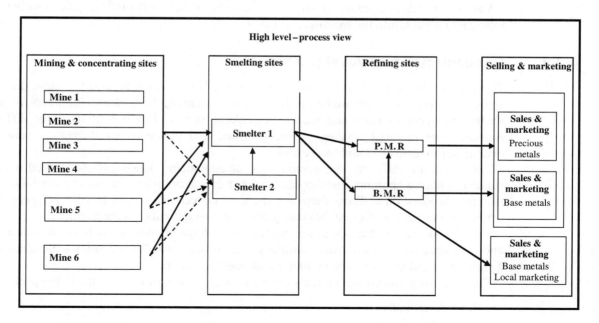

High-level business process overview

A.3.2   High-level metallurgical processes

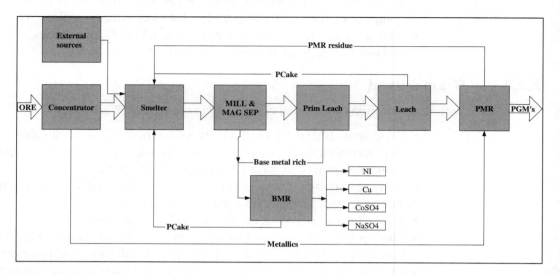

High-level metallurgical processes

PGMs are mined and the run-off-mine (ROM) ore are then processed through a benefication plant where it is crushed, milled and separated from the waste material and concentrated to the correct levels of metal content. The high PGM grade concentrate (metallics) is sent to the pure metals refinery (PMR) directly.

The other concentrate is smelted in a furnace, and the matte from the furnaces is sent to the BMR or pure metals refinery (PMR). In the BMR the base metals are separated from the PGMs through milling and magnetic separation and a leaching process.

The base metals are then refined and sold.

The PGMs are sent to the PMR where the PGM metals are refined through a series of batch chemical processes into saleable product.

Various residue streams from other processes are returned to the smelter for introduction back into the metallurgical process.

A.3.3   Concentrator high-level process

The feed to the concentrator comes from three different sources, and the three different source materials are each treated in its own benefication plant. Ore is trammed from the shafts and opencast mine and loaded in bunkers, one set for each of the three different sources. The material is then crushed, milled and concentrated through a series of cyclones, flotation cells and thickeners.

The plants use different configurations of the equipment, dependant on the feed material content to ensure optimum metal concentration for each ore type. The waste material or final tails are disposed on a slimes dam, and the final concentrate is sent to silos per plant from where it is despatched. Despatches take place over a weighbridge in the plant.

Samples are taken throughout the process on a frequent/continuous basis in concentrate and tails streams with automatic sample cutters or by hand. The samples are collected on a shift basis and tested by the laboratory for metals content.

The following reagents/chemicals/consumables are bought in and used in the process:

- Steel balls
- Caustic soda

- Depressant
- Frother
- Copper sulphate
- SIBX
- Xanthate Senkol
- IMP4.

All reagents/consumables/chemicals are weighed at the stores on receipt, before being delivered to the plant.

A.3.4   Metals accounting/reporting process

Daily report

Early in the morning, data is captured from logsheets, weighbridge tickets and laboratory spreadsheets into a daily/monthly report in Excel format. Plant throughput and efficiency indicators are calculated in the spreadsheet for review purposes.

Other information such as downtime and dam levels is also reported on a daily basis.

Plant metal accounting (PMA)

Most of the same information captured on the daily/monthly report spreadsheet is also captured in the PMA later during the day. Laboratory information is entered into the PMA daily. Fire assay information is captured to predict PGM metal balance on a daily basis. Once the official PGM content per metal from the nickel sulphide results are available (approximately five days after sample being taken), these results replace the fire assay results in the PMA.

This information is stored in an SQL database on a server. Data is exported from the database into an Excel spreadsheet where the calculation and transformation of the data takes place in order to generate the official month to date (MTD) report.

Platinum division evaluation (PDE)

Ore weights received by the concentrater plant and official production declared on the PMA is stored in the PDE database. The PDE database is used to verify and reconcile data (quality and quantity) from the different mines and plants (as depicted in error! reference source not found) in the division.

Month end

At month end, all nickel sulphide results need to be completed and available before the final month-end report can be generated. Consumable stock and usage are recorded daily on the logsheets, but only captured on the PMA and reconciled at month end.

A.3.5   Ore receipt process

The ore is weighed at the mine before being trammed to the concentrator. The weights are captured in an Excel spreadsheet and e-mailed to the concentrator on a daily basis. The ore is loaded into the concentrator bunkers per source/plant. The bunkers have level measurement equipment that is used with tramming figures for daily stock/consumption/throughput calculation.

A.3.6   Metals despatch process

Material is despatched separately per plant. Material is loaded and weighed on a weighbridge and a weighbridge ticket generated for each load containing the lot number. The weighbridge ticket information is manually captured on the PMA. Only despatched product is officially declared as production.

Concentrates and metallics are despatched in day-lots. Lot numbers are assigned daily, with the following naming convention YYYY-DDD-NN where Y = year, D = day of year and N = truck number.

Each concentrate despatch is sampled and the samples tested per lot. Metallics are also sampled per despatch and a weekly weighted composite sample is tested. Samples are also taken and tested at the PMR, and a PMR sample is sent to the concentrater laboratory and tested. The results from these tests are compared and an agreement is reached on the metal content. A control sample is also sent to ARC labs on a regular basis for verification.

A.3.7   Laboratory process

Samples are received by the laboratory from the process plants, despatch and the mine. Process samples are received per shift and made up into a day composite. The composite samples are tested using the fire-assay method and or size analysis. The fire-assay provides a total PGM content result, from which platinum, palladium, gold, etc. content is calculated using ratios.

Despatch samples (concentrate and metallics) are made into day or lot composite samples and tested daily using the nickel sulphide method (on ICP equipment) for PGM content per element. Base metals are tested for using XRF equipment.

A week-weighted composite is made of the metallics samples. This weekly composite metallics result is compared to the result obtained by the PMR, and the laboratory also test the composite sample made by PMR on receipt of the metallics.

The mines send geology (drill-core) samples to the lab that are tested for PGM's (using the fire assay method) and base metals. If the results indicate PGM content of above 3 g/ton, the sample is tested on the ICP.

Results are reported in MS Excel format and the file/s placed on the network from where they are accessed by the users.

A.3.8   Current systems

The following systems are in use:

- WinCC SCADA for plant monitoring and control
- In SQL database for historical information
- SAP R/3
- Expert system for mill control and optimization
- Custom-developed plant metal accounting (PMA) system on SQL database
- MS Excel is being used extensively for metals accounting and reporting.

A.3.9   Other findings

The following other findings were made:

- Plant running is monitored by the SCADA system for critical equipment, but no functionality exists to capture reasons for downtime. The downtime is also not analyzed to identify a sequence of events in order to raise alarms or generate work-orders.

- No event logging is done on the historian, such as start/stop, trip, alarm.
- No condition monitoring is done.
- The SAP R/3 installation runs on an Oracle 8I database.
- Dell hardware is used as standard.

A.4   Practical exercise 4

Using the information below, draw a proposed project lifecycle with an indication of the parties involved at each stage.
Hints: Big bang versus phased approach
  Objectives

  Is there a defined roadmap/vision?
  What influence will 'corporate' have?
  When should technology supplier be involved and how?
  When do we involve the employees/users and how?
  When do we pin down deliverables?
  When/how do we motivate the project?

PoliCo is part of the ChemPetro group, and consist of PoliCo and PurePhos plants. The business in its current form is still a new entity that started with the selling of PoliCo by the MetalCo group. ChemPetro took over the business interests of PoliCo from MetalCo and that of PurePhos from PhosCo to combine the two plants into one business unit under ChemPetro.

PurePhos plant and PoliCo is situated in different towns, and each plant functions as a separate entity. The business unit management function is situated on the PoliCo site. The PurePhos plant produces one product, purified sodium phosphate (PSP) that is the main raw material of the PoliCo plant. The PoliCo plant produces sodium tripolyphosphate (STPP) that is used mainly in the manufacturing of detergents.

The PurePhos plant is mainly a continuous chemical processing plant using liquid–liquid extraction columns for purification and a chemical reactor to convert phosphoric acid to sodium phosphate. The PoliCo plant is a more mechanical process using spray dryers, kilns, filters, crushers and mills to convert the PurePhos solution into saleable STPP product.

In terms of information technology, the two plants were historically almost inverse in their implementation philosophy. The PurePhos plant has a SAP R/3 ERP system in place on the business management level while PoliCo has very little on this level. PoliCo plant has a well-developed SCADA system on the process control level while PurePhos has only recently moved to PLC and digital technology on this level. The only commonality between the two plants is on the operations management level where both plants rely on manual systems, Excel spreadsheets and Access applications for effective operational management of the plants.

The current challenge for PoliCo management is to integrate the two businesses with diverse cultures and business processes into a single business unit.

PoliCo has a business requirement for total integrated systems to support its operations. The systems must automatically and accurately perform monitoring, control, coordination, scheduling, optimization and reporting. PoliCo's overall systems vision is that every part of the integrated process, as well as essential support functions, operate as a focussed unit, striving together to satisfy customer requirements in terms of quality and service at the lowest possible production cost. The systems must also be scalable, flexible and complement current initiatives within the ChemPetro group.

The current PoliCo requirement is to find the best suitable solution to fully integrated business, operations, and process control systems. The integration of both plants, coupled with automated monitoring and control, will lead to more efficient processing and the ability to streamline total operations and optimize processes. Amongst others, the integrated systems requirements include:

- Common operational data repository
- Process control (PLC and SCADA) and production data repository for the PurePhos plant
- Common business management system (SAP R/3) for both plants
- Quality information sharing and availability
- Planning and scheduling (operational and supply chain)
- Raw material, work in process and final product inventory tracking and management
- Inventory reconciliation
- Transport and dispatch planning
- Real-time or near real-time data exchange
- Operational reporting.

A.5   Practical exercise 5

The company *described below decided to implement a totally integrated solution. Develop a process to assist the company in their selection of vendor/s package/s to enable the integration vision.*
Hints: Best of breed versus best of suite

> *Do they have a defined roadmap/vision?*
> *How will we ensure they do not buy 'vaporware' or 'slideware'?*
> *How do we keep the selection process fair/unbiased?*
> *When do we involve technology providers/vendors?*

Company C has the following systems in place:

- An aging ERP/financial system

  - data-entry intensive
  - extensive work-arounds in use where system is inflexible or system functionality has been forgotten
  - system blamed for every problem ever occurring in the plant
  - business comes to a standstill for one week per month as month-end reports are generated.

- A combination of SCADA systems from four different suppliers with ages between 2 and 10 years

  - Some easily integratable, some much more difficult.

- A stand-alone maintenance system

  - Intensive data entry
  - Usually three to four days out-of-date.

- Custom developed recipe management system

  - Stand-alone
  - MS Access
  - Out-of-date documentation.

- An advanced expert process control system in one manufacturing department but nothing in similar departments elsewhere.
- MS Excel reporting systems
  - Personalized systems reporting the same data but according to needs of individuals
  - Little integration – Large unwieldy spreadsheets where integration has been attempted.

A.6   Practical exercise 6

Identify/define the lessons learned by the team below and develop a process that can be implemented in future to prevent similar problems. Indicate in terms of activities and time how this process relates to the normal project life-cycle.

As any rookie starship officer just out of the Academy will tell you, installation of a new software system requires the implementation of changes to the current system to improve productivity or efficiency. What these geniuses or champions of new software systems normally fail to mention is that it will not be all red dwarfs and roses. The change includes not only using a new software system, but also new methods and work practices. Initially, productivity and efficiency will more than likely go full reverse thrust, even with Mr. Spock in charge. Why is this so, and what should a Starship captain do to reduce the impact? The answer captain, must be out there, and its *not* boldly going where no man has gone before! Captain Kirk with his vast experience in human nature, says you must anticipate and be prepared for potential problems that may occur. A captain must be aware that he is asking people to do things differently. Unless they are involved early in the process, they may not understand what is required of them. People establish 'zones of stability'. These 'zones' can be interpreted in different ways. A captain might say things like 'Scotty is a creature of habit', 'McCoy is accustomed to doing it this way' or 'In engineering we've always done it that way'. As commander, a captain is responsible for his crew accepting new ideas, for involving them and making them understand the changes that will take place and the benefits to them and others.

When installing a new system, you also install change. Problems can occur both in the technical and 'how to' aspects of the change, and in the behavioral aspect. Spock will tell you that of the two, technical problems are easier to recognize and solve. The technical and 'how to' aspects are by no means simple to resolve though, and are strongly interrelated with the behavioral aspects. If a captain takes responsibility for the behavior of his people, if he starts to effectively steer their behavior along constructive courses, only then can technical problems be resolved easily.

General behavior on the part of officers, support personnel and crew will determine the time needed to work on the technical bugs normally found within a new system. Spock found that the installation uncovered problems that previously existed but were thought to have been resolved, or problems that were not discovered before the installation. An excerpt from his ships log reveals some of the problems uncovered by the new system.

Large backlogs of work have been uncovered and nobody knows how these were not noticed before. Some officers even refuse to acknowledge that they actually do exist. Some sections fail to meet deadlines, creating problems for other sections that rely on their output. This is due to productivity differences that were not apparent before, and system usage and training inefficiency after installation. Some officers are also weaker in command, and where the old system hid this, the new system highlights it.

The system also highlights the sections where Safety or quality is sub-standard, and officers and personnel alike obviously do not like having their weaknesses exposed. This in turn is leading to behavioural problems in officers and personnel.

Captain Kirk was not impressed when he received this report, and he was really worried when he received the reports from McCoy and Sulu.

Reports from Dr McCoy and Lt Sulu revealed a number of support personnel and crew behavioral problems after the system installation. Sickbay visits and work apathy suddenly increased, personnel slowed down the work process as a reaction to the new system, not on an organized basis, but nevertheless contributing to the failure to meet objectives.

Personnel also became careless and produced at substandard safety and productivity levels, often opposing the new system and verbally expressing resentment of the new system.

Starting to take notice of the effect of the system on his crew, Captain Kirk noticed changes in the effectiveness of his junior officers. Due to nervousness or because they are operating with a new system, officers failed to follow or adhere to normal deadlines, procedures and work sequencing, and failed to react to unusual circumstances as they had previously done.

Officers also temporarily forgot their responsibilities for safety and quality. For a reconnaissance vessel, this was obviously of major concern, and Kirk recorded how failure to fulfill their responsibilities was expressed.

Some officers are criticising the system or personnel rather than actively participating in the correction of problems, thus contributing to low morale. Others are unwilling to participate or support the installation, and rather than adopting the system and data as their own, refer to it as "Kirks baby". Even some of the science officers, chosen for their logic and scientific approach to problems, are failing to analyse or determine the facts before jumping to conclusions. Some of the weaker officers have expressed their own weakness by fighting the system, its principles and basic data, rather than doing what the system calls for, taking what appears to be the least distasteful way out. This is often done deliberately to show personnel that the officer is on their side. Generally, most officers are blaming anything that goes wrong on the new system, using the system as a scapegoat for current problems and in some cases even things that have occurred in the past. It has been reported that problems that have occurred have not been discussed within the section with superior officers, but have been voiced to other sections and officers, increasing distrust and animosity between crewmembers, sections and officers. These sometimes minor issues have thus, due to action not being taken by superior officers who are unaware of them, become major problems.

So how did Kirk resolve the situation? What did he do to make the new system a success? To understand the remedy and be able to prevent the same thing from happening again, Kirk found it necessary to investigate and understand the pre-installation misconceptions of the officers. An installation post-mortem was done and Kirk and his team uncovered some of these misconceptions.

The senior officers of all sections were involved in the system design and planning. They thus did not deem it necessary to involve the lower level personnel that were going to use the system. The 'tricks of the trade' used by the personnel to make the old system work were not taken into account. As a result the new system had most of the same deficiencies as the old system.

The senior officers thought that all their subordinates understood the benefits of the system for the ship, but it appeared that some of the officers were not completely committed to the change required to make the new system work. This was especially true for officers not directly benefiting from the installation, but whose sections had to provide inputs into the system on a regular basis. As a result, the effort they spent on making the system work was directly proportional to the perceived benefit to their section.

As the system design and development required long and exhausting hours of work, the officers accepted that their workload would decrease as soon as the system was installed. They thought that they would be able to take a well-deserved rest a week after implementation. What they did not take into account is the disruption and accompanying reduction in productivity as result of the installation. It took them a while to recognize their mistake, losing the opportunity to decrease the effect of the installation. Spock depicted this with the following graph, once again confirming Kirk's belief that he is too logical for his own good.

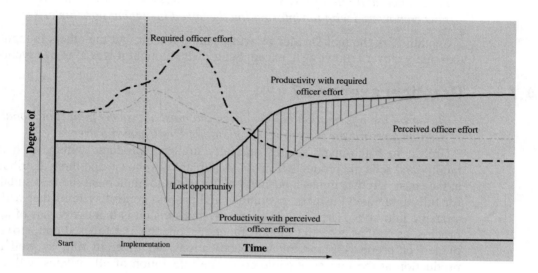

Sulu related to the graph immediately, commenting that the team from Omega-5 that helped to design, configure and install the system left the enterprise soon after the installation. It would have helped to have them around for a longer period, as they knew the system and would have provided additional hands and guidance. As the installation meant changes in work procedures and made things easier for personnel, they assumed the system would do everything. They were not ready to adapt to the changed practices.

The officers themselves assumed the same thing, and were not ready to follow up on the functions not addressed by the system. The credits saved by sending the Omega-5 team back early did not even start to make up for in subsequent productivity losses. They were not around to guide the system users and officers about their responsibilities in terms of the system use, and how to interact with the system. Sulu felt that the system was not used properly from the start and any system is only as good as its usage. Captain Kirk thought this is a very valid point, and made a mental note, 'If the system is not used properly, the benefits will not be there, sounds logical!'.

The officers accepted that personnel will follow orders and use the system immediately, unfortunately this did not happen. Even after training, system usage was low, until officers became actively involved. Dr McCoy explained why this happened.

During any installation junior officers and personnel will have to be taken through 3 Phases. The first thing that must be achieved is "Compliance". During this phase an officer needs to instill the discipline to adopt the new procedures and methods totally and completely. Lots of re-explaining will be required and it may take considerable time and sometimes even disciplinary action before compliance is achieved. Personnel will usually merely comply, often mechanically, and as quickly as possible. This is fine, as compliance is necessary before the junior officer or person can move onto the next phase "Understanding"

Once personnel are complying, their officer can teach them to understand the value of the information obtained from the system. Once personnel understand the meaning and effect of what they are doing, they progress to the third and final phase 'Usage'

Now that personnel are recording meaningful and comprehensive information, they can start to use it in their day-to-day activities. Usage involves reacting to problems that cause safety risks, lost time and process inefficiencies, events can now be anticipated and often prevented. Crises will become less frequent, as personnel are now empowered with the information they need to combat possible crises.

Captain Kirk thought Dr McCoy was pushing it a bit, but the others thought he was an extremely clever fellow, well, except Scotty who thought it was a lot of psycho-babble.

A.7   Practical exercise 7

Based on the information below, develop a master systems plan (functional solution architecture) for the integrated solution that UretChem requires.

UretChem batch-manufactures specialty urethane products within its manufacturing facility and sells the products from its local/plant warehouse and three distribution centers at the coast. UretChem has a business requirement to implement the best suitable solution for fully integrated business, operations and process control systems that will support its strategic initiatives. UretChem's overall systems vision is that every part of an integrated process, as well as essential support functions, operate as a focused unit, striving together to deliver products according to specifications, on time, in full, as well as ensuring production at the lowest possible cost. The integration of all systems will lead to more efficient plant processing and provide the ability to streamline total operations and optimize processes.

A.7.1   Process description

Raw materials are received in the raw material warehouse (UD) and stored in specific bin locations. On receipt of a picking slip, full drums of the required raw materials are issued to production.

Drums are moved by forklifts onto roller conveyers, from where it is placed on the scales next to the respective vessels. Production operators pump these raw materials directly from the drums into the vessel according to the sequence specified on the blendsheet. Part drums left after loading are stored in the production area and not returned to UD. The blending procedure may include heating, cooling, recycling and/or degassing. After the blending the product in the vessel, a sample is taken to the quality control (QC) laboratory for testing.

The quality controller performs tests that are specified for the type of product blended. Should material adjustments to the batch be required, a new picking slip is generated and additional raw materials are received from UD and/or the part drums are used. These materials are loaded into the vessel, blended and another sample is submitted for testing. This procedure is repeated until an *approved* or *reject* status is given to the final product.

When a product is approved, the operator pumps the product out of the vessel into the drums that are specified on the blendsheet. The drums are standing on a pallet on the scale next to the vessel. Labels with the product name and stock code, batch number, weight and any additional QC information is printed and placed on the drums. These drums are now moved to the transfer area where the packaging and labeling is inspected before transferal to the final product warehouse (UC).

In the case where a product is rejected by QC, the operator empties the vessel into 210 liter drums and a reject label with the products name, stock code and weight is placed on the drums. Reject material is transferred to the quarantine store (UQ) from where it will be reworked into similar blends.

Some of the final products sold by UretChem are not blended by the plant. These products are purchased in 210 liter drums and repacked into 25 kg or 5 kg packs according to customer requirements. The UretChem plant consists of eight vessels and three stainless steel intermediate bulk containers (IBCs) used for blending.

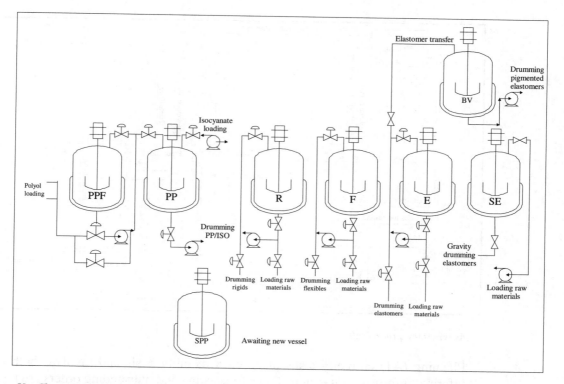

UretChem material flow

Due to the different composition and functionality of the various types of polyurethane products produced at Company CU, vessels are dedicated to specific types of polyurethane. The allocation of vessels is as follows:

Allocation of vessels

Vessel Name	Polyurethane Blended
PPF	Polyol raw material for prepolymer production
PP	Prepolymer and isocyanate
SPP	Prepolymer and isocyanate (currently being replaced)

Vessel Name	Polyurethane Blended
R	Rigid foams
F	Flexible foams
E	Elastomers
SE	Elastomers
BV	Blending vessel for the pigmenting of elastomers
IBCs	Manually blending of various small batches

A.7.2   As-is business process

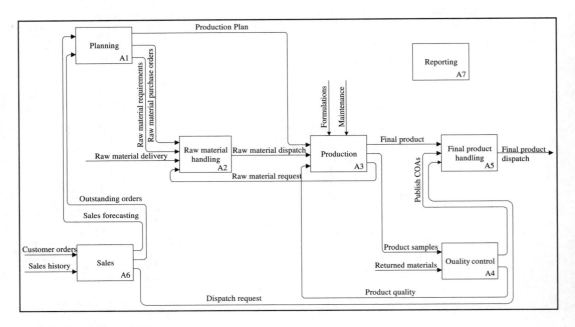

As-is business process (A0)

Planning (A1) currently takes place on a monthly basis and involves both formal and informal information transfers. Sales forecasting and outstanding orders are typical inputs that are received from the sales division (A6) and are used to determine the raw material requirements. Coordination of these inputs and consolidation of the plant capacity leads to the generation of the production plan.

Before production of chemicals can take place the raw material handling (A2) ensures that the purchased raw materials are received, stored and that the correct type and quantity of raw materials are dispatched to the production area when required for blending.

The production (A3) of polyurethane chemicals is a batch driven process where the respective raw materials are weighed and loaded into vessels according to the formulated recipes. Some of the blends require heating, cooling or degassing during the blending phase. Different vessels are dedicated to the various types of polyurethane chemicals, like elastomers, rigids, flexibles and prepolymers.

Samples of the blended products are send to QC (A4) where the product quality and any additions are determined. The Certificates of Analysis (COA) are made accessible once final products are approved. Quality control is also responsible for the handling and rework of returned and rejected products.

Final products are transferred to final product handling (A5) from where it is dispatched to customers on receival of dispatch requests/customer orders from the sales division (A6). Impact and Microsoft Excel are extensively used as the reporting system (A7). Some information is viewed in Impact and manually captured into Excel Spreadsheets in order to acquire the desired reports. The Citect SCADA system has a limited use on reporting or plant management.

A.7.3   Current systems in use

Impact

Impact is a small to mid-range business management system operating on the ERP level. The following modules are implemented at UretChem:

- Work-in-progress module
- Inventory control module
- Sales order module
- Purchase order module
- Requirements planning module (MRPII).

Proteus

Proteus is the QC system currently in use at UretChem. Functionality includes the following:

- Blend management (list of final products, generate batch numbers, picking slip and blendsheets and OH/NCO calculations)
- Technical sales and development (capture new and rework formulations, QC specifications, blending and packing instructions)
- Quality control (records batch results, manual approve/reject, list test methods)
- Reports (COA, formulation, progress, raw materials).

Citect SCADA

The Citect SCADA is used to monitor and control the plant systems. Pumps, valves and stirrers are controlled by PLCs, with Citect as the graphical user front end. Field equipment like scales and operator panels interface directly to the Citect system. Currently the plant control and recipe execution are manual processes and the Citect system is not used to automate batch processing. The recipe handling on the Citect system is developed in VB code and Cicode.

Others

Apart from the systems above, various Excel spreadsheets are used for planning, forecasting and reporting.

Deficiencies

The major deficiencies that should receive top priority in UretChem business are that of systems integration and total quality and stock control. A number of operations require determination of parameters on a MS Excel spreadsheet, manual input of the parameters, reading of information from separate systems and input of information back into Excel for analysis. Due to time constraints of these manual inputs, reports are mainly generated on a monthly basis and therefore management decisions are based on old data.

The total quality control system does not operate as a focused unit that integrates to the essential support functions. Recipes are currently captured in both Proteus and Impact, which leads to errors with the adjustment of these recipes.

The major areas that should receive attention in this regard include:

Planning

- Sales forecasting is not received in advance, which leaves little time for raw material and production planning.
- Production plan changes almost daily, this complicates the planning of raw material purchasing.
- No training has been given on the use of the Impact MRPII system.
- Changes in formulations are not always communicated to raw material planning.

Raw material handling

- OH and NCO values not captured on delivery of raw materials.
- No proper verification of the quality of raw materials received.
- Raw material transfer procedures are not followed properly.
- Lack of system integration hampers accurate stock control.

Production

- No daily visibility of production status to management.
- Production plan available to all departments not on a live basis.
- No OH and NCO values available for Citect to calculate ratio of raw materials needed.
- No feedback loops on critical valves on the plant. This may lead to substantial product losses.
- Maintenance of plant equipment is not planned or logged.
- No means to measure downtime.
- No stock control of materials used for cleaning or flushing of pumps and pipelines.
- No weight control on the filling of individual drums, only spot checked by QC.
- Blendsheets and picking slips are generated for planned batches before verification of raw material availability. As a result batch numbers are generated that are not always used. This could be prevented by having proper standard operating procedures (SOPs) in place of the production planner.
- Batch numbers, Blendsheets and picking slips are generated in Proteus, which is no longer a stable operating system.
- Blendsheet should include the type of drum required for packing and make provision for more operator input on times.
- No SOPs for operators.
- No accurate measure of operator attendance.
- No means of measuring operator performance accurately.
- No existing procedure for the flushing of vessels.

- Part drums are weighed, but the weight not communicated to the production planner in order to keep track of losses and accurate stock.
- The trial kitting functionality in Impact is not used by production.
- During the trial kitting procedure, the raw materials are not allocated per lot numbers out of specific bins. This leads to ageing stock.
- Trial kitting can only allocate raw materials in one warehouse. Currently WP and UD are different warehouses.
- Raw materials are not always booked out in advance to provide adequate time for materials that should be placed in the waterbath before blending.
- Raw materials that were issued to production are only booked out once the blend is completed and ready for transfer to UC. This results in inaccurate raw material availability.
- No proper procedure for the handling of reject batches by Impact.

Quality control

- No daily visibility of product quality to management.
- No tracking of adjustments made to blends.
- No proper logging of rework of reject materials.
- No labels for sample cups.
- No unique number for the logging of samples.
- Currently no control of lab consumables.
- Manual control of retained samples.
- Manual logging of sample detail in QC logbook.
- No means of measuring QC performance.
- Generates manual picking slip for batch adjustments.
- No interface between Proteus and Citect for production instructions from QC.
- Slow functionality of Proteus.

Final product handling

- No proper procedure for transfer of raw materials stored in UC.
- The dispatch of raw materials is not captured immediately by Impact, which affects the credibility of the final product stock levels.

Sales

- Inadequate visibility of updated production plan.
- No visibility of conversion cost calculations.
- Difficult to measure forecast accuracy. No report configured by Impact for this requirement.
- No notifications if sales forecast is lower than the sales budget.
- Manual allocation of payments received.
- Inadequate training on Impact.
- Has to service customers on old stock data.
- No notification if order has not been delivered.
- Price increases have to be changed manually, printed and posted to each customer.

Appendix B

Model answers

B.1   Practical exercise 1

Tools at various levels

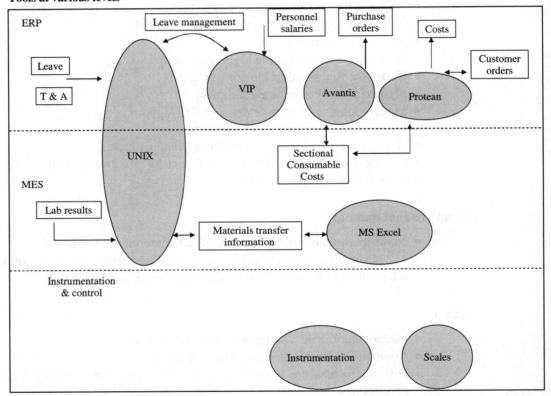

B.1.1   Opportunities

The following top level opportunities were identified for the PS plant:

- No process automation tools (PLC and SCADA) are used, all manufacturing is done manually.
- Batch processes are controlled manually.

- Scales are not integrated to any system, leading to duplicate data capturing.
- Security process is not integrated with anything.
- No integrated maintenance system in place (spares, resource planning, etc.).
- No historical real-time process data is available for analysis and optimization studies.
- QTL results are not available electronically for verification, analysis and progress tracking.
- The metals accounting system should be improved in order to determine the metals deportment within the PS plant on a real-time basis. Accurate volumes from the plant and tight integration with a lab information module will be required. By doing this the plant personnel will be able to see all work-in-progress (WIP) metals quantities for their own as well as upstream plants.
- An improved metals accounting system with the ability to automatically generate reports and calculate metal content from lab results and transfers will reduce the need for duplicate entering of data and rework of figures. This will free some time for the plant foreman and chemist to do process improvement and training.
- Online (automated) process monitoring and control will enable better control of the KIFs, providing access to statistical analyses, identification of out-of-control conditions and possible process automation opportunities. This includes measurement of weight, volume, consumption, temperature, pH, time, flow-rate, mV and sg.
- The online data can be used to generate alarms, and the history can be captured for later analyses.
- Better coordination between process plants can be achieved if abstracted information suited for use between processes, plants and services is provided online. This includes batch/metals transfer forecasts, progress between steps, weights, last reported results and raw materials availability. This will ensure information availability and visibility between the plants and supporting processes and services. It will ensure uninterrupted flow of materials thus driving down cost.
- A system to identify/flag and report exceptions or deviations from standards, forecasts or plans at different intervals in the process will ensure that exceptions are focused on. Currently it is required to work through much data to identify deviations. If done online it will ensure quick response to process deviations. This includes metals transfers, process time, pH, quantity, weight, assay, ratio, volume, addition rate, temperature exceptions and unavailability of materials, services or capacity.
- A system enabling statistical analyses of process performance data, the creation of automatic reports and the automatic update of transfer forecasts will assist in reducing time spent on less productive activities. Rather than spending time generating schedules and reports the foreman can be on the plant floor preventing problems and deviations or can be working on process improvement.
- The duplication and effort for data capturing within the plant must be reduced. This includes transfer data, sample data and laboratory results. Data must be captured once and then shared between plants and service processes. This can be achieved using bar-coding and bar-code readers for concentrates, caustic-fused recycles, chemicals and samples with direct scale inputs into the system at transfer points.

B.2 Practical exercise 2

Information flow diagram

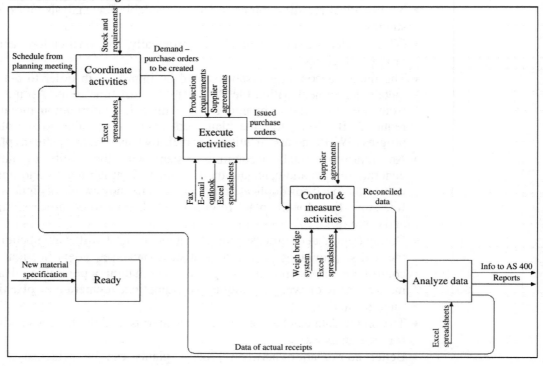

Information flow diagram of raw material handling

B.2.1 Opportunities

From the analyses opportunities were identified in the coordinate, execute, process, analyze and ready elements of the workflow. The following top level opportunities were identified for RM:

- There is an opportunity to eliminate duplicate data capturing. An integrated system should be in place where data is only captured once, preferable at the source thereof. Such a system should include automatic reporting and built in checks and balances. Currently data is being captured at the weighbridge from where manual reports are compiled that are then manually re-entered on the AS 400 and on an Excel spreadsheet. In this process, time is lost and data integrity is seriously affected. Various reconciliation steps are also required to manage these separate systems, which would not be necessary if the systems are integrated and have build-in checks.
- The raw materials system should be integrated with other operational systems. This will make information directly available from the LIMS for reporting of laboratory results and from the furnaces for indicating the furnaces requirements in advance. It will once again ensure the availability of accurate information and various departments will work from the same data, eliminating discrepancies between the figures and results from different departments.

- Built-in business rules and practices will help deliver accurate accounting/reconciliation figures. Standard definition, procedures and standard units of measure will provide uniformity across functional departments. e.g. a standard or adjusted figure will be used for accounting/reconciliation purposes that will only use figures that exclude slag or moisture.
- Get the same infrastructure platform for RM and all the suppliers. The standard, e.g. MS Outlook, will allow uniform communication and using of standard purchase orders.
- Very little feedback of information takes place to improve processes.
- Very few controls are in place to ensure correct handling and disposition of materials.
- Very little to no information is available to ensure that processes are ready for the next steps, such as materials, equipment, utilities, etc. No interaction between ready and coordinate processes.

B.3 Practical exercise 3

S95 model

Enterprice	Site	Area	Storage Areas & Production Units & Process Cells	Storage Units & Process Units	Equipment Modules	Control Modules
Corporation Z	Mine 6 Mine 5 Mine 4 Mine 3 Mine 2 Mine 1	Shaft 1 Shaft 2 Opencast Concentrator	S1 bunker S2 bunker OC bunker S1 Concentrator S2 Concentrator OC Concentrator S1 conc Storage S2 conc Storage OC conc Storage Slimes dams	S1 conc Silo S2 Metallics silo S2 conc Silo S2 Metallics silo OC conc Silo Dam 1 Dam 2		
	Smelter 1 Smelter 2	Receival area Smelter area Despatch area Waste area	Conc.Storage area Smelter 2-1 Smelter 2-2 Granulation unit Separation unit BMR Storage area PMR storage area Waste bunker1			
	BMR	Receival area Mill area Leach area Despatch area	Matte Storage area Mill 1 Mill 2 Megnetic separator Primarty leach unit Ni Unit Cu Unit CoSo4 unit NaSo4 unit Secondary leach Ni Storage Cu Storage CoSo4 Storage NaSo4 Storage PGM Storage	Ingot silo Dust silo		
	PMR	Calcine Primary Separation Precious Metals Other Metals Strong rooms	Process cells Process cells Process cells	Process units Process units Process units	Equipment modules Equipment modules Equipment modules	Control Modules Control Modules Control Modules

B.3.1 Functions of interest

Product inventory control
Quality assurance
Maintenance management

Production control
Product shipping administration
Production scheduling
Material and energy control
Product cost accounting.

B.3.2   Opportunities

The following high-level functional architecture encapsulates the opportunities for improvement:

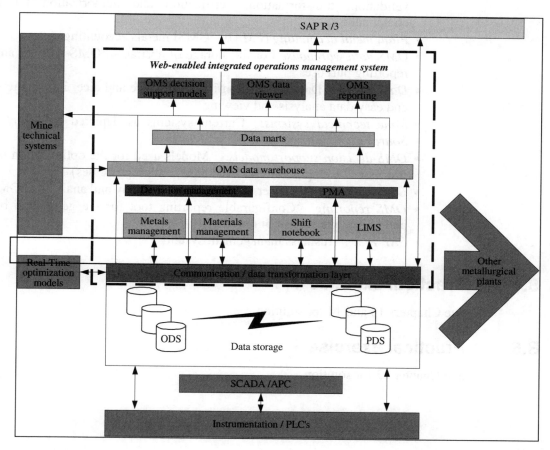

Illustrative conceptual architecture

- *Instrumentation/PLCs:*   Flow and density meters, scales and weighbridges
- *SCADA/APC* Real-time plant control and advanced process control (APC) or expert systems
- *Operational data storage (ODS):*   Data store for relational and transactional data
- *Process data store (PDS):*   Data historian for real-time process data
- *Real-time optimization models:*   Process modeling and optimization tools for the optimization of process control loops
- *Communication/data transformation layer:* A layer of data restructuring applications ensuring communications between various systems and components
- *Metals management:*   OMS component handling ore receipt, metals tracking, metals inventory and metals despatch

- *Materials management:*   OMS component handling chemicals/consumables receipt, inventory and consumption
- *Shift notebook:*   OMS component handling shift instructions, checklist configuration, generation and tracking, manual data capturing, downtime reason recording and shift reporting
- *Laboratory information management system (LIMS):*   OMS component handling sample notification, tracking, testing (including automatic interface to instruments) and results publishing
- *Deviation management* OMS component handling all business rules, data validation, transformation, verification and authorization for all OMS components
- *Plant metal accounting (PMA):*   GMSI metals accounting system
- *OMS data warehouse* For the storage of abstracted OMS data for analysis and reporting purposes
- *Data marts:*   Data 'cubes' configured to 'slice and dice' information for quick and easy data analysis and viewing
- *Mine technical systems:*   Current systems as depicted in *Error! Reference Source not found.*
- *OMS decision support models:*   Models used for the optimization of business processes (as opposed to material benefication processes)
- *OMS data viewer:*   User interface for the display and analysis of OMS data
- *OMS reporting:*   Configurable reporting tool for the generation of standard and ad-hoc generation of reports
- *SAP R/3:*   Business management system.

B.4   Practical exercise 4

See Chapters 13 and 14 for solution.

B.5   Practical exercise 5

See Chapter 13 for solution.

B.6   Practical exercise 6

See Chapter 15 for the change process.
The activities and how they relate to normal project life cycle is indicated below.

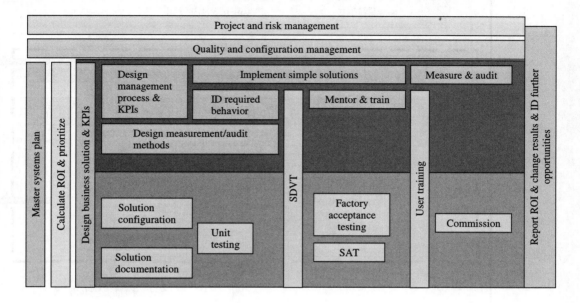

B.7   Practical exercise 7

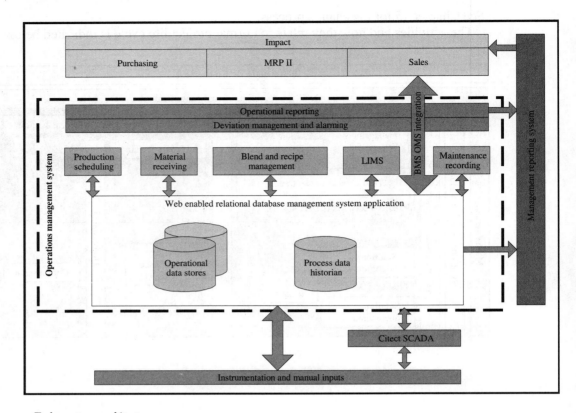

To-be systems architecture

This section specifies the business function requirements of each system module as depicted in the figure above:

B.7.1   Production scheduling

The production scheduling module will interface with Impact in order to download information needed for production scheduling and upload information needed by Impact during of after scheduling. This module will directly communicate to Impact's business rules layer, thus allowing the appropriate and efficient data validation on a business system level. The following business actions are the responsibility of the Production Scheduling module:

- Production reconciliation
- Production scheduling
- Batch number generation
- OH/NCO calculation
- Work-order generation
- Raw material stock transfer scheduling.

B.7.2   Recipe management

The recipe management module will be the master recipe management system and upload active recipes to Impact to ensure the availability of the active formulations on a central system. The recipe management module is responsible for the following:

- Formulation management
- Blending recipe management.

B.7.3   Material receiving

The material receipt module will interface with Impact in order to download information needed for raw material receipt and upload information to Impact after receipt. This module makes provision for the functionality to capture raw material details and execute handling instructions, not available by Impact. The material receipt module is responsible for the following:

- Raw material detail list
- Capturing of raw material details during receipt
- Placing raw material on hold in Impact for quality verification.

B.7.4   Blending management

This module will be responsible for all blend control and execution functionality, interfacing with Impact and Citect where required.

- Blend initiation
- Ensuring that the raw material kit issue is executed on time
- Total batch delay calculations
- Control of blending process
- Raw material utilization reconciliation
- Down packing control
- Yield calculations
- Ensuring product losses are booked as expense issues
- Interfacing with Impact for packed product issue
- Control of packaging inspection.

B.7.5   Laboratory information management (LIMS)

The laboratory Information Management system is responsible for the following:

- Raw material quality verification
- Handling instructions on non-conforming raw materials
- Product quality verification
- Quality control instructions to blend management
- Certificate of Analysis (COA) generation.

B.7.6   Maintenance recording

The Maintenance module is interfacing with blend management and is responsible for:

- Yearly maintenance plan
- Monthly maintenance schedule
- Actual maintenance recording.

B.7.7　Impact

Impact is responsible for the following functionality within the BMS/OMS integration.

- Sold product reconciliation
- Sales forecast validation
- Production plan generation
- Raw material requirements consolidation
- Raw material purchase order scheduling
- Trial kitting
- Raw material allocation to blends
- Raw material stock transfer
- Raw material dispatch to production
- Kit issue of raw materials for blends
- Specific raw material issue of additional raw material utilized
- Packed product issue
- Expense issue of product losses.

B.7.8　Citect

Citect is responsible for the following functionality within the OMS:

- Control of plant equipment
- Control of plant parameters
- Signals to blend management system
- Generation of plant equipment alarms.

B.7.9　Process data historian

The data historian will capture and store all real-time process data for analysis and reporting purposes.

Appendix C

Glossary of terms

3PL	Third-party logistics
4GE	Fourth-generation environment
4GL	Fourth-generation language
ABC	Activity-based costing
AI	Artificial intelligence
ALM	Asset life cycle management
AMR	Advanced manufacturing research
ANSI	American National Standards Institute
API	Application programing interface
APS	Advanced planning and scheduling
ARIS	Architecture of integrated information systems
ASCII	American Standard Code for Information Interchange
ASP	Application service provider
ATP	Available to promise
ATS	Available to sell
B2B	Business to business
B2C	Business to customer
BMT	Business modeling tools or business modeling techniques
BOE	Bill of equipment
BOL	Bill of lading
BOM	Bill of material
BPM	Business process management
BPR	Business process re-engineering
BTO	Build to order
CAD	Computer-aided design
CAM	Computer-aided manufacturing
CAPA	Corrective and preventative actions
CAPP	Computer-aided process planning
CAS Number	A number assigned to specific chemicals by the Chemical Abstracts Service
CASE	Computer-aided software engineering
CBM	Condition-based monitoring
CCR	Capacity Constrained Resource
CD	Cross Direction
CEO	Chief executive officer

CFO	Chief financial officer
CGMP	Current good manufacturing practices
CIM	Computer integrated manufacturing
CISC	Complex instruction set computer
CM	Customer management
CMM	Collaborative manufacturing management
CMMS	Collaborative manufacturing management systems, also Computerized maintenance management systems
CNC	Computer numerical control
COA	Certificate of analysis
COGS	Cost of goods sold
COMMS	Customer-oriented manufacturing management system
CPAS	Collaborative process automation systems.
CPFR	Collaborative planning, forecasting and replenishment
CPG	Consumer packaged goods
CPU	Central processing unit
CR	Critical ratio
CRM	Customer relation management
CRP	Capacity requirements planning also continuous replenishment planning
CSC	Customer service center
CSF	Critical success factors
CSR	Customer service representative
CSV	Comma separated values
CTP	Capability to promise
CVO	Contract variation order
CWIP	Current work in process
DBA	Database administrator
DBMS	Database management system
DBR	Drum-buffer rop
DCOM	Distributed component object model
DCS	Distributed control systems
DDE	Dynamic data exchange also direct data entry
DDX	Digital data exchange
DFD	Data flow diagrams
DM	Dimension models
DMC	Dynamic material control
DMS	Document management system
DNC	Distributed numerical control
DRP	Distribution resource planning
DSS	Decision support system
EAI	Enterprise application integration
EAM	Enterprise asset management
EBIT	Earnings before interest & tax
eBPO	Electronic business process optimization
EBR	Electronic business records
ECP	Engineering change proposal
E-CRM	Electronic customer relation management
EDI	Electronic data interchange
EDMS	Electronic document management system

EFT	Electronic funds transfer
EH&S	Environment health and safety
EOQ	Economic order quantity
EPCM	Engineering procurement and contract management
ER	Entity relationship
ERA	Electronic remittance advice
ERP	Enterprise resource planning
FAD	Function allocation diagram
FAT	Factory acceptance testing
FBD	Function block diagram
FCS	Finite capacity scheduling
FDM	Factory talk data model
FIFO	First in first out method
FMEA	Failure mode, effects and analysis
FMECA	Failure mode, effects and criticality analysis
FOB	Free on board
FORTRAN	FORmula TRANslation
FTE	Full-time equivalents
FTP	File transfer protocol
GAAP	General accepted accounting practices
GMP	Good manufacturing practice
GSM	Global system for mobile communication
GUI	Graphical user interface
HCS	Hybrid control systems
HDI	Help desk institute
HMI	Human machine interface
HMS	Holonic manufacturing system
HTML	Hypertext Markup Language
HTTP	HyperText Transfer Protocol
I/O	Input/output.
IEC	International electro-technical commission
IHS	Industrial health and safety
IIS	Industrial information system
IL	Instruction list
IP	Internet protocol
IPEC	Initiate, plan, execute and control
IPP	Inter process planning
ISA	Instrumentation systems and automation society/ organization
ISDN	Integrated services digital network
ISO	International Standards Organization
IT	Information technology
JAB	Java automation bus
JIT	Just in time
KIF	Key influencing factors
KPI	Key performance indicators
LAN	Local area network
LCC	Life cycle cost
LD	Ladder diagram
LIFO	Last in first out
LIMS	Laboratory information management system

MAP	Manufacturing automation protocol
MES	Manufacturing execution system
MESA	Manufacturing enterprise solution association
MIS	Manufacturing information system
MMI	Man machine interface.
Modem	MODulator-DEModulator
MOMS	Manufacturing operations management system
MPI	Manufacturing profitability index
MPS	Master production schedule
MRO	Maintenance repair and operations
MRP	Material requirements planning
MRPII	Manufacturing resource planning
MSDS	Material safety data sheet
MTBF	Mean time between failures
MTTR	Mean time to repair
MTU	Master terminal unit
NEMA	National Electrical Manufacturing Association
NIIIP	National industrial information infrastructure protocols
NNTP	Network news transfer protocol
NPD	New product developments
NPI	New product introductions
NSF	National Science Foundation
OBDS	On board data system
OCAPS	Out-of-control action plans
OCP	Oracle certified professional
ODBC	Open database connectivity
ODD	Operational detail design
ODS	Open data services also operational data store
OEE	Overall equipment effectiveness
OLAP	Online analytical processing
OLEDB	Object linking and embedding database
OMS	Operations management system
OOS	Out of stock
OPC	Open process control
ORB	Object request broker
ORM	Object role modeling
OSI	Open systems interconnect
P&PE	Product and process engineering
PAC	Process automation and control
PAS	Process automation system
PBIT	Profit before interest and tax
PC	Personal computer
PDA	Personal digital assistant
PDM	Product data management
PLC	Programmable logic controllers
PLM	Product life cycle management
PM	Preventative maintenance
PMC	Process monitoring and control
POC	Process operation control
POD	Proof of delivery

POS	Point of sale
POSIX	Portable operating system interface for computer environments
PTP	Profitable to promise
QA	Quality assurance
QC	Quality control
QCR	Quality control report
QMS	Quality management system
R&D	Research and development
RAD	Rapid application development
RAID	Redundant arrays of independent disks also redundant arrays of inexpensive disks
RAM	Random access memory also reliability and maintainability
RAS	Random access storage also remote access service also Reliability, availability and serviceability
RD&E	Research, development and engineering
RDMS	Recipe distribution management system
RFI	Request for information
RFP	Request for proposal
RFQ	Request for quotation
RFT	Rich text format
RISC	Reduced instruction set computer
ROA	Return on assets
ROCE	Return on capital employed
ROI	Return on investment
ROM	Read only memory
RONA	Return on net assets
RTD	Real-time dispatcher
RTDB	Real-time database
RTKBS	Real-time knowledge-based systems
RTU	Remote terminal unit
S&T	Subsistence and travel
SADT	Structured analysis and design technique
SAT	Site acceptance testing
SCADA	Supervisory control and data acquisition
SCE	Supply-chain execution
SCM	Supply-chain management
SCO	Supply-chain optimization
SCP	Supply-chain planning also support center practices
SCS	Supply-chain solution
S-DBR	Simplified drum-buffer rope
SDVT	Software design verification testing
SFA	Sales force automation
SFC	Sequential function chart also sequential flow chart
SIM	Science interface module
SKU	Stock keeping unit
SMART	Solutions for mes-adaptable replicable technology also specific, measurable, achievable, realistic, time-based
SMTP	Simple mail transfer protocol
SOP	Standard operating procedures
SP	Strategic planning

SPC	Statistical process control
SQC	Statistical quality control
SQL	Structured query language
SRM	Supplier relationship management
SRPT	Shortest remaining processing time
SSM	Sales and service management
SSPA	Software support professionals association
ST	Structured text
Step-NC	Step numerical control
STIS	Sample tracking and inventory system
SWIP	Standard work in process
T & A	Time and attendance
TCO	Total cost of ownership
TCP	Transmission control protocol
TDD	Technical detail design
TOC	Theory of constraints
TPM	Total productive maintenance
TQM	Total quality management
UML	Unified modeling language
URL	Uniform resource locator
URS	User requirement specification
VACD	Value-added chain diagram
VAN	Value-added network
VAT	Value-added tax
VBR	Value-based return
VMI	Vendor-managed inventory
WAN	Wide area network
WBS	Work breakdown structure
WIP	Work in process also work in progress
WMS	Warehouse management systems
WOM	Work order management.
WYSBYGI	What you see before you get it
WYSIWYG	What you see is what you get
XML	Extensible markup language

Appendix D
Bibliography

Websites

www.amrresearch.com
www.allair.com
www.invensys.com
www.entegreat.com
www.mesa.org
www.aspentech.com
www.advancedmanufacturing.com
www.aberdeen.com
www.ac.com
www.sequencia.com
www.sun.com
www.siemens.com/e-business
www.noosh.com
www.ARCweb.com
www.hmssoftware.ca
www.wonderware.com
www.andrewscg.com
www.hsesystems.com
www.hse-global.com
www.mySAP.com
www.microsoft.com
www.swift-computing.com
www.mmsonline.com
www.here4business.co.uk
www.edistribution.com
www.imaginewms.com
www.macola.com
www.exactsoftware.com
www.elance.com
www.skyva.de
www.interwavetech.com
www.vista-control.com
www.hytec-electronics.co.uk

www.kenonics.com
www.isa.org
www.it-analysis.com
www.cimagenovasoft.com
www.global-business.net
www.tenrox.com
www.relexsoftware.com
www.dencoplans.com
www.b-sources.com
www.coda.com
www.ebizmarketserver.com
www.yancy.org
www.xyntekinc.com
www.sofsol.com
www.manufacturing.net
www.thesupplychain.com
www.apics.org
www.s95.info
www.rockwellautomation.com
www.standishgroup.com
www.supply-chain.org
www.pera.net
www.gartner.com

Books

1. Supply Chain Management: For Global Competitiveness
   Sahay, Dr B.S.1st edition, McMillan.
2. Supply Chain Management: Strategy, Planning and Operation
   Chopra Sunil; Meindl, Peter, 1st edition, Prentice Hall.
3. Supply Chain Management Workbook, Harrison Francis, 2002.
4. Supply Chain Management In The Internet Age, S.K. Sharma, ICFAI press, 2002.
5. Quality Planning and Analysis, From Product Development through Use – 4th edition by J.M. Juran and Frank M. Gryna. McGraw-Hill.
6. UML Distilled, by Martin Fowler, 1st edition, Pearson Education.
7. UML Explained, by Kendall Scott, 1st edition, Pearson Education.
8. E Business Organizing For Success, Promod M Mantravadi, ICFAI press, 2002.
9. Perry's Chemical Engineering Handbook, 4th edition, McGraw-Hill.
10. Industrial Management & Operations research by Prof. K.K. Ahuja, 4th edition, Khanna Publication.
11. Advanced Cost & Management Accounting: V.K. Saxena and S.D. Vashist, 4th edition, Sultan Chand & Sons.
12. ANSI/ISA-95.00.01-2000 Enterprise Control System Integration – Part 1: Models And Terminology, ISBN: 1-55617-727-5.
13. ANSI/ISA-95.00.01-2000 Enterprise Control System Integration – Part 2: Object Model Attributes, ISBN: 1-55617-773-9.
14. A Reference Model for Computer Integrated Manufacturing, Theodore Joseph Williams, Purdue Research Foundation, Instrument Society of America, 1989.
15. Supply Chain Operations Reference (SCOR) Model, Supply Chain Council, 2000.

16. ANSI/ISA-S88.01-1995 Batch Control – Part 1: Models And Terminology, ISBN: 1-55617-562-0.
17. ANSI/ISA-S88.01-1995 Batch Control – Part 2: Data Structures and Guidelines for Languages.

White papers/newsletters/catalogs/research papers

1. SIBRO ^{TM} – e-procurement tools by Orbis Online Inc., 2002.
2. White papers of Keops Altech Technology Information Systems.
3. A Tutorial on the SP95 Enterprise/Control Integration Standard by Dennis Brandl Director, Sequencia Corporation.
4. Future Manufacturing IT Architectures by Roddy Martin, AMR Research, Boston.
5. Beyond Six Sigma E-Manufacturing in the dot.com era – May 2001 by Allen Yurko, Chief Executive, Invensys plc.
6. What is E-Manufacturing and how does it work? June 2002, EnteGreat, Inc.
7. Integrate ERP with Control systems using the S95 model, 2002, EnteGreat, Inc.
8. Linking the Plant Floor to the Enterprise: The Benefits & Pitfalls by Brian L. Harkins, Elliott S. Middleton and David A. Mushin, Aspen Technology, Inc.1999.
9. IDEFO as a standard for Function Modeling- Dec 1993, the Computer Systems Laboratory of the National Institute of Standards and Technology (NIST).
10. ERP – MES INTEGRATION: CLOSING THE GAP by Rogier van de Ree, Delaware Computing Nederland B.V., Nieuwegein, the Netherlands.
11. White Paper: Supply Chain Management by SerCom Solutions.
12. Five Steps to an eSynchronized Supply Chain by Andrew Berger, Andersen Consulting.
13. White papers on Supply Chain Management by Supply-Chain Council Inc., Pittsburgh, Pennsylvania, USA 15238.
14. E-assurance by Paul Bahnisch and Brett Miller, KPMG Adelaide.
15. S88.01 – The Standard for Flexible Manufacturing and Batch Control by Dennis Brandl, Sequencia Corporation.
16. SCADA TALK by Ian Wiese Industrial Computing, December 1999.
17. Measuring Supply-Chain Performance by Kevin P. O'Brien, Cap Gemini Ernst and Young.
18. The World of The Theory of Constraints, by Vicky Mabin and Steven Balderstone, St. Lucie Press, 1999.
19. Introduction to the Unified Modeling Language by Terry Quatrani, UML Evangelist. Rational Developer Network, 2001.
20. UML: Unified Modeling Language by Bill Raduchel, Sun Microsystems.
21. Paperless Online E-Procurement Solution by Sun Microsystems, 2002.
22. Strategic Sourcing and e-Procurement – Which One, or Both? By Ernest G. Gabbard, Allegheny Technologies Incorporated, Pittsburgh.
23. E-Logistics & E-Fulfillment: Beyond The "Buy" Button by Deborah L. Bayles, BridgeCommerce, Inc.
24. Inbound Logistics: A Catalyst for Supplier Collaboration By Adrian Gonzalez, 2002 Arc Insights 2002.
25. E-Industrial Services for planning and optimization of production and logistics systems through an example of simulation by Dipl.-Ing. Tom-David Graupner, Fraunhofer Institute for Production technology and Automation.
26. Product Catalogues of Ellipsys, Inc.
27. Product Catalogues of Amplicon Liveline.
28. Distribution Planning and Control:An Experimental Comparison of DRP and Order Point Replenishment Strategies by S. T. Enns and Pattita Suwanruji, University of Calgary, Canada.

29. Real-Time Dispatching Reduces Cycle Time by Ying-Jen Chen, Yeaun-Jyh Su, Ming-Shing Hong, Ivan Wang, Macronix International Co., Ltd, Taiwan, Semiconductor International, 2000.
30. Modeling Events in Object-Process Methodology and in Statecharts Iris Reinhartz-Berger by Arnon Sturm, Dov Dori, Technion – Israel Institute of Technology.
31. Tracking and Genealogy by Mike Cudemo – Interwave Technology, 2002.
32. Making more money with Finite Capacity Scheduling by Michael D. Novels, The CIMulation Centre, UK.
33. Following the Path from FCS to APS to SCM with Preactor, 1999 Preactor Inter-national Ltd.
34. An Introduction to Just-In-Time (JIT) Manufacturing by Debasish N. Mallick, Carlson School of Management, University of Minnesota, 2002.
35. WMS Planning, Design & Procurement by John M. Hill, Esync International, Ohio.
36. Collaborative Manufacturing Management Strategies, John Moore and Ed Basset, ARC Advisory Group, October 2001.
37. MESA White Paper 1: The Benefits of MES: A Report from the Field, 1997.
38. MESA White Paper 2: MES Functionalities and ERP to MES Data Flow Possibilities, 1997.
39. MESA White Paper 3: Controls definition and MES to Controls Data Flow Possibilities, 2000.
40. MESA White Paper 4: MES Software Evaluation/Selection, 1996.
41. MESA White Paper 5: Execution Driven Manufacturing Management for Competitive Advantage, 1997.
42. MESA White Paper 6: MES Explained – A High Level Vision, 1997.
43. MESA White Paper 7: Justifying MES: A Business Case Methodology, 2000.

Index

THIS BOOK WAS DEVELOPED BY IDC TECHNOLOGIES

WHO ARE WE?

IDC Technologies is internationally acknowledged as the premier provider of practical, technical training for engineers and technicians.

We specialise in the fields of electrical systems, industrial data communications, telecommunications, automation & control, mechanical engineering, chemical and civil engineering, and are continually adding to our portfolio of over 60 different workshops. Our instructors are highly respected in their fields of expertise and in the last ten years have trained over 50,000 engineers, scientists and technicians.

With offices conveniently located worldwide, IDC Technologies has an enthusiastic team of professional engineers, technicians and support staff who are committed to providing the highest quality of training and consultancy.

TECHNICAL WORKSHOPS

TRAINING THAT WORKS

We deliver engineering and technology training that will maximise your business goals. In today's competitive environment, you require training that will help you and your organisation to achieve its goals and produce a large return on investment. With our "Training that Works" objective you and your organisation will:

- Get job-related skills that you need to achieve your business goals
- Improve the operation and design of your equipment and plant
- Improve your troubleshooting abilities
- Sharpen your competitive edge
- Boost morale and retain valuable staff
- Save time and money

EXPERT INSTRUCTORS

We search the world for good quality instructors who have three outstanding attributes:

1. Expert knowledge and experience – of the course topic
2. Superb training abilities – to ensure the know-how is transferred effectively and quickly to you in a practical hands-on way
3. Listening skills – they listen carefully to the needs of the participants and want to ensure that you benefit from the experience

Each and every instructor is evaluated by the delegates and we assess the presentation after each class to ensure that the instructor stays on track in presenting outstanding courses.

HANDS-ON APPROACH TO TRAINING

All IDC Technologies workshops include practical, hands-on sessions where the delegates are given the opportunity to apply in practice the theory they have learnt.

REFERENCE MATERIALS

A fully illustrated workshop book with hundreds of pages of tables, charts, figures and handy hints, plus considerable reference material is provided FREE of charge to each delegate.

ACCREDITATION AND CONTINUING EDUCATION

Satisfactory completion of all IDC workshops satisfies the requirements of the International Association for Continuing Education and Training for the award of 1.4 Continuing Education Units.

IDC workshops also satisfy criteria for Continuing Professional Development according to the requirements of the Institution of Electrical Engineers and Institution of Measurement and Control in the UK, Institution of Engineers in Australia, Institution of Engineers New Zealand, and others.

CERTIFICATE OF ATTENDANCE

Each delegate receives a Certificate of Attendance documenting their experience.

100% MONEY BACK GUARANTEE

IDC Technologies' engineers have put considerable time and experience into ensuring that you gain maximum value from each workshop. If by lunch time of the first day you decide that the workshop is not appropriate for your requirements, please let us know so that we can arrange a 100% refund of your fee.

ONSITE WORKSHOPS

All IDC Technologies Training Workshops are available on an on-site basis, presented at the venue of your choice, saving delegates travel time and expenses, thus providing your company with even greater savings.

OFFICE LOCATIONS

AUSTRALIA • CANADA • IRELAND • NEW ZEALAND • SINGAPORE • SOUTH AFRICA • UNITED KINGDOM • UNITED STATES

idc@idc-online.com • www.idc-online.com

Visit our Website for FREE Pocket Guides

IDC Technologies produce a set of 4 Pocket Guides used by thousands of engineers and technicians worldwide.

Vol. 1 - ELECTRONICS
Vol. 2 - ELECTRICAL
Vol. 3 - COMMUNICATIONS
Vol. 4 - INSTRUMENTATION

To download a **FREE copy** of these internationally best selling pocket guides go to:
www.idc-online.com/freedownload/